Cases on Technology Innovation:
Entrepreneurial Successes and Pitfalls

S. Ann Becker
Florida Institute of Technology, USA

Robert E. Niebuhr
Florida Institute of Technology, USA

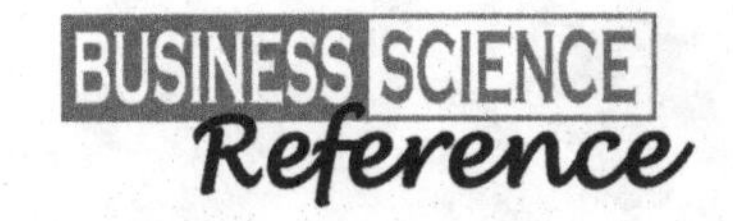

BUSINESS SCIENCE REFERENCE
Hershey · New York

Director of Editorial Content: Kristin Klinger
Director of Book Publications: Julia Mosemann
Acquisitions Editor: Lindsay Johnston
Development Editor: Christine Bufton
Publishing Assistant: Devvin Earnest, Keith Glazewski
Typesetter: Michael Brehm
Production Editor: Jamie Snavely
Cover Design: Lisa Tosheff
Printed at: Yurchak Printing Inc.

Published in the United States of America by
Business Science Reference (an imprint of IGI Global)
701 E. Chocolate Avenue
Hershey PA 17033
Tel: 717-533-8845
Fax: 717-533-8661
E-mail: cust@igi-global.com
Web site: http://www.igi-global.com/reference

Library of Congress Cataloging-in-Publication Data

Cases on technology innovation : entrepreneurial successes and pitfalls / S. Ann Becker and Robert E. Niebuhr, editors.
p. cm.
Includes bibliographical references and index.
Summary: "This book presents cases on theory, research, and practice in the areas of technology transfer, innovation, and commercialization, offering illustrations and examples of entrepreneurial successes and pitfalls in university, industry, government, and international settings"--Provided by publisher.
ISBN 978-1-61520-609-4 (hbk.) -- ISBN 978-1-61520-610-0 (ebook) 1. Technology transfer--Case studies. 2. Technological innovations--Case studies. 3. Research, Industrial--Case studies. 4. Entrepreneurship--Case studies. I. Becker, S. Ann. II. Niebuhr, Robert E. III. Title.

HC79.T4.C385 2010
338'.064--dc22

2009039911

British Cataloguing in Publication Data
A Cataloguing in Publication record for this book is available from the British Library.

Editorial Advisory Board

Table of Contents

Detailed Table of Contents

Section 1
University, Community, and Institutional Involvement in Technology Innovation

Section 1 focuses on case studies that involve universities and institutions in technology innovation, transfer, and commercialization; as well as, university and community efforts to promote entrepreneurship. Alan Collier, University of Otago, and Fang Zhao, American University of Sharjah, present their research on North American university performance in technology transfer and innovation. Nicholas Maynard, RAND Corporation, presents research on Thai and Malaysian science and technology institutions showing that institutional and policy reform process is directly influenced by regional activities as countries seek to match their regional peers for technology development. Shirley Ann Becker, Bob Keimer, and Tim Muth, Florida Tech, describe a framework for university and community collaboration in providing entrepreneurial technical assistance services. Each chapter is briefly summarized.

This chapter reports on case studies of four North American universities engaged in technology transfer and commercialization. The literature and case studies permitted an understanding of the characteristics possessed by universities and university technology transfer offices that appear to be successful in technology transfer and commercialization. Fourteen characteristics, or institutional enablers, are identified and analyzed in order to determine which among these characteristics have greater influence in the success of technology transfer offices. The chapter concludes that universities with superior-performing technology transfer offices possess two factors in common. First, the university President and other executives concerned in commercialization have to believe in it and make a genuine commitment to its success. Second, the technology transfer office has to be led by an individual who possesses several attributes: the ability and willingness to work within the university structure; the ability to be both an entrepreneur and a manager; the ability to see what is happening in technology transfer and commercialization as it evolves and matures; and to be a leader of people and business.

A country's national technology strategies can be an important contributor to economic development through its support of technology adoption and by advancing the national technology capacity. The development of a domestic information and communications technology (ICT) sector within a developing country requires the creation of specialized institutions that carefully coordinate their initiatives with the private sector. This case study research of Thai and Malaysian science and technology (S&T) institutions shows that this institutional and policy reform process is directly influenced by regional activities, as countries seek to match their regional peers for technology development. This effort to support ICT utilization requires governments to rapidly alter their policy goals and initiatives in response to shifts in technologies, global market demand, international investment, and local workforce capabilities.

Small businesses are viewed as the backbone of America and integral in the recovery of any economic downturn. Creative approaches to university and community collaboration are being explored to achieve high rates of success in launching, sustaining, and growing small businesses. One such approach, the Entrepreneurial Training Services (ETS) program, is being studied by Sci-Tech University as a means of technology innovation and regional economic development. The ETS Program has several unique features including: the entrenchment of a large number of adults in the program, an intensive training approach that is implemented in a short time frame, personalized mentoring offered to each entrepreneur in the program, and the leveraging of resources with a large, diverse group of community partners. The case profiles the region using Strengths, Weaknesses, Opportunities, and Threats (SWOT) Analysis, identifies an ETS framework on which the program is based and explains the process of implementation. The case concludes with challenges facing the university and local community in offering the ETS Program to a large and diverse group of entrepreneurs. It also summarizes benefits and successes from initial implementation efforts.

Section 2
Organizations, People, Processes, and Paradigms in Technology Innovation

Section 2 focuses on organizations, people, and processes as related to technology innovation and entrepreneurship. Michael Workman, Florida Tech, contrasts various theoretical perspectives on how innovations are adopted and shaped by organizational processes and structure. Francisco Chia Cua, University of Otago, presents a case study on the process by which a new enterprise system is introduced and adapted by a university. Brian O'Flaherty and John O'Donoghue, University of College Cork, explore the application of the Lead-user method in the development of medical applications based on Wireless Sensor Network (WSN) technology by three independent research teams. Roman Boutellier, Mareika Heinzen, and Marta Raus, ETH Zurich, contribute to the research in diffusion and adoption of innovation using science progress and the interplay of science and technology as dominant concepts. Arvind Karunakaran, Jingwen He, Sandeep Purao, and Brian Cameron, The Pennsylvania State University, describe a study on small to medium enterprises (SMEs) at different growth stages and levels of maturity (with respect to their information systems) and with different perceptions of the usefulness of information systems. Suryadeo Vinay Kissoon, RMIT University, introduces the CTIO (Concern-Task-Interaction-Outcome) Cycle as a means of studying team member interaction using face-to-face and virtual interaction media in retail banking.

The literature on technology innovation adoption and diffusion is vast. This chapter organizes and summarizes some of the major perspectives from this body of literature, contrasting various theoretical perspectives on how innovations are adopted and shaped by organizational processes and structure. The author first introduces the technology acceptance model, and innovation diffusion theory; and then categorizes viewpoints about organizational innovativeness. Drawing from this framework, the case study background introduces adaptive structuration theory, redefining some of its conceptual relationships in "structuration agency theory," putting primacy on the actions of agents and the means by which they operate through and around institutional structures. The author then presents a case study example of an expert decision support system, and concludes with a discussion of implications for managers and entrepreneurs.

The common structured procurement process of the Request for Information (RFI), Request for Proposal (RFP), and Business Case Development (BCD) is thought to establish ties with the right vendors and to strengthen relationships among other stakeholders. This single-case study gathered information through archival documents, observations, and in-depth interviews and examined whether RFI-RFP-BCP processes fostered favourable relationships with vendors. The study revealed certain disadvantages of the process.

This case study explores the application of the Lead-user method in the development of medical applications based on Wireless Sensor Network (WSN) technology by three independent research teams. This exercise produced surprising results, with the emergence of diverse WSN technology product concepts applied to Geriatric Falls Detection & Analysis, Sport Cardiac Screening and Critical Care Vital signs within accident and emergency environments. This case highlights the segmented nature of medical areas and the difficulty in applying a generic WSN technology to meet the functional requirements of the broader individual medical domains. It questions the appropriateness of applying 'total' highly functional technologies broadly across highly specialised niche medical areas.

This chapter explores the concept of paradigms, science, and technology in the context of information technology (IT). Therefore, the linear model of Francis Bacon and Thomas Kuhn's notion of scientific paradigms are reviewed. This review reveals that the linear model has to be advanced, and supports the adoption of Kuhnian ideas from science to technology. As IT paradigms transform business processes, a five-level concept is introduced for deriving managerial implications and guidelines. Within the case of e-customs, a European-funded project tries to ease border security and control by adopting a common standardized e-customs solution across the public sector in Europe. The rise of the IT paradigm within customs and its effect on business operations will be explained. This chapter contributes to the research in diffusion and adoption of innovation using science progress and the interplay of science and technology as dominant concepts.

This case describes two small to medium enterprises which are located within the same region and sharing the broad industry sector but at a different "growth stage" perceive the role of Information Systems differently. The authors describe how these two firms, at different growth stages and at different levels of maturity with respect to their information systems, perceive the usefulness of information systems differently. They extend the interpretations to discuss sub-sections within SMEs, which are at different stages of growth, and how the nature of information systems' risks is likely to differ depending on these growth stages. The authors emphasize the importance of owner/manager's "sensemaking of risks" as a key variable that influences the demarcation between entrepreneurs and small business owners, beyond the oft-discussed variables such as "achievement motivation," "risk-taking propensity," and "preference for innovation." This case concludes with the proposition that SMEs should not be considered as unitary entities; and suggest that there are likely to be different varieties of risks that SMEs face, and suggest the growth stage and organizational filters as key determinants of the owner/managers' understanding of these risks.

This chapter introduces the CTIO (Concern-Task-Interaction-Outcome) Cycle as a means of studying team member interaction using face-to-face and virtual interaction media in retail banking. The type of interaction is discussed in terms of different conceptual cycles having a linkage in the framing of the CTIO Cycle. In the past, routine teamwork using face-to-face communication was important. Today, with emerging technologies for retail banking organizations, teamwork through virtual communication has been gaining importance for increased productivity.

Section 3
Innovations in Information and Communication Technologies and Software Systems

Section 3 focuses on innovations in information and communications technology and software systems. These chapters are case-driven within the context of a government, industry, or business scenario. Divakaran Liginlal, Carnegie Mellon University, Lara Khansa, Virginia Polytechnic and State University, and Jeffrey P. Landry, University of South Alabama, describe the entrepreneurial vision and business model of Wikimedia and the success and challenges of its Web 2.0 innovations, the wiki and Wikipedia. Shirley Ann Becker, Florida Tech, focuses on the business development of a social network targeting older adults with chronic health conditions and family and friend acting in caregiver roles. L-F Pau, Copenhagen Business School and Rotterdam School of Management, provides a case on how usage-led technology innovation leads to entrepreneurial successes and pitfalls. Sherif Kamel, The American University in Cairo, takes an in-depth look at how information and communications technology (ICT) has improved the quality and range of services offered by the Egyptian National Post Organization (ENPO), while asserting the magnitude of its impact on the country's emergence as a competitor in today's global postal market. Desai Narasimhalu, Singapore Management University, presents a case on redefining medical tourism through the use of technology. Biswatosh Saha, Indian Institute of Management Culcutta, describes entrepreneurship as a temporal evolution of the creation and control over assets. Carolyn Fausnaugh and Mary Helen McCay offer a case involving collaboration between an entrepreneur and a university for market and scientific research to establish commercial viability of an invention.

This chapter describes the entrepreneurial vision and business model of Wikimedia, particularly the successes and challenges of its innovations, the wiki and Wikipedia. The case study first traces the history of how Wikimedia was founded, as such providing a rich descriptive background, using information obtained from scholarly news sources and websites. This historical overview is followed by a description

Preface

The *Cases in Technology Innovation: Entrepreneurial Successes and Pitfalls* is a compilation of theory, research, and practice in the areas of technology transfer, innovation, and commercialization. The book also contains illustrations and examples of entrepreneurial successes and pitfalls in university, industry, government, and international settings. The book is divided into three sections each of which is composed of chapters associated with a central theme. These sections include: *University, Community, and Institution Involvement in Technology Innovation and Entrepreneurship; Organizations, People, and Processes in Technology Innovation; and Innovations in Information and Communication Technologies.*

SECTION 1: UNIVERSITY, COMMUNITY, AND INSTITUTION INVOLVEMENT IN TECHNOLOGY INNOVATION

Alan Collier, University of Otago, and Fang Zhao, American University of Sharjah, present in their chapter titled, "Case Studies of North American University Performance in Technology Transfer and Commercialization," a comprehensive overview of the characteristics of universities that appear to be successful in technology transfer and commercialization. They identify fourteen characteristics, or what they call "institutional enablers," and analyze them for influences in the success of university technology transfer offices. Their findings identify common factors among universities with superior-performing technology transfer offices. One such factor is high-level administration, inclusive of the university president, having a strong commitment to technology transfer and commercialization. Another factor is leadership in working within the university structure; whereby, leaders have entrepreneurial and management abilities and stay abreast of what is happening in technology transfer and commercialization as it evolves and matures.

Collier and Zhao report on case studies covering four North American universities, as they develop benchmarks for which university technology transfer and commercialization performance can be measured. They conducted interviews and compiled data about the universities and the environment. Their data sources included the universities, venture capitalists, and consultants in the field. The authors provide a comprehensive discussion on their data finding in general and regulatory environments, commercialization structure, process, and incentives, industry links, intellectual property, entrepreneurial culture, and commercialization office performance. The authors conclude their chapter with the challenges facing North American Universities.

Nicholas Maynard, RAND Corporation, explains in the chapter titled, "The Evolution of ICT Institutions in Thailand and Malaysia," that a country's national technology strategies can be an important contributor to economic development through its support of technology adoption and by advancing the national technology capacity. Maynard points out that the development of a domestic information and communications technology (ICT) sector within a developing country requires the creation of specialized institutions that carefully coordinate their initiatives with the private sector. The author presents research on Thai and Malaysian science and technology (S&T) institutions showing that institutional and policy reform process is directly influenced by regional activities as countries seek to match their regional peers for technology development. The author discusses ICT utilization as requiring governments to rapidly alter their policy goals and initiatives in response to shifts in technologies, global market demand, international investment, and local workforce capabilities.

S. Ann Becker, Bob Keimer, and Tim Muth, Florida Tech, describe in their chapter titled, "A Case on University and Community Collaboration: The Sci-Tech Entrepeneurial Training Services (ETS) Program," an entrepreneurial training program provided by a university to the regional community as a means of promoting technology innovation and economic development. In their chapter, they profile the regional economic environment as an impetus to build university and community relations for technology transfer and business development. They outline the ETS program and identify unique features in promoting university and community outreach to entrepreneurs in the region.

The authors conclude the chapter by identifying benefits and initial successes from implementation of the ETS program. The overall objective is to provide a basis for further study of university and community partnerships in providing regional entrepreneurs technical assistance services. The authors summarize the challenges facing the university and local community in offering the ETS Program to a large and diverse group of entrepreneurs.

SECTION 2: ORGANIZATIONS, PEOPLE, PROCESSES, AND PARADIGMS IN TECHNOLOGY INNOVATION

Michael Workman, Florida Tech, presents in the chapter titled, "Technology Innovation Adoption and Diffusion: A Contrast of Perspectives", various theoretical perspectives on how innovations are adopted and shaped by organizational processes and structure. The author reviews two seminal streams of innovation adoption theory through the introduction of the Technology Acceptance Model (TAM) and the Diffusion of Innovations Theory. He then categorizes the major theoretical perspectives on how innovations develop from, or are shaped by, organizational processes and structures (referred to as organizational innovativeness).

Workman points out that technology innovation adoption and diffusion have been actively researched. But, given the vast body of literature it can be difficult to determine under what circumstances innovations are adopted and diffused and what factors may lead to resistance. In addressing this, he organizes major streams of theory and perspectives on innovation and adoption and diffusion. The author uses structuration agency as a framework for a case study to show how actors and structures play out in the adoption and diffusion of an innovative technology.

Francisco Chia Cua, University of Otago, presents in the chapter titled, "The Challenge of a Corporate Matchmaker," the process by which a new enterprise system is introduced and adapted by a university. The author uses Everett Rogers' Diffusion of Innovations (DOI) theory as the primary model for evaluation criteria during the matchmaking phase. Cua describes how deploying new enterprise systems or replacing old ones requires problem-solving intervention under conditions of incomplete information.

The case study illustrates how innovations depend on reasoned action for success. Cua points out that "reasoning" is often based on a subjective set of beliefs and motives held by the executives sponsor, the opinion leaders, and other supporters of the innovation. These beliefs, motives and other assumptions are called a "mindset," and the case study showed that they play a bigger role in procurement than the rigid structure of the process would suggest. Cua describes a proactive mindset among those involved in the innovation process as a positive influence for fostering relationships with opinion leaders, change agents, project team members, and other stakeholders. The same mindset influences some vendors to develop relationships with prospective customers and to understand more thoroughly their needs. Successful matchmaking depends on an alignment of these mindsets.

Brian O'Flaherty and John O'Donoghue, University of College Cork, explore in their case study, "The Development of Emerging Medical Devices - The Lead-User Method in Practice," the application of the Lead-user method in the development of medical applications based on Wireless Sensor Network (WSN) technology. The authors point out that the Lead-user process has been successfully adopted within a diverse range of application domains such as development of medical equipment technology, medical infection control devices in 3M, weblog technology, and extreme sports communities. The authors chose the Lead-User process to help guide each of three student teams in developing potentially successful commercial products or services.

The authors describe how student research teams, utilizing the Lead-User process, produced surprising results in the emergence of diverse WSN technology product concepts applied to Geriatric Falls Detection and Analysis, Sport Cardiac Screening, and Critical Care Vital Signs within accident and emergency environments. The authors highlight in their case study the segmented nature of medical areas and the difficulty in applying a generic WSN technology to meet the functional requirements of individual medical domains. The authors point out that the Lead-user method is useful in teaching technology entrepreneurship, as it sensitizes the students to alternative sources of innovation and encourages them to interact with domain experts in niche areas.

Roman Boutellier, Mareike Heinzen, and Marta Raus, ETH Zurich, explore in their chapter titled, Paradigms, Science, and Technology - The Case of E-Customs," the concept of paradigms, science, and technology in the context of information technology (IT). They review the linear model of Francis Bacon and Thomas Kuhn's notion of scientific paradigms recommending that the linear model be advanced. The authors introduce a five-level concept for deriving managerial implications and guidelines taking into account that IT paradigms transform business processes. The chapter contributes to the diffusion and adoption of innovation using science progress and the interplay of science and technology as dominant concepts.

The chapter introduces a case on e-customs, a European-funded project that tries to ease border security and control by adopting a common standardized e-customs solution across the public sector in Europe. The authors provide an overview of the rise of the IT paradigm within customs and its effect on business operations. The technological progress of the adoption of a common standardized e-customs system in Europe is explained. The authors discuss resistance in adopting e-customs. They point out that although the paradigm has changed, the culture has not and that an information technological paradigm shift from customs to e-customs is irrevocable.

Arvind Karunakaran, Jingwen He, Sandeep Purao, and Brian Cameron, The Pennsylvania State University, focus in their chapter titled, "Growth Trajectories of SMEs and the Sensemaking of IT Risks - A Comparative Case Study," on two small to medium enterprises (SMEs) located in the same region and sharing an industry sector. The authors describe how these two firms, at different growth stages and at different levels of maturity with respect to their information systems, perceive the usefulness of information systems differently. The authors discuss sub-sections within SMEs, which are at different stages of growth, and how the nature of information systems' risks is likely to differ depending on these growth stages. They emphasize the importance of the owner or manager's "sensemaking of risks" as a key variable that influences the demarcation between entrepreneurs and small business owners, beyond the oft-discussed variables such as "achievement motivation," "risk-taking propensity," and "preference for innovation."

The authors discuss the proposition that SMEs should not be considered as unitary entities. They point out that Risk Management studies, within the IS/IT stream, should not ignore "organizational context" and move beyond the development of ideal frameworks for abstract organizations. They recommend focusing on contextual and structural dimensions such as, enterprise size and age, growth rate, formalization, centralization, and number of organizational levels, among others. The authors suggest that there are likely to be different varieties of risks that SMEs face, and also suggest growth stage and organizational filters as key determinants of the owner or managers' understanding of these risks.

Suryadeo Vinay Kissoon, RMIT University, introduces in the chapter titled, "Use of the Concern-Task-Interaction-Outcome (CTIO) Cycle for Virtual Teamwork," the CTIO Cycle as a means of studying team member interaction using face-to-face and virtual interaction media in retail banking. Kissoon discusses the importance of teams using virtual communication for increased productivity in retail banking organizations. The author also discusses the type of team interaction in terms of conceptual cycles in the framework of the CTIO Cycle.

Kissoon addresses different problem-solving cycles, each of which relates to the mode of interaction medium (whether face-to-face or virtual) used by team members, facilitators, or managers to resolve problems in the workplace. The author focuses on understanding the relationship between direct (face-to-face) and virtual interaction variables and as they relate to retail banking trends in hybrid teams and virtual group networks. The author discusses the use of virtual team interactions in data life cycles linkages as gaining importance from perspectives of data and information quality. Kissoon identifies current trends in the triangulation of continuous improvement, routine teamwork, and virtual teamwork to support retail banking organizations in achieving efficiencies in performance.

SECTION 3: INNOVATIONS IN INFORMATION AND COMMUNICATION TECHNOLOGIES AND SOFTWARE SYSTEMS.

Divakaran Liginlal, Carnegie Mellon University, Lara Khansa, Virginia Polytechnic, and Jeffrey P. Landry, University of South Alabama, describe in their case study titled, "Collaboration, Innovation, and Value Creation – The Case of Wikimedia's Emergence as the Center for Collaborative Content," the entrepreneurial vision and business model of Wikimedia, particularly the successes and challenges of its innovations, the wiki, and Wikipedia. They compare the Wikimedia business model to other Internet business models inclusive of Knol and Google's open encyclopedia. The authors use a modified version of Weill and Vitale's model schematics, which prove useful in visualizing the flows of information, resources, and revenues among Wikimedia's contributors and consumers. The authors discuss the use of Kaplan and Norton's Balanced Scorecard to analyze the value generated by Wikimedia's business model.

The authors use the case as a means of exploring the wiki model from societal and ethical perspectives. Based on the principles of collaborative innovation, self-organization, democratization, and leadership by merit, the authors discuss how wikis can generate value for businesses. They also discuss innovativeness in the collaborative philosophy of Wikimedia as both a contributor to success and a challenge to credibility. The case explores this concept, its controversy, and the associated ramifications to society, along with an illustrative example of its use for collaborative work in a funded academic research project.

S. Ann Becker, Florida Tech, examines in the chapter titled, "Social Networking for Distance Caregiving and Aging in Place: A Case on Web 2.0 Technologies for Virtual Support," the business development process for a social network targeting older adults and unpaid caregivers. Becker proposes technology for seniors to share daily entries on health with family and friend caregivers some of whom may be geographically distant.

A focal area of the case is Web site design in terms of usability by older adults. The author points out that many Web sites, inclusive of social media sites, meet the online needs of younger adult users. However, some do not take into account usability needs of older adults in terms of vision, cognition, and motor skills associated with normal aging. The case provides an overview of these factors to be considered in the design of a social network Web site. The case concludes with a discussion of challenges facing online startups given rapid changes in technology, minimal barriers to market entry, and a near saturation point for Web sites with social networking capabilities.

L-F Pau, Copenhagen Business School and Rotterdam School of Management, presents in the chapter titled, "Case 'Mobile-INTEGRAL'" , a case on how a multinational company specializing in machinery goods uses high technology in its field support and mandated safety solutions to migrate its customer relationships into partnerships of growing scope and with new revenue streams. The key technologies are in-situ equipment monitoring and wireless communications. The key management ingredients are top management's understanding and respect for operational issues. The history of the case illustrates the importance of the strategic choice of the in-house vs. in-sourced nature of the needed technical expertise and of a gradual deployment compatible with the fast technology evolution.

Sherif Kamel, The American University in Cairo, presents in the chapter titled, "The Egyptian National Post Organization: Past, Present and Future - The Transformational Process Using ICT," an in-depth look at how information and communications technology (ICT) has improved the quality and range of services offered by the Egyptian National Post Organization (ENPO), while asserting the magnitude of its impact on the country's emergence as a competitor in today's global postal market. Kamel points out that the international postal sector over the last 20 years has changed drastically due to several forces, including globalization, changing technology, greater demands for efficient services and market liberalization. He explains that for Egypt, keeping up with the changing atmosphere in the global market meant investing in information and communication technology.

Kamel describes how Egypt has been gradually building its information society since the mid 1980s, adapting its strategy and approaches to the evolution of the global ICT sector. The steps taken included supplying accurate and timely information, encouraging private investment, formulating effective economic reforms, improving productivity, providing programs for lifelong learning, making public services more efficient, improving health care, optimizing the use of natural resources and protecting competition.

Desai Narasimhalu, Singapore Management University, describes in the chapter titled, "Redefining Medical Tourism," a case on the use of information and communications technology (ICT) in the medical tourism industry. The author explains that medical tourism is a term coined by tour agencies to facilitate travel across international borders to get either affordable or specialized healthcare. Narasimhalu describes challenges in the medical tourism industry inclusive of quality control. Several innovative uses of ICT (e.g., iMedical Butler) are introduced as a means of serving an international market.

Biswatosh Saha, Indian Institute of Management Culcutta, describes in the chapter, "Institutional Innovation and Entrepreneurial Deployment of a Software Product: Case of Financial Technologies Group in India," entrepreneurship as a temporal evolution of the creation and control over assets. The author explains that the value of an asset lies in its transactional relations with other assets in the ecosystem. Saha introduces the chapter by explaining how assets generated by entrepreneurs derive value as part of wider ecosystems, which in turn can be viewed as an architecture of inter-related and inter-connected assets.

Saha outlines a case on a software firm called Financial Technologies (FT). It is argued that the deployment of financial trading software, as a product in brokerage houses in the emerging securities trading ecosystem in India by FT, hastened institutionalization of new rules governing transactions embedded in the software design. The case explains how FT implicitly collaborated with the regulator and other ecosystem participants who coordinated the innovation in design of the ecosystem. The software firm went on to expand the market for its own products (trading software) by incubating exchange ventures. The author explains that this was achieved through a strategy of spawning of linked subsidiaries that led to both a growth of the trading ecosystem and further entrenchment of the innovated ecosystem.

Acknowledgment

The editors would like to acknowledge the help of all persons involved in the collaboration and review of this book. The authors did a remarkable job of ensuring high quality chapters were created. Many served as reviewers for chapters written by other authors. Thanks to all of you who provided constructive and comprehensive reviews. The editors extend a note of thanks to the IGI Global staff for guidance and professional support throughout the process. A special thanks to family and friends for their unfailing encouragement and patience.

S. Ann Becker
Florida Institute of Technology, USA

Robert E. Niebuhr
Florida Institute of Technology, USA

Section 1
University, Community, and Institution Involvement in Technology Innovation

Chapter 1
Case Studies of North American University Performance in Technology Transfer and Commercialization

Alan Collier
University of Otago, New Zealand

Fang Zhao
American University of Sharjah, UAE

ABSTRACT

This chapter reports on case studies of four North American universities engaged in technology transfer and commercialization. The literature and case studies permitted an understanding of the characteristics possessed by universities and university technology transfer offices that appear to be successful in technology transfer and commercialization. Fourteen characteristics, or institutional enablers, are identified and analyzed in order to determine which among these characteristics have greater influence in the success of technology transfer offices. The chapter concludes that universities with superior-performing technology transfer offices possess two factors in common. First, the university President and other executives concerned in commercialization have to believe in it and make a genuine commitment to its success. Second, the technology transfer office has to be led by an individual who possesses several attributes: the ability and willingness to work within the university structure; the ability to be both an entrepreneur and a manager; the ability to see what is happening in technology transfer and commercialization as it evolves and matures; and to be a leader of people and business.

INTRODUCTION

This chapter reports on case studies covering four North American universities (two in the United States and two in Canada) examined by the authors as they developed benchmarks against which to measure university technology transfer and commercialization (TT&C) performance in other countries. While there has been a significant increase in research into university TT&C since around 1990, there remains a need to develop a coherent and

DOI: 10.4018/978-1-61520-609-4.ch001

broad understanding of the factors that define performance in this environment. This is not to say that fine work has not been done in certain metrics – we use some of the work by the likes of the Association of University Technology Managers (AUTM) in our analysis. But we felt that there remained a need to gain a greater understanding of the range of factors: TT&C processes; Technology Transfer Office (TTO) structure; incentives among researchers; the role of venture capital; and many others, to develop a full appreciation of what made TT&C work in some universities and, from this, to begin defining a model.

Our case studies are developed from interviews conducted at the universities and from data about the universities and their environment from a wide range of reliable sources that included interviews with venture capitalists, consultants in the TT&C field, and written data. Our case studies are of the type called 'explanatory' (sometimes also called 'causal'), because they present: "...data bearing on cause-effect relationships – explaining how events happened" (Yin, 2003, p.5)[1].

We have structured this chapter into five further sections. Section 2 provides background to the environment in which universities engage in TT&C and offers some definitions. Section 3 describes the role of universities in national innovation, identifies barriers facing universities participating in TT&C, and develops fourteen environmental factors likely to affect university TT&C performance. Section 4 contains the case descriptions. It discusses the North American environment and then examines the case study universities under the fourteen environmental factors identified in Section 3. Section 5 discusses the challenges facing North American universities, identifies key factors relevant to TT&C performance, and identifies some actions that universities could take to improve TT&C performance. Section 6 draws together the threads of the analysis and offers some conclusions.

BACKGROUND

The Importance of Universities in Technology Innovation

Publicly-funded research, of which universities represent a significant proportion, is a vital and continuing ingredient in innovations introduced by industry and has been for many years. As it concerns the United States, Narin, *et al.* (1997, p.317) opined that "[a]mong both scientists and economists it is widely accepted that public science - scientific research that is performed in academic and governmental research institutions and supported by governmental and charitable agencies - is a driving force behind high technology and economic growth". They concluded (p.340) that "... public science plays an essential role in supporting U.S. industry, across all the science-linked areas of industry, amongst companies large and small, and is a fundamental pillar of the advance of U.S. technology".

There is strong evidence that university research in the U.S. has been important to a number of innovative developments in seven major industries[2] during the period 1986-1994. Mansfield (1998, p.773) reported that: "Innovations that could not have been developed (without substantial delay) in the absence of recent academic research accounted for over 5% of the total sales of all major firms". And, at p.775: "over 10% of the new products and processes introduced in these industries could not have been developed (without substantial delay) in the absence of recent academic research". He also observed that university research has led to cost savings among the same major firms: "Innovations that could not have been developed (without substantial delay) in the absence of recent academic research resulted in cost savings of about 2%". He further noted that, between his two surveys (in 1991 and 1998), in the later one, "... there was a decrease in the average time lag between academic research results and the first commercial introduction of new products and processes based on these results…" from 7 years to 6 years. Results consistent with those of Mansfield are reported by Cohen, *et al.* (2002), particularly noting the importance of university research to the pharmaceutical industry. But they also made the observation that universities are important not only in ground-breaking discoveries, but in research that aids industrial innovation in many industries and for all sizes of companies. They reported that almost one-third of industrial research and development projects from their sample made use of research findings from publicly-funded research and that knowledge originating from public research is often conveyed through consulting and other informal channels, although their results are tempered by the finding that other sources of knowledge, typically customers and competitors, are, in many cases, more important than the public research component.

Classifying Research Commercialization

Introduction

Many, and possibly most, commentators and researchers on commercialization tend not to discriminate between the various fields of research and the modes used in diffusing research results. For example, on the desirability of patenting, economics researchers often examine patents in aggregate and usually assume that the results of their analysis may be applied generally. However, an understanding of the commercialization of university research requires that it be segmented in two principal ways: by broad field of research; and by mode of diffusion. The need to segment university research in this way arises because, first, commercializing outputs from different fields of research (see next section) has distinctly different characteristics, and what applies to one may not (and often doesn't) apply to the others; while, second, the mode of dissemination used in relation to the outputs of the research will almost always have an impact on the costs, complexity and speed of getting the research results into the public domain.

Field of Research

University research may be categorized into three broad fields:

- Medical and health sciences and biological sciences (called here life sciences);
- Science disciplines (in particular information and communications technology, and other forms of engineering); and
- All other disciplines (including humanities, arts and social sciences, or HASS).

In general terms, life sciences account for about half the TT&C results published. For example, of the "100 innovations from academic research to real-world application" published by the Association of University Technology Managers (AUTM, 2007a), 48 involve medicine, pharmacology, health and biotechnology (rising to 52 if veterinary science is included), 16 involve electronics and IT (rising to 31 if all engineering disciplines are included), 8 deal with food and agriculture, while there are only 5 in humanities, arts and social sciences. Similar results can be found in other countries[3]. One of the reasons for the preponderance of life sciences in commercialization is the historically strong link between universities (and other institutes of medical research) and industry.

As a result of the extent of university-industry involvement in life sciences, this category of research generally tends to produce a return to universities greater than that in other research fields. While universities may not participate in life sciences research based on the revenue that it can generate, this difference is a factor that often distinguishes life sciences commercialization from that in other fields.

The next basis for this division of research outcomes can be seen in the market cycle. Life sciences discoveries generally take many years between discovery and market presence – ten to fifteen years is not unusual, largely as a result of strict regulatory requirements. For example, in 2006 Gardasil[4] reached the markets of the first world after having been patented in 1991. Innovations in the field of communications and information technology, on the other hand, are becoming measured in months rather than years. In many cases the issue of a patent for developments in communications and IT is not crucial in taking a discovery to market. The reason for this is that, by the time a patent is granted, the commercial life of the discovery may well have expired. In this case, any infringing conduct by an unlicensed user during the life of a product may be better settled by court-ordered damages rather than a prohibition against any continuing infringing conduct. In the case of humanities, arts and social sciences, patents and registered intellectual property rights are of limited or no use at all. Ideas originating from the HASS fields are unlikely to be protected to any significant extent by registering an interest – indeed, most ideas are simply not amenable to registration[5].

Finally, life sciences, in particular, is generally subject to a rigorous approvals and regulatory regime which other disciplines are not. For example, getting new medical technologies (such as pharmaceuticals) through regulatory approvals and into the market can cost in the order of one billion dollars (Goozner, 2004) and take in the order of fifteen or more years. This rigorous regime distinguishes life sciences from all other fields of research.

Analysis of patent data from Massachusetts Institute of Technology (MIT) suggests that the optimum means of commercializing from a robust patent is by licensing. University inventor commercialization is generally done as a result of some limitation in the patent (or, maybe, where a patent is not the ideal protection) and is a second-best option which achieves lower financial results. As Shane (2002, p.135) concluded from his data analysis:

"The results provide evidence that university inventors become entrepreneurs because of failures in the market for knowledge, suggesting that inventor entrepreneurship is a second-best solution to the commercialization of new technology. This view stands in contrast to the perspective of most of the entrepreneurship literature (and the popular press), which argues that independent entrepreneurship is a better mechanism for university technology commercialization than commercialization by established firms. This difference is important because theories in which independent entrepreneurship is considered the best approach to technology commercialization yield different implications from theories in which independent entrepreneurship is considered a second best approach".

Mode of Dissemination

The research commercialization process may be conveniently divided into four types, namely: knowledge diffusion, knowledge production, knowledge relationships, and knowledge engagement (Howard, 2005). Within these four categories there are numerous 'output indicators', of which only three involve TT&C dissemination: patenting and licensing (and income streams from them); spin-off company formation; and contract research and consulting. Other commercial activities such as university-industry joint ventures (Howard, 2005), and sporting teams and identity licensing (Bok, 2003), do not involve a TTO, and are excluded from the meaning of research commercialization for the purpose of this chapter. Also excluded are the broader non-commercial activities included in the so-called "third stream" activities of universities discussed by authors such as Hatakenaka (2005).

Table 1. Characteristics of commercialization activities

Research Commercialization Activities	Risk	Effort Required to Commercialize	Need for TTO/ Commercialization Office	Need for Early-stage Capital	Need for Venture Capital	Period Before Returns	Potential Returns
Consultancy	Minimal	Small to medium	None – legal assistance for contract	No	No	Short (months)	Based on time and skill
Contract Research	Some, but low	Small to medium	None – legal assistance for contract	No	No	Short (usually months)	Based on time and skill and special resources
Licensing	Generally low	Generally medium to high	Desirable: IP, contract, management	Often needed for proof-of-concept and a little beyond	No	Short-medium (months to years)	Variable. A great majority yield low results with some major exceptions
Spin-off Company	Increasing	Generally high: long-term commitment needed	Highly desirable	Almost always needed	May be needed	Usually long (years)	Variable. Many will fail, but some can be huge

Source: Authors

Thus, for the purposes of this chapter, there are four components to the university repertoire of research commercialization: consulting, contract research, licensing, and spin-off companies. These four components are consistent with the activities observed by the authors in the case studies.

In order to undertake a cogent analysis of university commercialization activities it is helpful to understand the practical activities in which universities engage. The activities which constitute commercialization, as it is practised in most universities, may be described under four headings, as shown in the left-hand column of Table 1. Table 1 shows not only a list of commercialization activities, but also a summary of the main characteristics of each activity.

Academic faculty members have been undertaking consultancy and contract research for many years, often before a TTO was contemplated. Consultancy and contract research generally offer little commercial risk to the staff or university, can be done usually without prejudice to other academic duties and provide reasonable and immediate cash returns to the service provider. Generally speaking, neither consulting nor contract research have any special capital needs outside those normally required for research. Universities in all advanced economies contemplate participation in consulting and the pursuit of private commercial activities by academic faculty by granting time-off within working hours to pursue such activities. Consulting and contract research would continue to flourish in most universities whether or not a TTO existed. In the United States and most of Canada universities assume no role in private faculty consulting and commercial arrangements while in some countries, such as Australia, TTOs often play no role in contract research (it usually being handled by an Office of Research or some similar entity). In many universities, and all the North American case study universities reported here, TT&C handled by a TTO embraces only the last two activities of this list: licensing intellectual property, and spin-off companies.

Research in industry has changed its character in recent decades. Many large company-owned research laboratories are gone, replaced by an emphasis on co-operation amongst companies and between companies and research entities like universities[6]. This is evident in every advanced economy and many industries. A manifestation of this in Canada, the Network of Centers of Excellence often involves several universities and several companies with some financing from government. Industry, or at least that part of it participating in the centers, obtains benefits from this more open approach to research, in particular reduced costs and improved access to research generated by the best investigators in the field.

It must be kept in perspective that only a small proportion of the output of research conducted by universities goes through a TTO. In the U.S., figures suggest that the proportion of university research transferred by way of patent does not exceed 7% at MIT, a premier commercializing university (Agrawal and Henderson, 2002, p. 45), and this may well be an overestimate.

THE ROLE OF UNIVERSITIES

The Role of Universities in National Innovation

The seminal work on the evolving role of universities in the national economy is by Etzkowitz and Leydesdorff (2001) who described the triple helix of university-industry-government relations. There are a number of consequences arising from the perspective they propound, in particular the expanding importance of the "knowledge sector in relation to the political and economic structure of the larger society" (p.155), and the re-structuring of each participant in the triple helix, such as increasing numbers of research centers in universities and strategic alliances amongst companies (p.156). The resulting increasing interactions among the three participants has led universities, in the view of some authors (Gunasekara, 2005, p.526), "… to initiate and drive agglomeration through knowledge capitalization projects, often with government and industry support".

One of the many ways that universities stimulate economic activity is through spin-off companies. But there are real gaps in understanding the extent to which spin-off companies benefit the economy, as described by O'Shea, *et al.* (2004, p.26), when commenting on the literature on this topic they said:

"... much of the literature has focused on a single university or on a very small number of institutions making it hard to draw any generalizations...As a result, the conclusions of much of the current research concerning university spinout performance may not be generalisable to other settings".

Heavy investment in university research in Europe apparently has not led to a commensurate increase in commercialization (Audretsch and Lehmann, 2005). These authors, using Germany as an example, concluded that (p.343): "Investments in German universities have ranked among the highest in Europe. Still, the ensuing commercialization has been disappointing". The correlation between investment and result is not always clear.

Japan, which has a strong industrial base, has only relatively recently begun to promote the university-industry link through formal structures such as the TTO, and the evidence suggests that it will take many years before this link will be functionally effective (Collins and Wakoh, 2000).

The Australian Productivity Commission (2007, finding 7.1) found that:

"... [the] policy framework for universities should be focused on maximizing the social return from public investment in R&D through the transfer, diffusion and utilization of knowledge and technology. The pursuit of financial returns from the sale or licensing of intellectual property, and the creation of university spin-off companies, while important pathways in their own right, should not be to the detriment of this overarching objective".

Barriers to TT&C Facing Universities

There are many factors that can constitute barriers to the transfer and commercialization of research from universities. These range from factors within the control of universities, to matters of government policy and legislation, through to the economic environment, social culture and other exogenous factors. A number of studies have identified barriers to research commercialization, and each seems to have prepared its own list of barriers.

In terms of barriers and challenges, the Lambert Report (Lambert, 2003) made some valuable points in relation to the U.K., which are equally relevant to North America:

- "The best forms of knowledge transfer involve human interaction…" (p.31). This is a major theme of the report and emphasizes the importance of individual relationships in successful commercialization;
- "… a lack of clarity over the ownership of IP in research collaborations" (p.4);
- "… the variable quality of [university] technology transfer offices" (p.5);
- "… too much emphasis on developing university spinouts…and not enough on licensing technology to industry" (p.5);
- "Universities are playing an increasingly important role in regional economic developments…" (p.5);
- "… proximity matters when it comes to business-university collaboration" (p.6), particularly when dealing with Small-Medium Enterprises (SMEs);
- "Business is critical of what it sees as the slow-moving, bureaucratic and risk-averse style of university management" (p.6), although it noted there have been improvements in recent years.

A paper by the Canadian Advisory Committee on Science and Technology (ACST, 1999), identified four principal barriers that prevent Canadian universities from achieving their full potential in commercialization:

- the absence of a coherent national university intellectual property policy (p.19);
- a lack of support and resources in university commercialization offices (p.21);
- uncompetitive business conditions in terms of taxation, Executive Share Option Plans (ESOPs) and the ability to invest pension funds (pp.21-22); and
- low levels of investment in university research (pp.22-23).

A paper by Adamson (2004) identified three key elements that need to be fixed in order that universities have effective commercialization:

- implement sound and effective commercialization processes;
- employ the right people in the commercialization office; and
- spread the rewards, including to the commercialization office staff, in order to create the right motivation.

A report by the Australian Government (DEST, 2003) on national and international research-business linkages identified three frequently reported concerns (p.5):

- cultural and operational differences between the public and private sectors, which can impede collaboration;
- a lack of visibility of Australian research and development to international players; and
- a limited capacity of small and medium sized firms to connect with Australia's science, engineering and technology base.

Other issues identified in the literature for difficulties experienced in some countries by universities engaging in TT&C include (ACST, 1999; Ferris, 2001; Zhao, 2004):

- poor university governance;
- the lack of a coherent framework describing ownership of intellectual property;
- the lack of clear ownership of intellectual property generated by universities;
- a culture of risk aversion;
- a tradition of trade barriers for the purpose of job protection, leading to a self-defeating culture of reliance on protection rather than innovation;
- a pervasive cultural cringe – things are always done better elsewhere;
- a culture of social egalitarianism leading to an aversion to individual success (except in sport) – the so-called tall poppy syndrome.

Just as it was identified earlier that different types of research may need different approaches to TT&C, there may well be barriers that are particular to individual types of research. The most obvious area of research with potential for distinctiveness is life sciences because of its particularly high regulatory barriers and long lead times to market.

There are also the barriers that arise from industry, such as a lack of understanding by industry of the process of academic discovery, cultural and gender barriers between academe and industry, conflict over financial incentives, conflicts of interest, insufficient university resources, and insufficient university time (Sobol and Newell, 2003). There is also an arguable position that the continuing obligation of universities to a common good may limit their ability to commercialize (Argyres and Liebeskind, 1998).

At the same time, commercialization through intellectual property licensing and spin-off companies, which normally involves a registered interest in intellectual property, has to invoke formal procedures. The difficulty here from the academic faculty viewpoint is the effort required to make disclosures versus the benefits obtained. Many academics view the disclosure process as unnecessarily long and involved and of peripheral relevance to them, particularly given that they can obtain better career returns by engaging in other activities such as publication. A further important factor at work is the perceived efficiency of a TTO in handling disclosures and its success at commercialization. The balance of the equation in favor of commercialization can depend critically on individual TTO performance, as demonstrated by successful TTOs. Owen-Smith and Powell (2001) attributed the variations in commercialization success to faculty perceptions of the benefits of TT&C, the quality of the TTO, and the institution (through its history, environment, capacity and reputation) as a collective enterprise. Owen-Smith and Powell (2001, pp.112) found that: "… inconvenient or frustrating interactions with TTOs may be enough to convince ambivalent inventors that the benefits of IP protection do not outweigh the costs", but that success, on the other hand, is self-reinforcing (p.113):

"Where faculty are highly aware of other's successes, prestige is associated with commercial success. When academic and commercial rewards are linked, incentives to patent are enhanced. In this kind of setting, frustrations with the patent process may be overcome by the general positive reputation of the multiple benefits of IP protection and even ambivalent inventors may begin to disclose".

From this summary and analysis a list of barriers and challenges that confront universities attempting to undertake commercialization can be developed, as follows:

- Linkages and relationships between the three principal classes of actor: researchers, industry and financiers (usually identified as the most important factor);

- University legislation, policies and procedures (intellectual property policies; academic promotion policies; risk management; managing conflicts of interest; and human resources policy rigidity limiting opportunities for university-industry staff exchange);
- University support for commercialization through such activities as establishing effective frameworks within which commercialization operates as well as mechanisms to identify, capture, protect, disseminate and exploit the ideas created through research;
- Ensuring that the costs of commercialization are realistic compared to the benefits obtained, financial and otherwise;
- Ensuring that the historical mission of the university to encourage the free-flow of information is not compromised, consistent with the need to ensure that the community obtains maximum benefit from novel ideas with commercial potential;
- The size and scale of research and industry clusters extant in each field of research present within individual countries;
- The relative lack of entrepreneurial and commercialization skills and propensity amongst university academic faculty;
- The lack of capital available for the commercialization of ideas, particularly during early-stage development;
- A fiscal regime that may not be supportive of commercialization, particularly given the mobile nature of capital;
- An intellectual property regime (covering legislation and registration of intellectual property) that may be complex and expensive and does not always ensure clear-cut ownership of intellectual property;
- A National Innovation System (NIS) that is less supportive of TT&C or which promotes research in fields where local industry may lack the capacity to absorb the volume of new ideas generated;
- A commercialization regime that is oriented towards research-intensive universities and offers little recognition to the diversity found among university missions and capabilities; and
- A local industry that lacks depth, may be risk-averse and, in some important areas, foreign-owned and inclined to seek innovations in their home countries.

From this list of barriers the authors developed fourteen environmental factors which are used to characterize the case study universities in order to allow comparisons and contrasts to be drawn between universities individually and within particular countries:

- The regulatory environment;
- The character of the university;
- The TT&C structure employed by the university;
- The TT&C processes used by the university;
- The TT&C performance of the university;
- The TT&C performance of the TTO (as distinct from the university);
- Knowledge and awareness of TT&C among faculty members and researchers;
- The incentives offered by universities for faculty members and researchers to engage in TT&C;
- The nature and strength of the linkages between the university and industry;
- The impact technology parks and incubators on a university's TT&C performance;
- The impact of intellectual property rules;
- The availability of financing for development of ideas – both at the early stage and later venture capital finance;
- The entrepreneurial culture of the university and its impact on TT&C; and

- The role played by intermediaries in promoting and supporting TT&C.

CASE DESCRIPTIONS

Case Study Universities

The four university case studies reported in this chapter were prepared in 2007, and are identified in the following analysis by the following terms, with quotes from staff of the various TTOs similarly identified:

- United States Private University (U.S. Private University)
- United States Public University (U.S. Public University)
- Canadian University (Canada Uni1)
- Canadian University (Canada Uni2)

The U.S. universities are differentiated by their respective public/private status, however each is a large and well-regarded institution: one each from the east and west coasts. Canadian universities are generally public institutions, and both Canada Uni1 and Canada Uni2 are eminent public universities offering a full range of course options. The Canadian universities are located respectively on the east and west coasts.

Additional opinions were offered by a range of experts in the field of university TT&C and are identified where they occur in this chapter.

Common Themes

There is a view that universities in developed economies are changing and becoming more "entrepreneurial". For example, the school of thought identified by Etzkowitz and others in which commercialization is often characterized as the 'third mission', or a part of a 'triple helix' of activities embracing government, industry and universities (Etzkowitz and Leydesdorff, 2001).

While this is generally true, our research has demonstrated that there is no 'typical' university commercialization model, and there is a great range in the way that TT&C tasks are completed by universities. For example, one university, Imperial College London, not only spins-off companies from the university, but has spun-off its TTO to form Imperial Innovations plc which is listed on the London Stock Exchange[7].

Some governments provide assistance to TTOs in the form of finance for commercialization such as in the U.K. under University Challenge funding. In the U.S., on the other hand, most funding comes from private sources with the encouragement of government through schemes such as the Small Business Innovation Research Program (SBIR) and the Small Business Technology Transfer Program (STTR) (explained below).

Some universities have committed substantial capital funds through the invention-innovation cycle from early- to mid-stage development through to venture capital to bringing an idea to market. For example, the University of Arizona, which claims the largest student body in the U.S., has substantial venture capital available through its commercialization company, Arizona Technology Enterprises LLC, of up to $10 million at the proof-of-technology stage alone. Some universities commercialize through wholly-owned companies, while most do it through unincorporated entities that form part of the university itself.

One persistent theme in university TT&C is the dominance of life sciences (Vitale, 2004), the other significant fields being physical sciences and engineering, and Information and Communications Technology (ICT). This was noted above, but it is also evident from the case studies. For example, a prominent United States venture capitalist noted that "[we] would rather have … a biotech company or an IT company because the payoffs are currently bigger, whether you're selling the company" or an Initial Public Offering (IPO), or whatever, "because the valuations, whatever multiple you are using: cash flow, earnings, revenue, or whatever, are higher in IT and biotech. The reason for that is that those kind of companies have been able to be grow faster". Similarly, U.S. Public University reported that while life sciences has declined in the past five years as the major source of inventions, it still represents about 50% of disclosures, with the majority of the balance being physical sciences, particularly IT and engineering. And this trend is confirmed by the experience of Canadian Uni2 where all eleven professional business managers in the TTO are in three fields: life sciences, physical sciences & engineering, and ICT.

There are probably four major reasons why U.S. universities have led the world in TT&C. The first relates to the ability of U.S. universities, certainly the private universities, to fund TT&C activities because of their substantial financial strength arising from a history of philanthropy for the benefit of universities, a factor that continues to the present day. As well as providing a high degree of financial independence, this endows universities with a distinctive entrepreneurial streak. The second reason is the consistency of national public policy as it concerns university TT&C. Whether or not the *Bayh-Dole Act*[8] creates an ideal policy, and whether or not it needs amendment to adapt to changing circumstances:

"... [one] of the great strengths of Bayh-Dole was that it provided no additional funding, therefore it never needed re-authorization... and so we've had in the U.S. the benefit of twenty-seven years now of consistency of policy" (U.S. Private University).

The third reason is the substantial financial support provided to universities from federal Government agencies, especially those in defense, health and energy, while the fourth is through the consistent and highly supportive policies of the federal Government through the SBIR and STTR programs.

A fifth, and somewhat less corporeal, reason for the success of university TT&C in the U.S. is the culture of entrepreneurialism that permeates the society. While not unique to the United States, it is a particularly potent element of the national psyche and colors much public and private decision-making.

North American Universities

Universities in the U.S. have become the *de facto*, worldwide standard for what Clark (1998) termed the *entrepreneurial university*[9]. The apogee of research commercialization in North America is probably represented by two particular universities in the United States: MIT and Stanford. These two universities (among others) were successfully commercializing research through licensing[10] and spin-off companies for decades before the introduction of the *Bayh-Dole Act* in 1980, and still represent international best practice in commercialization. The universities of Columbia and California are also notable as commercialization leaders among North American universities and provide some useful lessons, while the Wisconsin Alumni Research Foundation (WARF) remains one of, if not the, premier exemplar of life sciences commercialization (although it now does much more) in North America since its foundation in 1925. There are also individual universities that have developed leading attributes as a result of community inspiration, such as AZTE (Arizona Technology Enterprises LLC), a wholly-owned subsidiary of the Arizona State University Foundation that is responsible for technology commercialization from the University of Arizona, and which includes substantial associated funding for ideas throughout the process, from the laboratory through to industry acceptance.

The following sections of this chapter describe the characteristics, practices and experience of the four universities noted above: two in the United States and two in Canada[11]. They also synthesize information obtained during other interviews in the United States as well as from literature and public sources. To put North American practice into perspective, there is first a description of the environment in which the case study universities operate.

General Environment

The United States has held world technological pre-eminence since at least the end of the Second World War but, largely because of the globalization of industry, its marginal advantage is becoming muted. This loss of leadership is recognized by many eminent bodies in the United States such as The National Academies of Science (NAS, *et al.*, 2006) and the Domestic Policy Council (2006). It has caused the United States considerable concern which it is addressing through actions such as the American Competitiveness Initiative which was announced by the President of the United States in his State of the Union address in 2006[12].

According to the Carnegie Foundation for the Advancement of Teaching[13], the United States has 284 universities among a total 4,386 post-secondary institutions. Because of inconsistency in the use of names such as university, college, institute and the like in the U.S., the actual number of institutions equivalent to a university in other countries would be much higher than 284. For example, there are approximately 214 university members of the National Association of State Universities and Land-Grant Colleges (NASULGC) and 430 university and college members of the American Association of State Colleges and Universities (AASCU), the dominant university representative entities in the U.S., apart from the Association of American Universities (AAU). There is overlapping membership amongst these organizations, so that there are certainly well in excess of 400 universities as such in the U.S., with a further large number of specialist medical, law and other colleges offering undergraduate and graduate training and conducting research.

According to membership data prepared by the Association of University Technology Managers (AUTM) there are approximately 326 universities or colleges of higher learning that have at least one person who is a member of AUTM. This represents a large proportion of "universities" with staff who are members of AUTM but, if taken across the spectrum of all post-secondary education institutions, it is quite a low proportion (around 7%)[14]. At least in part this low proportion may be attributed to the fact that there are a large number of colleges and post-secondary institutions in the United States that are not actively engaged in post-graduate education or research but, even allowing for this, the number appears surprisingly low. For comparison, some 49 Canadian universities employ technology transfer staff who are members of AUTM. This means that of Canada's approximately 92 universities[15], over half appear to be actively engaged in technology transfer[16]. In fact, since 1998 all but the smallest Canadian universities and university colleges have had active TTOs (Fisher and Atkinson-Grosjean, 2002).

Turning to research intensive universities, the AAU represents sixty United States and two Canadian universities and focuses on issues that are important to research-intensive universities. Sixty research-intensive universities out of a population of approximately 400 universities represent a proportion of 15%. But whether we are dealing with 400 universities or 60 universities in the U.S. that are concerned with technology transfer, or some number in-between, the number of universities active in the technology transfer environment in the United States is substantial.

AUTM provides the best industry figures available for the United States, and these show the relative importance of technology transfer to the United States economy. For fiscal year 2005, the latest information available, in the United States alone, the contribution to the economy is estimated to be at least (AUTM, 2007b):

- $42 billion plus in R&D expenditures at U.S. academic centers;
- 4,932 new licenses signed in 2005;
- 28,349 current, active licenses;
- 527 new products introduced into the market in 2005;
- 3,641 in the 8 years from FY98 through FY05. That is 1.25 new products based on academic inventions introduced every single day over the last 8 years;
- 628 new spin-off companies created in 2005. That is 1.7 new companies every day of the year. Each is based on what is hoped to be a platform of academic technology that will address market needs through the application of invested money by well-paid employees;
- 5,171 new spin-offs since 1980. That is more than one company every two days during 9,133 days of innovation.

But the results are highly concentrated in a limited number of institutions. For example, of the approximately $1,593 million in licensing income reported for the 2005 year (AUTM, 2007b, Data Appendix) across 153 universities, almost half ($720 million) is attributable to two universities, with 70% attributable to eleven universities, and 78% to sixteen universities[17]. While, clearly, a large number of universities did not participate in the AUTM survey, the available data show that license income is highly concentrated. Thirteen of the sixteen universities that account for 78% of licensing income are member universities of the AAU.

Commercializable research results are highly concentrated in one field as well. For two years about 1996-1997 AUTM collected data on the proportion of commercialization in various fields and "they found that about two-thirds of the activity and about eighty percent of the income was from life sciences" (U.S. Private University). This is reflected in the research funding available in the U.S.: "... the NIH has a $26 billion extra-mural research budget, NSF has $5 billion...just follow the funding". The DoE and Defense budgets, while large, are not easily allocated into categories (U.S. Private University[18]).

AUTM data for Canada disclosed that for fiscal year 2005, the latest information available, the contribution of university TT&C to the Canadian economy from the thirty-six institutions that responded to the survey is estimated to be at least (AUTM, 2007c):

- Over $4.2 billion in R&D expenditures at 36 Canadian institutions;
- 565 new licenses and options signed in 2005;
- 1,433 invention disclosures and 685 patent applications (an increase of 20% over 2004) in 2005;
- 37 new spin-off companies created in 2005;
- License income of $52,863,816.

As in the United States, Canadian results are highly concentrated in a limited number of institutions. For example, of the approximately $4,234 million in research expenditure reported for the 2005 year (AUTM, 2007c, Data Appendix) across 36 institutions, almost two-thirds ($2,764 million) is attributable to nine universities. Of the $52.863 million in license income reported, over 50% is attributed to just two universities[19], while five institutions are responsible for over 70%. While, again, a large number of universities did not participate in the AUTM survey, the available data show that most indicia are highly concentrated. Similarly, the fields in which Canadian research is producing results are concentrated. At Canadian Uni1, life sciences accounts for 58% of disclosures and 51% of spin-off companies with the balance in physical sciences and IT (Canadian Uni1) which is typical of universities actively involved in life sciences research.

Canadian members of AUTM noted a common problem faced by TTOs in developed economies (AUTM, 2007c, p.12):

"...more than 50% of the personnel working [in Canadian TTOs] have five years or more experience. Yet on the other hand, retirements and other changes of employment status have recently caused upheavals in some offices. Anecdotal information continues to point to a challenge in attracting personnel well-suited to the peculiarities of the profession".

Industry experienced in dealing with the best United States universities believes that "... [it] is absolutely crucial, in today's world, for a university that does a lot of R&D, to establish a good tech transfer office" (U.S. Management Consultant Interview).

Despite the financial power of the large corporations in the United States and the impact of sponsored research from industry, the government still supplies something like 70% of funding for United States university research (U.S. Private University) which means that the government, in particular the federal Government, exercises great control over the conduct of research.

Regulatory Environment

A number of Acts and instruments intrude into the U.S. university research commercialization environment, the most important of these being federal legislation, and include the *Bayh-Dole Act*, the Small Business Innovation Research (SBIR) Program and Small Business Technology Transfer (STTR) Program[20].

As it concerns this chapter, the principal change brought about by the *Bayh-Dole Act* in 1980 was to vest title in intellectual property arising from federally-funded research in the university that undertook the research subject to certain conditions being met. This altered the previous position where, subject to agreement to the contrary, the federal agency that funded the research (mainly Defense, the Department of Energy, the National Science Foundation and the National Institutes of Health) took title to resultant intellectual property. Prior to the enactment of Bayh-Dole approximately seventy-five universities[21], including Stanford and MIT, had negotiated agreements[22] by which the university took title in the intellectual property developed, so that they were already engaged in research commercialization prior to the *Bayh-Dole Act*. Nonetheless, the Act had the effect of encouraging more U.S. universities to establish TTOs as evidenced by the statistics compiled by AUTM – there being some 21 TTOs prior to 1980, and at least 152 by 2005 (AUTM, 2007b).

The *Bayh-Dole Act* also had the intent of encouraging the exploitation of a university's intellectual property through small business. According to AUTM (2003), it has had this effect such that by 2003 United States universities, hospitals and research institutes reported 65.3% of executed licenses and options were with start-up and small companies combined, and 34.6% were with large companies, as compared with 66.6% and 33.4%, respectively in 2002.

According to the U.S. Small Business Administration (SBA) having the responsibility for implementing the SBIR program:

"[the] statutory purpose of the SBIR program is to strengthen the role of innovative small business concerns (SBCs) in federally-funded research or research and development (R/R&D). Specific program purposes are to: (1) stimulate technological innovation; (2) use small business to meet Federal R/R&D needs; (3) foster and encourage participation by socially and economically disadvantaged SBCs, and by SBCs that are 51 percent owned and controlled by women, in technological innovation; and (4) increase private sector of innovations derived from Federal R/R&D, thereby increasing competition, productivity and economic growth"[23].

The SBIR commenced in 1982 and was reauthorized to continue until 2008. It is open to all for-profit, American-owned, and independently operated businesses with 500 or fewer employees. Federal agencies are required to set aside 2.5% of their extramural budget for domestic small business concerns to engage in Research/ Research and Development (R/R&D) that has the potential for commercialization. Under the SBIR program, a company identifies research it needs to have done and writes a 25-page proposal. If the proposal is accepted, the company will receive up to $100,000 for Phase 1. The company can subcontract up to 30% of Phase 1 funds to research departments, including university departments. If Phase 1 research goes well, the agency may invite the team to submit a Phase 2 proposal. Phase 2 proposals cover two years and can provide up to $750,000. Subcontractors such as universities can share equally in these grants. Phase 3 contracts are also available. This money comes from private companies or other government agencies[24].

The importance of the SBIR program to biotechnology development in the U.S. was highlighted in evidence cited by Audretsch (2001). He noted that the SBIR program at the National Institutes for Health for 1999 alone exceeded $300 million. Similarly, the United States Department of Defense used the SBIR program to fund biotechnology, and invested over $240 million between 1983 and 1997. He said (p.13): "There is compelling evidence that the SBIR Program has had a positive impact on developing the U.S. biotechnology industry…".

But the SBIR Program is not limited to biotechnology alone, and has been described as the world's largest seed capital fund and the secret to a great deal of the United States' eminence in high technology through support for young ventures (Connell, 2006).

In addition, the United States Administration has made an Executive Order (Number 13329) the purpose of which "is to ensure that Federal agencies assist the private sector in its manufacturing innovation efforts"[25].

Character of the University

It was noted earlier that a significant majority of the universities responsible for the bulk of licensing income from research commercialization are member universities of the AAU. This result is not surprising given that AAU members comprise most of the United States' research-intensive universities, and suggests that a benchmark model of United States universities should embrace many of the characteristics of member universities of the AAU. This conclusion is subject to the important exception that some universities not members of this group are recognized as good at research and commercialization, such as shown by Tornatzky, *et al.*, (2002)[26].

There are apparent differences between public and private universities in both their attitude and the degree of support that they provide to TT&C. Private universities appear more inclined to see TT&C as a desirable activity while public universities are not always prepared to embrace TT&C with enthusiasm. As U.S. Public University said: "…we give [faculty members] the freedom not to be commercial. I think [our university] has a bit of a history there, that there is some real concern on campus even right now at the level of commercialization that we do" (U.S. Public University). This general attitude is reflected, too, in the availability of funding through the university to support early-stage and later-stage development of discoveries and inventions although, in the case of U.S. Public University, being located in California, with access to some of the world's deepest pools of venture capital, this is unlikely to present an insurmountable problem.

The first, and arguably the key, criterion against which universities were assessed as successful in TT&C in the Tornatzky, *et al.* (2002) study of commercializing universities in the U.S. was the Mission, Vision and Goal Statements. For without genuine belief in the importance of TT&C as a core value of the university, it is unlikely to develop; MIT had its Vannevar Bush and Karl Taylor Compton, and Stanford its Frank Terman. This view is no less recognized by the Canada Foundation for Innovation (2002, p.11) when it said "… success is…affected by the level of commitment from the universities themselves to commercialization". North American universities that are successful in TT&C genuinely believe in its importance to the community and the university. The next criterion for success is the establishment of a TT&C environment to which the university adheres for a period of many years, most likely measured in decades, to allow the faculty members, industry and TTO staff to develop relationships and processes that work and become understood by all participants. U.S. Private University observed that MIT is a unique institution - it has no medical school but is one of the best performing TT&C universities in the United States. The Director of this private university TTO said: "…[MIT has] had the same management team running technology transfer for twenty years, so they adopt a consistent set of philosophies and principles and investment. I think MIT is a unique institution" (U.S. Private University). And probably the third most important criterion is an adequate volume of high-quality research.

There is also an apparent generational change of attitude and approach to university TT&C, where younger researchers are more inclined to seek external, non-institutional, funding and are more willing to engage in TT&C.

Commercialization Structure

A common characteristic among North American TTOs is their diversity of structure. The Canadian Prime Minister's Advisory Council on Science and Technology, ACST (1999, p.11) reported that Canada had:

"...a wide range of organizational models for operating offices. Some are owned and operated by the university; some are owned by the university but managed by arm's length corporations whose activities are guided by boards of directors; some models involve a hybrid whereby innovation responsibilities are shared between in-house expertise and outside experts; and some universities collaborate in designing shared infrastructures while others establish their own infrastructure".

It went on to say: "Each model has merit and each university requires the flexibility to endorse the model that best meets its unique circumstances".

U.S. Public University provides a typical example of a TTO at a United States university. The university created one office to provide a "one-stop shop" for industry research partners to interact with the campus. This office reports to the Vice-Chancellor for Research and consists of two groups: one dealing with technology licensing (12 employees), and one with industry alliances (7 employees). Prior to 2004, there were two separate offices: a technology licensing office; and a sponsored projects office.

"One of the problems that we noticed, just a couple of years ago, basically, when we reorganized under one office, was that the Sponsored Projects Office really focused mostly on federal grants", such as NIH and NSF grants, "and they didn't service the industry-sponsored research as well as they could have". The common management of the two "helps a little bit in terms of keeping closer contact between [the two offices], with the expectation, of course, that those sponsored research projects tend to generate significant licensing opportunities for the university because they are more directed towards ways in which particular companies are interested" (U.S. Public University).

Similarly at U.S. Private University which employs 14 people in TT&C, almost half (6) are in technology licensing. The first TT&C function established by the university was the Venture Fund:

"... which was started in late 1974, first investment 1975; and technology transfer was added a couple of years later; that was it until 1997 when we added entrepreneurial assistance. Incubation was going on elsewhere in the university and was not brought under our control until 2006, and Corporate Business Development was about a year ago, too" (U.S. Private University).

Even large sophisticated universities such as Canadian Uni2, which began commercialization activities in the 1980s, have struggled to find the optimum TT&C model. For about twenty-five years until 2006 it undertook commercialization through a separate but wholly-owned company. For most of its life this company had a tenuous financial existence because it was expected to become self-financing through its commercialization activities. When it was clear that this was not going to happen, the company was given a degree of financial independence in 2000 through the university extending a $10 million line of credit. When, by 2006, the line of credit had been exhausted, the university elected to terminate the mandate to commercialize research held by the company and return the function in-house under the management of the Vice-President (Research) (Canadian Uni2). This is an example of the difficulties experienced by universities as they work to find the best working model for their particular circumstances, and the fact that one model is unlikely to work in all circumstances.

Canadian Uni1 TTO is a unit within the office of the Vice-President Research. The TTO commenced as a part of the Office of Research Services in the 1980s and has always been a unit of the university. In the mid-1980s this office extended its brief to include industry liaison with the aid of 5 years' funding from the Provincial Government. At the time of formation it was expected that this Office would be self-funding within 5 years. In 1987 the TTO was separated from the Office for Research Services as a separate unit. The university has a wholly-owned for-profit company for the purpose of allowing the TTO to do things that require a corporate structure such as limiting liability in some cases, permitting access to some federal funding programs available to industry (but not universities, as such), and to undertake some hands-on business incubation prior to spinning-out particular companies (Canadian Uni1).

Of the 151 United States universities that responded to the most recent AUTM survey (AUTM, 2007b, p.18), half reported five or fewer full-time staff, with one-third reporting three or fewer[27]. The mode in respect of staff employed in the TTO was six, with a range between one and fifty, reinforcing the grouping towards the smaller office. Only two TTOs reported a staff number greater than fifty.

Most North American universities operate on the model of a single office where all the TT&C resources are located centrally. This approach is reflected in (or maybe results from) the relatively small number of staff at most North American TTOs. The typical number of employees at a United States university TTO is six, a number too small to accommodate any practical alternative structure.

Where the function is not outsourced there are two practical alternative models to the central office where the size of the TTO permits it: the "hub and spoke" model and the "front-office/back-office" model. The hub and spoke model involves having a central resource of capabilities in such areas as intellectual property management and contract negotiation (the hub), supplemented by business managers embedded in university research areas on a permanent basis (the spokes). Because it embeds TTO staff with researchers across the range of university technologies, the hub and spoke model requires a large number of staff, which means that it is an option only open to universities with substantial research expenditure and the number of disclosures to justify the cost. There has been considerable debate in North America[28] on the issue of structure with no general consensus as to the optimum model for any given circumstance. Some, like the University of Michigan, operate a spoke only in one or a few academic departments. Each of the U.S. universities interviewed for this study operate principally along the central office model.

The front-office/back-office (FOBO) model used by the TTO at Canadian Uni1 is a variant of the hub and spoke model. The university tried and subsequently abandoned the hub and spoke model; the university said that:

"We tried for a number of years with the hub and spoke where we would have an individual down at various sites and what we found was that that doesn't work very well. It doesn't work very well because the individual is isolated from everybody. He is not part of the home base, and a lot of what people learn here they learn around the water cooler..." (Canadian Uni1).

The FOBO model involves central resources in the back office, like the hub and spoke, but employs groups of co-located business managers that specialize in particular fields. For example, the life sciences front-office is located at one hospital affiliated to Canadian Uni1, but works with all other affiliated hospitals. In this way the university believes the group obtains the benefits of working together as a group (rather than being embedded with the researchers as in the hub and spoke model) while still being identified by the target researchers as a specialized group to which they can relate. As the university said:

"...around 2001 we set up a formal hospital office, and that's were we played on the front-office/back-office component. And the front office down there at that time had six [employees] now has seven or eight people in it, and that creates a large-enough group that they have enough synergy amongst themselves to be self-supporting, and yet they still have to come back here in terms of finance, patent management, contract management, internal legal services, communication, all of that back-office stuff" (Canadian Uni1).

Staff at larger North American TTOs generally possess outstanding academic qualifications. At U.S. Public University, life sciences licensing officers are all PhDs, while licensing officers in physical sciences possess more of a business background with some MBAs and patent experience. Canadian Uni2 TTO staff all possess technical degrees, with all life and physical science staff possessing PhDs, while at U.S. Private University, of the thirteen professional staff involved in licensing, six hold doctorate degrees (and another an MD), three hold MBAs (including two of the PhDs), and two hold Master's degrees, while two are paralegals. Similarly at Canadian Uni1, where the thirty-five TTO staff include nine PhDs, five MBAs, two lawyers, and one accountant, with all other professional staff holding a technical qualification including some up to Masters level. Staff possessing this level of training and experience are difficult to find in the numbers needed and are difficult to retain. They are also an expensive resource, and universities experience challenges in remunerating staff in some cases, particularly where the TTO is a unit within the university structure and obliged to pay salaries on a par with other faculty members and staff. This is in contrast to TTOs that have been structured as companies where, generally, management has greater flexibility in the amount and way it can remunerate staff. Each of the North American case study universities commented on the difficulty of appropriately paying staff and the attraction, in this regard, of having a commercialization company rather than a TTO within the structure of the university. The challenge presented by the structure chosen was expressed by Canadian Uni2 when it said:

"With the internalization of the group, it really takes a set of functions that could probably be better managed with the flexibility to provide incentive instead of putting people on a salary grading, and I think the impact of that will be that we will have, as long as we have commercialization and tech transfer responsibilities internalized within the university, we will have a lot of churn and we will be in a constant training mode with people who come in, learn the business and then spin-out to be part of the opportunities... Paying people who are very entrepreneurial on an incentive basis is very important and we don't have that flexibility" (Canadian Uni2).

One critical weakness regularly identified by most TTOs in their skill-set is industry experience. Canadian Uni1 said succinctly: "I think an area where we would be light in terms of qualifications is in deep industry experience" (Canadian Uni1). And it is not sufficient to fill this gap with a person who has simply "worked in industry". When Canadian Uni1 TTO employed a person out of big industry he obviously floundered in the environment of a small-medium entrepreneurial business which is characteristic of most TTOs.

The difficulties attending the structures presently used by TTOs in North America invite consideration of alternatives. The paradigm is changing, particularly with the emergence of open systems innovation and the other pressures arising from government desires to use research results to enhance economic development and address the grand challenges arising from the likes of homeland security, health and environmental concerns. This leads to pressure on TTOs to behave less like technology transfer offices and more like offices of innovation and commercialization, a role for which they are often ill-equipped largely because only a handful of staff have sufficient skills in industry. What is needed may be a fresh view – to see the TTOs as a part of the university's research enterprise.

The future involves an acknowledgement of the new and different roles facing TTOs. An example of a new intermediary is the TTO Director at the University of North Carolina (UNC) who has, basically a title that includes technology transfer, commercialization and economic development.

A continuing conundrum for universities in Canada is the source of funding for the TTO. Most TTOs in Canada (as in the United States) will not show an accounting profit and have to be subsidized in some way. Even TTOs on their way to profit within research-intensive universities take many years after their foundation to achieve profitability, and therefore need funding to support them in their formative years. By way of example, Canadian Uni2 has re-formed its TTO and is working towards demonstrating its profitability. As the TTO Director said:

"...[because] we have had, traditionally, to draw the resources made available for commercialization in [our Province] from the general revenues of the university, there is an argument that that's taken out of the classroom, particularly when commercialization wasn't as focused-upon as it has been. There's a lagging funding for commercialization activities relative to government policy orientation on it, and we're drawing from a very weak base... we're quite below the national average on per-student funding" (Canadian Uni2).

This demonstrates the difficulty in funding TT&C activities, particularly in the early years. Governments expect universities to perform TT&C activities for the common good including stimulating regional and national economic development, even when TTOs are financially marginal or unprofitable, while supplying no specific funding for the activity. Universities either take this from endowment (where it exists and is permitted) or recurrent funding. Most government recurrent funding is for teaching so that the university may have to appropriate some of these funds to the purposes of TT&C, something that is made more difficult when, as in the case of Canadian Uni2, it already receives comparatively low levels of funding for teaching.

In order to remedy the perceived deficiency in TTO staffing numbers and skill levels in Canada, the ACST (1999, Recommendation #3) recommended that, in addition to current spending on technology transfer, "... [the] federal government should invest new and additional resources to strengthen the commercialization of universities in an amount equal to 5 percent of its investment in university research...". Such an amount would equate to $50 million in 1999 dollars (being 5% of the $1 billion invested in research by the federal government), a significant sum.

Commercialization Process

Just as there are a variety of TTO structures among North American universities, there are differences in the way that TT&C is performed. Factors affecting the process include whether the university owns the intellectual property (not an issue in the United States in the case of federally-funded research, but still an issue in some Canadian universities), the availability of university early-stage funding, the size of the university, the size and nature of the medical and life sciences schools (life sciences represents the majority of licensing, and the presence of a medical school can affect the ability of a university to conduct some development such as clinical trials), the presence and use of intermediary firms, the availability of venture capital, and the preferred route for commercialization (licensing or spin-off).

TT&C starts with disclosure to the TTO by a faculty member of a relevant invention or discovery. While researchers are often obliged by university rules to make disclosure of discoveries and inventions, generally there appears little concern on the part of TTOs in the event that disclosure is not made and a faculty member chooses instead to publish, with the exception of sponsored research where the results may not be the faculty member's alone to make public.

Each disclosure is then subject to an examination, test or gate of some description to determine whether it has commercial potential to justify the expenditure of effort and money to patent. At U.S. Public University, when deciding whether and how an invention should be commercialized following disclosure to the office: "... we look at a number of aspects, [the first] would be a technical evaluation of the disclosure in terms of 'is it patentable?', 'is there a lot of prior art?', 'what potential value do you see, just as a patent?'" (U.S. Public University). This decision is made by the licensing officer in conjunction with the faculty member. Conversely, U.S. Private University will usually file a patent upon disclosure in order to preserve a position and then wait for some months before making a decision, based on further discussions with the faculty member on whether the patent should be pursued.

Canadian Uni2 has a rigorous process for life sciences in particular:

"We have been accepting, over the past few years, only 10% of the invention disclosures that we see. So our screening process has a fairly fine grid, and when we accept them, we think they have a fairly good chance for commercialization - it is only these that are going to go to the review panels" (Canadian Uni2).

Canadian Uni2 has a large proportion of life sciences disclosures because it has an associated teaching hospital with faculty cross-appointments between the university and hospital. Canada also has a large proportion of young life sciences companies, a major reason being that spin-offs tend to be favored in Canada because of the general lack of receptor companies.

Canadian technology transfer professionals and scholars are of the view that commercializing university research should rest in the first instance with the university (Canada Foundation for Innovation, 2002, p.20).

Canadian Uni1 prefers the licensing route for commercialization over a spin-off where this is practical:

"A spin-off is going to take to close to ten times the amount of work and effort, [while], in terms of the return, we're not doing this based on financial return as our sole indicator, so the spin-off has economic return, it has political return and, in fact, it does have, in many cases, a very strong financial return" (Canadian Uni1 Interview, 2007).

On the other hand, Canada is said to have a policy favoring the establishment of spin-off companies over licensing (Riddle, 2004). Clayman and Holbrook (2003) analyzed data on university spin-off companies since 1995 and reported that Canada appears to be successful in keeping spin-off companies going, with a 73% survival rate. They also reported that biotechnology is the major focus of spin-off companies, at 52% of all spin-offs.

But not all academics embrace TT&C uncritically, and there is some disquiet about United States universities engaging in ever-more commercialization. Press and Washburn (2000) expressed alarm that sponsored research is putting at risk the paramount value of higher education - disinterested inquiry, while universities are operating ever more like for-profit companies. Nelson (2001) expressed concern at the potential compromise commercialization is imposing on open science. On the other hand, Thursby and Thursby (2002) suggested that the increasing emphasis on TT&C in United States universities has brought about a culture change rather than compromised research integrity: "... to an increased willingness of faculty and administrators to license and increased business reliance on external R&D rather than a shift in faculty research" (p.90).

Every university involved in TT&C seems to have developed its own commercialization process. The University of Virginia Patent Foundation is a leading example of an institution that employs a sophisticated technology transfer process. Its operating manual sets a benchmark for commercialization processes (UVAPF, 2004) that provides a useful benchmark against which to test models involving ideas created otherwise than by contract research. At the same time, it is interesting to observe that this process is quite similar to that necessary for any ordinary business decision. For example, two of its steps are as important in any business decision as they are in TT&C: market analysis (or market research – that is: is this idea valuable; to whom is it valuable; and to what extent?); and marketing (who wants it; how do they want it packaged; how should it be priced; and so on?). Even in sponsored research, where the industry partner is known, marketing remains essential because a researcher needs to have located the industry partner *before* the research begins, and should have established agreements on matters such as how risks and rewards will be shared.

Universities have frequently approached TT&C somewhat opportunistically, often looking at each new idea as a discrete item to be commercialized. A more sophisticated approach involves scanning the totality of a university's intellectual property (and other intellectual property sources) in order to identify clusters of ideas that may have greater value than one idea alone. The University of California system, which operates 10 campuses, developed software to permit the collection and combination of intellectual property that was not, if left as one-off pieces of intellectual property, sufficiently broad or deep to be of commercial value. This permitted UC to see whether a more valuable, commercializable mass of intellectual property might be assembled from these smaller parts. It was found that there was potential benefit by grouping intellectual property that had not been evident in stand-alone intellectual property (Sime, 2004). This technique has also been used by the Virginia Polytechnic Institute and State University in conjunction with a major corporation (DuPont) in one particular technology area (nanotechnology) (Martin, *et al.*, 2004). However, presently at U.S. Public University, the TTO noted that the inability to assign intellectual property can complicate the process of trying to assemble different fragments of intellectual property in order to create a more useful bundle, although this rarely seems to be a difficulty experienced in practice (U.S. Public University).

Commercialization Performance

A common problem facing universities engaged in TT&C, governments and every other participant in the environment is to establish a set of metrics that accurately reflects a university's TT&C performance.

AUTM, probably the world leader as the representative body for university technology transfer professionals, has developed a range of metrics in order to assess the performance of universities in TT&C. These include input measures (such as the amount of R&D funding, and TTO staff numbers) and outputs (numbers and value of licenses, numbers of spin-off companies and their value, the number of resulting products introduced to the market) (AUTM, 2007b). But what constitutes suitable metrics is not a closed issue. It is almost a case of measuring what is available rather than what would best reflect reality. As the Canadian ACST (1999, p.14) said:

"... [the] success of a university commercialization office should not be measured by the number of licenses it negotiates or the number of spin-off firms it creates. Commercialization offices should endeavor to maximize the value of the companies which license their innovations and maximize the value of the companies they create. If they are successful in maximizing their clients' value, universities will maximize the economic and social returns to Canada as well as themselves".

This laudable proposition would appear to require, however, sophisticated economic and social measurement tools that may not be readily available.

Using data that are available, simply because they are available, as a proxy for commercialization performance could be self-defeating. Most United States TTOs do not create an excess of income over costs, but their value is greater than can be reflected in financial accounts alone. There are probably two principal measures that provide some useful insight into commercialization performance: measures of results, including income to the university; and current measures of productivity, such as the numbers of deals done and their continuing value. Metrics such as the number of spin-off companies may not be a valid measure unless it measured, for example, a sustainable company as expressed through the numbers of people employed in the spin-off company (U.S. Private University).

The TTO at U.S. Private University is not given specific performance targets. Instead, it is given a patent budget within which it must manage, and the manager informs the university how much he expects to make in the coming year. "As long as I make more money than I spend, they seem to be happy", reflecting an enlightened attitude that is concerned to make technology available to the community without the imposition of targets. This attitude reflects the thinking of the university President, and the management of the TTO believes that this is the appropriate way to manage commercialization (U.S. Private University). The same approach is evident at Canadian Uni2 where the university has not imposed specific performance targets on the TTO, although there is the possibility that these may be introduced later once it has been re-structured and operating for some time. Nonetheless, the TTO itself presently evaluates how it should measure its own performance. On the basis that some industry-consistent performance measures are used by AUTM there is a likelihood that, simply because they are available, these may form the basis of some future reporting obligation because they are likely to appeal to government funding agencies (Canadian Uni2).

The unsuitability of patents as an indicator of university TT&C performance is evident in the fact that the U.S. Patent and Trade Mark Office (USPTO) is some 600,000 patents behind which may take as long as 3-5 years to clear up. Second, the increasing use of patents as a tactical weapon is creating aversion to their use. In some fields business attempts to patent-build around a significant technology opportunity and then sue any apparent patent infringer, a fact that "is stifling the interest" in patenting genuine ideas.

United States universities do not generally take copyright to material produced by faculty members (with the exception of software) so that the likelihood of the university commercializing output from liberal arts and the HASS sector generally is probably remote. U.S. Public University reports that it has not seen any substantial potential for revenue generation from the liberal arts areas and does not presently see any likelihood that commercial ideas will emerge from this area (U.S. Public University).

Commercialization Office Performance

The ideal situation for all universities is to have a TTO that not only pays its way, but contributes substantial amounts of unencumbered income to the general purposes of the university. Riddle (2004, p.17) reported that in 2001 Canadian universities and research hospitals received $47.6 million in royalty revenue and incurred $28.5 million in TTO operating expenses. He said: "…[underlying] the aggregate data, is the fact that for many Canadian institutions the cost of running a commercialization office currently exceeds their revenues". This view is consistent with data compiled and reported by *The Times* Higher Education Supplement relevant to U.K. universities (Shepherd, 2006) in which it claimed that most U.K. university TTOs do not pay their way.

The experience of U.S. Public University, a "premier major research" U.S. university, was that:

"…In terms of our office, our expectation is that we will just about bring in enough in licensing to justify the office. Certainly, if you look at the overall activity in terms of the total license income that comes into the university, our office brings in more in licensing than we currently spend, but not that awful lot [more]" (U.S. Public University).

And this is after the TTO at such a sophisticated university has been operating for almost twenty years. In the opinion of an experienced technology transfer officer at U.S. Public University:

"If one were just to be starting up a new office at a university that didn't have one, or had a very weak one, I think you would have to expect at least five years, and maybe ten years before you could have an expectation that the office would be generating an income equal to its expenses…" (U.S. Public University).

A senior TTO officer who has been involved in collating and analyzing university TT&C data for AUTM over many years confirmed that a minority of United States TTOs generate income in excess of costs. He said:

"Even on the most expansive measure [by allocating all the income to the commercialization office]…only about two-thirds of U.S. institutions make more than they spend on technology commercialization." At the other extreme, where TTOs are attributed only the revenue that the university keeps, then about one-third of offices make more than they spend. It has been at this level for many years (U.S. Private University).

As to the criteria generally required for institutions to make money from commercialization, they are:

"... your office needs to have been around for at least...fifteen years, you need to have half-a-billion in research [expenditure], and you have to have twenty people in the office.... it took fifteen years at Stanford... And, when you look at individual institutions, there is an enormous difference between the mean and the median, and what that is saying is that a small number of institutions have got real lucky" (U.S. Private University).

Because TTOs are frequently seen by university management as cost centers, they are often given limited resources. This attitude often seems to persist even when the TTO is paying its way or is profitable. This may be attributable, in part, to the relatively long times it takes TTOs, including United States and Canadian TTOs, to become cash-flow positive. It can become difficult for a TTO to demonstrate a cash-flow positive position if it is subject to the regular re-structuring which many are. There is a risk, of course, that any TTO viewed as a cost centre and not as an essential component of the university-industry linkage may become subject to staff reductions or otherwise allowed to wither for want of capital. U.S. Public University TTO was established in 1991 and restructured in 2004, yet it remains relatively small with a total of fewer than twenty total staff for a university with a research budget exceeding $600 million per annum. The university may not be achieving its full TT&C potential with such a relatively small staff level. Additional staff in the Office has the potential to increase the amount of intellectual property licensed and income generated:

"Certainly in terms of maximizing the commercial return and to maximize out-licensing, we could use a lot more resources". The number of staff could be increased for this purpose because, at the moment "we end up spending most of our time on, really, the best technologies, and other technologies that might be able to be [commercialized]... we don't have the time to spend on them". Potentially, a doubling of staff could lead to a doubling of income (U.S. Public University).

A further potential constriction on the performance of the TTO identifiable in the quote above from U.S. Private University is this: there has to be significant triage done on technologies presented to the TTO because of the limited resources it possesses, encouraging TTO staff to work only on the technologies and with researchers that have apparently the greatest potential, perhaps prejudicing the prospects of commercializing otherwise worthy technology. This could lead not only to a loss in potential commercialization to the university, with the attendant loss of income, but could mean the loss of potentially useful technology to the community, while it could also lead to the demoralization of researchers who may, as a result of rejection by the TTO, become less inclined in the future to spend time making disclosures or attempting commercialization. It has the potential for technology to begin leaking from the university through other routes with consequent loss of income.

The financial performance of North American TTOs is also truncated by the fact that their income is derived principally from licensing, with an increasing amount derived from spin-off companies, but no income derived from faculty consulting assignments (which are generally treated as private arrangements) nor from sponsored research. For example, at Canadian Uni2 the TTO does not receive direct income from negotiating sponsored research, but is remunerated through general income paid by the university that recognizes negotiating sponsored research as one of the essential activities undertaken by the TTO for the university (Canadian Uni2).

There is a view prevalent in Canada that there is a shortage of people with the mix of skills needed in technology transfer (Canada Foundation for Innovation, 2002, p.22). Suitable staff are in short supply because they: "...require an in-depth understanding of the academic, financial and industrial sectors. They should possess an unusual combination of research, business, legal, interpersonal and communications skills" (ACST, 1999, p.11). Even assuming such people are available and prepared to work in a university, Canadian universities have difficulty in remunerating staff at an appropriate level, certainly where the TTO is a unit within the university wherein salary levels have to be comparable with other university staff as distinct from being a separate company where salary levels are more flexible (Canadian Uni2). Substantially the same problem faces United States universities when seeking TTO staff with the skills and experience needed to discharge the duties expected.

Knowledge and Awareness

University researchers are, generally speaking, not natural entrepreneurs. Their willingness to engage in TT&C may be enhanced by a number of factors such as rewards, but the first and most basic requirement to getting faculty members engaged in TT&C is educating them in what it is and how it is done. This is true no less in North America than anywhere else.

U.S. Public University has attempted training of academic faculty in TT&C, but: "... in the past, but we haven't been very successful. And I think the reason, typically, is that the faculty aren't interested, in general, unless they are specifically involved in something that they need it for, then they want it". Some of the most effective programs at U.S. Public University are run in the business school where they have courses in entrepreneurship and conduct business planning competitions. (U.S. Public University). Similarly, Canadian Uni2 conducts ad-hoc commercialization orientation sessions to groups of students and faculty members. It also actively tries to:

"... generate interest in intellectual property and commercialization by going out and meeting with the faculty and the chairs of the Departments and their faculty members to encourage invention disclosure and to encourage engagement with us. Because we undertake to represent them as their agent, we are very much in the mode of attracting them as clients to use our services".

While not done under the auspices of the TTO, the university Business School conducts a business planning competition and uses TTO cases as case studies for their Executive MBA students to develop business plans and prepare pitches to venture capitalists (Canadian Uni2).

Universities have generally embraced the concept of offering grants and prizes to encourage or support entrepreneurship and innovation. There are also close relationships between the TTO and academic departments in which TTO staff teach some entrepreneurship and business modules (for example, at U.S. Private University and Canadian Uni2).

Engagement in TT&C is also influenced by the prevailing university culture, and the United States has some outstanding examples of commercial acculturation in universities such as MIT and Stanford. It is a case of success breeding success where it is done well.

Incentives for Commercialization

It is likely that, in similar manner to company employees, all the right, title and interest in intellectual property generated by university-employed researchers is owned by the employer[29] but, unlike company employees, university-employed researchers are given a right to share in the profits from an invention. Largely as a result of the conditions imposed on universities by the *Bayh-Dole Act*, the general approach to sharing the benefits of commercialization with faculty members in North American universities is to grant them a proportion of income from licenses. Link and

Siegel (2005) reported that licensing activities among United States universities are improved by increasing pecuniary incentives to academic faculty members. U.S. Public University is typical in that it has a policy dating from approximately 1998 under which investigators receive 35% of license income calculated after the costs of patenting and other direct costs associated with the licensing process (but excluding the costs associated with the TTO). The balance goes to the university which allocates it: 15% to the Department, 10-15% to the State fund, and 25% to a general campus research fund. It is important to share the financial benefits of commercialization with academic faculty because:

"... most of the university technologies are such early stage that in order to have a successful commercialization, you really need the input and continued assistance of the faculty, and the only way to stimulate that is to give them some reason to do that".

Indeed, it is a requirement of the *Bayh-Dole Act*, "in order for the university to take title, it has to have some structure that [permits] some of the revenue to go back to the inventors..." (U.S. Public University).

At U.S. Public University:

"... [if] the faculty get no reward for patenting and commercialization they're not going to bother, because it doesn't help them as much with their career. Their tenure and their status within the community is [determined] primarily on their research [and] their publications; and whether they have a patent doesn't enter into it nearly as much..." (U.S. Public University).

Some universities offer royalties that vary between different disciplines, but generally grant around 30% of net royalties from an invention to the inventor, although both greater and lesser amounts apply in some institutions.

However, there is some evidence that financial remuneration of researchers is not a primary motivating factor. Harman (2002) cited evidence from his research that academic researchers in Australia have not received significant financial rewards from the commercialisation of their research. Only 5% of academics were able to increase their income by more than 20% of their basic income (typically around 70% achieved no increase) from "royalties from licensed patents, consulting and other similar means" (p.152). If the results had been cut more finely, it is highly likely that most increased income comes from consulting rather than registered intellectual property because it is the easiest, least complex and most immediate way for academic faculty to make additional income.

Colyvas *et al.* (2002) examined eleven university inventions from Columbia and Stanford Universities, and concluded that prospective financial incentives played little or no role in motivating faculty members to undertake the research projects. This is not quite the same as providing little or no incentive to commercialise the resulting invention, but speaks to the relatively low importance most academic researchers place on commercialisation when commencing research.

It may be that personal financial gain by researchers is not a major motivating factor encouraging researcher participation in research but is assuming greater importance in TT&C, although the data in this regard appear equivocal. If correct, this conclusion could arise because personal financial gain may be a greater factor in, for example, life sciences TT&C than some other areas, and may have more importance where TT&C has a longer history such as in some U.S. universities.

The increasing propensity for universities to take equity positions in companies in lieu of licensing royalties, as reported by Feldman, *et al.* (2002), raises the issue of remuneration of academic researchers when the potential income is more speculative (return on equity rather than a royalty stream). This may be seen as more attractive in some cases and by some faculty members, but it also raises the issue of whether and in what proportion equity should be shared between the university and faculty members. As a first-pass estimate researchers would, presumably, be entitled to one-third (or whatever the relevant proportion) of the equity received by the university, thus reflecting the proportion to which they would have been entitled had it been income, although there appears to be no reported research on this matter.

Unlike the United States, it is reported that Canadian universities do not permit academic staff who have obtained research funding to buy out their teaching obligations, that is, to pay for another to be employed to fulfill teaching duties. This may limit the willingness of Canadian academic researchers to engage in both research and commercialization, and was identified as one of the reasons underlying Canadian universities' relatively worse performance vis-à-vis the United States, in commercialization (ACST, 1999).

Linkages with Industry

Linkages with industry are fundamental to the performance of a university in TT&C and to the performance of a university TTO. For TT&C to work, effective linkages and relationships with industry, above all else, must exist. Yet research shows that it is consistently one of the greatest weaknesses in a university's efforts to promote TT&C (Lambert 2003). U.S. Public University conceded that this: "… is one area where we could do more". Apart from the industrial affiliates program, linkages with industry tend to be *ad-hoc*, although it depends on the Department involved, because: "… some of the Departments have [closer interaction], but our Office doesn't really manage it from that side…". This is due, at least in part, to the relatively small number of licensing staff in the TTO (U.S. Public University).

To attempt to overcome weaknesses in industrial linkages in the physical sciences, U.S. Public University has organized an industrial affiliates program. This program "… is to try … to cultivate that interaction". The university has a number of consortia:

"... particularly in the physical sciences. There are a couple of consortia that have twenty, thirty companies that come together; they pay some small annual fee to be part of this consortium, and then they get an option on any technology that gets developed [from the funding of the consortium], and that is quite useful in terms of interaction with industry".

Companies pay in the order $20,000 to $50,000 per year to participate in a consortium. The university has none of these in life sciences at present. Generally, where there is: "… a consortium, part of that usually is to have a yearly or semi-annual meeting where everybody that was funded by a consortium gives a report, talks about their research, and the companies all send representatives" (U.S. Public University).

United States universities generally encourage a close relationship with industry in order to stimulate sponsored research but also for the more altruistic reason of benefiting the community. An example of this at U.S. Private University was the establishment in the United States of the first industry-university research commercialization intermediary institution based on a European model. The centre: "… conducts applied research for local and international industry. Its mission as a non-profit institute is to develop next generation manufacturing technologies for industry based on emerging U.S. and European research" (U.S. Private University).

A major and continuing link between universities and industry is in sponsored research which often involves the university TTO in negotiation. "We do about $65 million per year worth of contract research… the contracts go through our office. We don't tend to count that in our metrics [but] we will, and it reflects much stronger basic industry relationships than people give us credit for" (Canadian Uni2).

Despite its many years in TT&C Canadian Uni2 is still working to develop a successful model for working with industry. Where previously industry linkages had been conducted in something of an *ad-hoc* manner, the TTO is now developing a more structured approach to dealing with industry using 'showcases' where industry is invited to presentations describing university research and driving connections through other university links such as the Development Office (links with alumni) (Canadian Uni2).

Canadian universities generate relatively more spin-offs and less in licensing revenue than comparable universities in the United States, a situation that is attributed by researchers to the fact, at least in part, of limited receptor industry capability for technology in Canada's private sector (Canada Foundation for Innovation, 2002; Canadian Uni2).

Canadian Uni2 TTO has not found SMEs to be suitable receptors for university research output. While SMEs are agile: "... [they] are not ready acceptors of early-stage new technology that takes a lot of development, which is typically what comes out of the university", nor do they have sufficient money. The university has not licensed technology successfully to a SME (Canadian Uni2).

The Canadian Government is clearly concerned to try to ensure that the benefits of research are retained in Canada, for example Recommendation #1 of the ACST (1999) report was:

"... [the] federal government should require an explicit commitment from all recipients of federal research funding that they will obtain the greatest possible benefit to Canada, whenever the results of their federally funded research are used for commercial gain".

Arguably, there are two key missing structural components in Canada's research network relevant to research commercialization: the first is corporate research laboratories; and the second is not-for-profit organizations that connect research with the market, like Battelle in the U.S., Fraunhofer in Germany and ITRI in Taiwan (Canada Foundation for Innovation, 2002).

The Canadian Government has made attempts (ACST, 1999) to encourage the participation of small businesses, including spin-off companies, by giving them preference in licensing university intellectual property, albeit dependent on finding appropriate businesses and equitable terms, which may constitute a considerable hurdle. In addition, there are concerns about the quality of spin-offs as suitable business receptors capable of exploiting university research output, as noted by Canadian Uni2:

"...my personal worries about spin-offs are the management team, the capitalization and the market access and how long it takes to build into the market. Licensing can solve the market issues, but there are other factors around the terms of the deal". Canada also has to face the fact that it is working in a global economy, which means that it may have to license to international companies, not only Canadian ones. "...we've got to change our mind frame" (Canadian Uni2)

One of the challenges in licensing software and other technologies to industry, particularly when it is high-volume low-margin technology, is the need for a rapid turn-around. Canadian Uni1 has addressed this through software it has developed for the purpose, a web-based application for marketing and licensing on-line.

Technology Parks and Incubators

There is a clear difference of opinion between advocates of the worth of incubators and technology parks and those that view them as unnecessary and being mainly real estate investments. Most of the leading U.S. universities identified by Tornatzky, *et al.* (2002) in their survey had an associated research park, incubator or similar facility, a factor that clearly impressed the authors; although in many cases, the facilities were quite young at the time of the survey and their longevity had yet to be proven. On the other hand, people who have been in the business of identifying and supporting young spin-off companies for many years appear somewhat more skeptical about their value. There is a view from an experienced venture capitalist that:

"You can't start a company with a manager. A company's got to be started by an entrepreneur. An entrepreneur doesn't really fit into an incubator. So I think all the incubators in the world are a waste of time. And, of course, I've spent a lot of time working with development people at the State level... They just want to do the right thing as they see it. They want to take the taxpayers' money and they want to build these incubators, and what they end up being is poor real estate investments. And the reason is that an entrepreneur doesn't work that way. I have never seen a successful incubator, whether it's funded by the State or whether it's funded by private [funds]. Entrepreneurs don't want to go into an incubator where there's a... telephone ... and a Xerox machine down the hall – they just don't want to do that, they want to be off in their own place. Incubators are just a waste of time and money".

This sentiment is consistent with the views found at both United States case study universities. While U.S. Public University has access to a number of technology parks and incubators and refers some potential start-ups to them: "... [my] opinion has been that it is very rarely successful in terms of economics – I'm not sure that it really pays... And I think there's a philosophical question about that... if [a] commercial idea is really commercial, it ought to pay for itself" (U.S. Public University). And while U.S. Private University has two large incubators, one for life sciences and one specializing in electronics, the fact is that:

"... universities have very high cost structures ... I think we're now [of the view] that they're nice to have for our own spin-outs for a period, but the President said to us: 'Why... do you bring external companies onto campus? We're never going to make money from them. The only reason to bring them on-campus is if they're going to become part of the intellectual climate of the university - [for example], take students and give them internships, talk about their business plans in the School of Management and stuff like that', so that's what we're doing" (U.S. Private University).

In Canada, the ACST (1999) is also skeptical about the results obtained from research parks and incubators. Indeed, the genesis of the financial difficulties experienced by the TTO at Canadian Uni2 lay, at least in part, in funding it provided to a university incubator:

"It was not successful simply because it wasn't properly funded and it wasn't fully occupied. We were trying to do it out of our existing budget, which included this line of credit and it consumed an enormous part of this line of credit, which was really not core business for us and consumed a lot of resources. I suspect it consumed more than anybody ever thought" (Canadian Uni2).

Academic literature seems to confirm the view that technology parks and incubators are not essential to new business success (Westhead, 1997; Siegel *et al.*, 2003; Peters *et al.*, 2004).

Intellectual Property

Arguably the greatest impact on U.S. university commercialization came about as a result of an Act supported by both sides of Congress, the *Bayh-Dole Act*[30]. Between 1979 and 1984, the number of patents issued annually to universities more than doubled (from 177 to 408) and more than doubled again to 1,208 between 1984 and 1989 (Mowery, *et al.*, 2001). It is continuing to grow arithmetically, reaching over 10,000 by 2005 (AUTM, 2007b). Over a similar duration the number of university technology transfer offices in the U.S. increased from 24 in 1980 to 150 by 2005 (AUTM, 2007b).

The principal effect of the *Bayh-Dole Act* was to permit universities to have perfect title in intellectual property developed as a result of federally-funded research at the university subject to certain conditions. Prior to this Act, title in intellectual property vested in the federal funding agency (subject to some notable exceptions involving Institutional Patent Agreements mentioned earlier), resulting in desultory exploitation rates which the Act was intended to turn-around (Sobol and Newell, 2003). Researchers such as Mowery, *et al.* (2001) postulated that the *Bayh-Dole Act* did not lead to an increase in university patenting, although later research by Shane (2004) suggested that it did in those fields where licensing is an effective method of technology transfer. On balance, the Act has resulted in increasing levels of exploitation of university-generated intellectual property, but it does come at some cost. A particular feature of the Act is that universities are not permitted to assign the title in intellectual property vested in them. One result of this has been that licensing (rather than spin-off company formation) has become the most common means of exploiting university ideas through industry in the United States, although, as U.S. Public University pointed out, while Bayh-Dole restricts some of the university's flexibility in this regard, it does not appear to cause a problem "... because the restrictions are... somewhat minor in terms of actually making a commercial deal" (U.S. Public University).

Notwithstanding the complexity of engaging in TT&C prior to the passage of the *Bayh-Dole Act*, many leading universities had been involved in research commercialization for some years by entering Institutional Patent Agreements with federal funding agencies permitting the universities to exploit intellectual property resulting from research. (U.S. Private University). And, according to Nelson (2001, p.14):

"... a significant increase in patenting and licensing was going to happen in any case. The passage of Bayh-Dole legitimated these trends, almost surely speeding them up and magnifying them... [but] in our view, the broad developments were inevitable, in the absence of policies and decisions to head them off, or to temper them".

Colyvas, *et al.* (2002, p.62) argued to the same effect:

"These two developments [new areas of life science research and a relaxation in what was patentable by the USPTO] were leading to increases in university patenting and licensing prior to the passage of Bayh-Dole, and in our view the principal effect of Bayh-Dole was to accelerate and magnify trends that already were occurring".

Consistent with this view, there is evidence that leading United States research universities such as Stanford and the University of California were increasing the size and scale of their TTOs before the enactment of the *Bayh-Dole Act* and that there were other more influential factors inspiring the growth of technology transfer (Mowery, *et al.* 2001). Indeed Mowery, *et al.* (2001, p.117) considered that the need for universities to grant exclusive patent rights to companies in order to develop and commercialize research results: "... flies in the face of the position that patents tend to restrict use of scientific and technological information, and that open publication facilitates wider use and application of such inventions and knowledge".

High rates of intellectual property exploitation are reported in institutions with an exceptional commercialization record. For example, it is reported that 60% of new patents at MIT are licensed within one year (Canada Foundation for Innovation, 2002, p.18). At the same time, however, while patents are an important method of university TT&C in the United States – probably more than anywhere else in the world – they still represent a relatively small proportion of the total intellectual output of a university. A potent example of this can be found at MIT, one of the United States' most active TT&C universities, where it is reported that patents represent approximately 7% of the knowledge transferred from its laboratories (Agrawal and Henderson, 2002).

It is a common condition of employment of academic faculty in the United States that faculty members are obliged to assign all right, title and interest in intellectual property they generate to the university. So, whether research is funded federally (and subject to the *Bayh-Dole Act*) or privately, the university ends up owning rights to intellectual property arising from faculty member involvement in research. Students and visitors are similarly obliged to assign their interest in intellectual property arising from sponsored research.

In countries where the rules are less rigid than those in the United States, and individual institutions have the right to develop their own intellectual property policies, there may be a lower rate of exploitation. For example, it was reported (Riddle, 2004, p.1) that only 68% of Canadian universities actively manage their intellectual property.

There are no national rules (laws or policy guidelines) in Canada affecting how universities may deal in intellectual property, with each university free to make its own rules as it sees fit, subject to any obligations arising from funding or other agreements with external parties such as industry partners. There have been previous attempts to establish a Canadian Intellectual Property Policy, such as that by the Expert Panel on the Commercialization of University Research (ACST, 1999), but these have not been adopted. In some ways this diversity of approach may disadvantage Canadian universities vis-à-vis United States universities because of the risk of uncertainty as to intellectual property title. Pressure from Canadian granting councils upon universities to adopt a uniform approach to intellectual property ownership reflects their concern on this issue. Support for a consistent approach to intellectual property amongst Canadian universities has been encouraged (Canada Foundation for Innovation, 2002). Canadian universities presently negotiate intellectual property issues with staff on a university-by-university basis as collective employment agreements come up for renewal. This position is unlikely to change without external pressure from relevant levels of government.

At Canadian Uni2 the rules prescribe joint ownership between the inventor and the university upon invention, after which the parties may elect to have the university exploit (whereupon the inventor receives 25% of the net income) or the inventor may exploit (and receive 75% of the net income). Not surprisingly, approximately 92% of university faculty members choose to exploit inventions personally (although half of these still engage the TTO to assist) notwithstanding that this requires them to undertake significant commercial activities themselves – with the attendant commercial risks that this may bring as well as the potential compromise to the faculty member's other activities as a result of the expenditure of time involved. To attempt to overcome this situation at Canadian Uni2 the university is proposing to offer inventors a large proportion of income (but less than 75%) providing the TTO undertakes all TT&C activities (Canadian Uni2).

While certainty as to who owns intellectual property at Canadian Uni2 has not usually been an issue, there have been occasions when a problem has emerged as a result of previously undisclosed interests. This has made the university "process careful" and it tries in all cases to have researchers assign all intellectual property to the university in the first instance, after which the university will assign it back to the inventor in the event that the inventor elects to exploit an invention personally (Canadian Uni2). Similar arrangements exist in other Canadian universities. While this overcomes some difficulties, it does not remedy the problem of the undisclosed interest while it also creates a rather onerous paper trail. Canadian Uni1 appears a little more concerned about the complications arising from intellectual property ownership uncertainty when venture capital is required:

"...common policy or common principles would be helpful, [but] the principles are not even there... intellectual property is rarely developed by an individual, it's developed by a team... [and] if you do it across multiple institutions it really gets kind of tricky, and if these are potentially investable opportunities, the VCs [shy] away from these because they [say] 'you can't show me clear title here'" (Canadian Uni1).

Another of the potential pitfalls in the Canadian *laissez-faire* approach is the risk of unfair outcomes arising from a power imbalance when many parties are involved in discoveries, as there usually is, such as between professors and graduate students, which becomes even more complex when multiple institutions are involved (Canadian Uni1).

But the Canadian approach is not all downside, because the flexibility of ownership of intellectual property offered to researchers has sometimes operated to their advantage by allowing researchers to negotiate with Canadian universities on the matter of ownership of research outcomes, making the Canadian approach attractive in some cases (Canadian Uni2).

There is ample evidence that Canadian universities and the federal Government do not intend to affect an individual academic's right to publish, whether or not some obligation to disclose a discovery to the university may exist (ACST, 1999).

There is also a difference in the intellectual property needs and demands of life sciences when compared with those of physical sciences and ICT, largely arising from differences in the market and their respective regulatory demands. Life science regulatory approvals can take many years so that time to market is almost always measured in years and sometimes decades, while physical sciences and ICT have market cycles generally between a few months and a few years. This means that patents are often less important in physical sciences and ICT (where time to market is often the crucial issue) than they are in life sciences:

"I think that the IT, engineering technology tends to be incremental rather than revolutionary – it is rarely patented or protected in any meaningful way by patent. It is always tied to the individual and so you can't just transfer it and let somebody else deal with it, it almost always has to go with the individual...". (Canadian Uni1).

While North America has resources to assist in connecting universities with each other and industry, it has proven difficult to combine patents amongst universities and it appears that it is rarely done among universities except when there is explicit collaboration on a research project. Parts of industry, however, do search university inventions and seek to combine ideas and patents in order to exploit combined intellectual property. U.S. Public University reported that it has had difficulty in packaging patents:

"... mainly because we have [the] inventor's interest in the technology being developed, and so we have to be careful in terms of combining technologies that we don't slight one inventor or make that inventor feel slighted because this other technology is being combined and maybe it's getting more of the revenue than he is getting, and 'of course his technology is more important than theirs'" (U.S. Public University).

It appears that university TTOs rarely, if ever, seek to combine their patents with those of other universities. The use of facilities such as on-line databases of ideas and patents appear to be exploited usually by industry but are rarely of interest to other university TTOs (U.S. Public University and U.S. Private University).

Early Stage Financing and Venture Capital

The journey between making a discovery or invention and getting the idea to market is rarely simple. The time taken to market will depend to some extent on the technology involved (life sciences, in particular, almost always takes longer), but every idea needs some development before it has been sufficiently proven to be marketable. This journey requires know-how and money, and it is the availability, or lack, of sufficient money to advance the idea that has a major influence on the time taken to get an idea to market. The significance of this and the difficulty in getting finance for the early stages of an idea was described elegantly by the TTO Director at Canadian Uni1 in the following way:

"The vast majority of the research that is undertaken at [our university] is government-grant funded academic research. Consequently, inventions that are disclosed to the [TTO] are typically at a very early stage along the development path to commercialization. Because of the high technical risk and long timeframe for developing products from such early-stage inventions it is often difficult for the [TTO] to attract the commercial partners (licensees or investors) that are required to develop the technology further towards the marketplace. Furthermore, traditional grant funding sources are no longer receptive to funding these inventions because their development has moved beyond the bounds of pure academic research. This leads to the unfortunate situation where many promising inventions are in danger of languishing between the realms of basic research and commercially viable technology, often due to the need for relatively small investments of development dollars. This disconnect is colloquially known as the 'Technology Funding Gap'" (Canadian Uni1).

This means that the funding needed to take the idea along the journey to market has to be found from some source that is not associated with the research itself.

There are two parts to funding the development of discoveries and inventions. The first, called here early-stage financing, involves providing funding at reasonably small levels–typically not exceeding $250,000 (but sometimes as high as $2 million) – to take an idea from bench or laboratory stage to proof-of-principle or a little beyond. It may be granted to a researcher in more than one tranche. The second, typically called venture capital (VC), involves funding at much higher levels either to get an idea to market, or to prove its worth to a company so that it may be sold-on. By this stage much of the technology risk has been removed, although some market risk remains. The amount of money involved in VC is as large as needed for the idea involved.

Early-stage financing of inventions and discoveries is provided by most leading North American universities, but not all. For example, U.S. Private University provides what may be viewed as a typical early-stage financing arrangement, involving two tiers of financing. The first tier enables researchers to generate commercially relevant data, reach key milestones, or develop a prototype in order to help bring raw technology to a mature enough state where it can be either licensed or form the basis of a new company. The awards are intended to accomplish specific applied research tasks. Typical awards will range from $25,000-$50,000, with a maximum award of $50,000. These will normally be one-time awards geared toward specific technology milestones. There are usually six such awards each year from a total amount available from the university's own money of up to $300,000 a year. These funds are recharged annually from other successful commercializations. The second tier helps faculty members start new companies based on technologies that they invented at the university. Each year 2-3 projects are selected for an investment by the university of $50,000-$200,000 which will form either debt or equity. The selected companies also benefit from assistance from the TTO Innovation and Entrepreneurship Group, with the intention of developing an independent, successful, self-sustaining company. Early-stage funds are sustained through injecting profits from other commercialization activities into them on an evergreen basis, ensuring that they continue (U.S. Private University).

The Canadian experience is similar to that of the United States in terms of the amounts and points at which funding is provided, although the government provides most of the funds which means that funding relies on continuing government interest in the sector.

Venture capital is characterized as being "early-stage, high-risk, high-return, long-term, high-technology investment" and may come from university or private funds. VC can be involved at any stage in the development process, including early-stage development, depending on the risk profile that the fund is prepared to accept, a decision that is contingent on the sophistication and experience of the fund managers. Private universities, such as U.S. Private University, often have a substantial fund at their disposal as a result of philanthropy and investment. U.S. Private University has a venture capital fund that invests in commercializable ideas, with particular emphasis on investing when university intellectual property is involved. The fund usually acts as a general partner or co-investor, investing alongside top tier venture capital funds as a strategic part of the investment syndicate. As a general partner the fund looks to investments in the range $2-5 million, while, as a co-investor, it will "typically invest between $500,000 and $1 million initially, depending upon the stage and capital requirements of the company. We act as a co-investor and require a separate larger venture fund to lead the round". U.S. Private University has about $40-50 million venture funds available on an evergreen basis which was created by the university with seed funding of about $11 million. The university is presently looking at dividing the investment functions handled by the TTO into two parts: a seed fund that the TTO will control that is not looking for a return; and some that will be managed for a return. Historically, the university has made much better returns on its investments in university spin-offs (U.S. Private University).

Notwithstanding the funds available to some universities, they:

"...struggle to pay for the patenting and for the office of technology transfer. And to make direct, dedicated, at-risk [investments] of our own money in these technologies, I think universities would sooner give up ninety-five percent of the economic benefit and limit their exposure, as well as limiting their up-side" (U.S. Private University).

The aggregate amount of venture capital available for investment depends on the incentives provided by government. Tax incentives are important for a private investor to be engaged in genuine venture capital, for example: capital gains tax, and R&D incentives. "Government can create more and better kinds of incentives... [government] should be trying to influence the way people behave – that's what government does...".

The need for private venture capital in the United States is very clear in the case of public universities where even, for example, U.S. Public University does not have internal funds to assist faculty members in forming spin-off companies. There are very strict conflict of interest guidelines which oblige faculty members to separate private interests from research obligations, as well as which the university does not permit the use of campus facilities to assist start-up companies.

These restrictions and limitations do not apply to private universities where greater flexibility generally exists.

Canada does not have the depth of venture capital that exists in the United States. Canadian Uni2 noted this deficiency not only in terms of cash but of management skills in the industry. The lack of venture capital is also evident in a lack of risk capital particularly at the early stages of development. For this, Canada is reliant almost entirely on publicly-sourced early stage funding which is available, for example, in Ontario through the Ontario Research Commercialization Program (ORCP). ORCP program funding is to be allocated over three years: physical sciences and IT getting $1.7 million; and life sciences approximately $2.7 million. Proof of principle funds of up to $120,000 can be allocated to individual ideas in three tranches, with first amounts typically around $10,000 to test an idea, such as performing a basic market review, a second tranche of $50,000 to make a prototype, with a second $50,000 available provided there is some matching fund from another source and with consultation with the Ministry. Based on its experience with venture capitalists, Canadian Uni1 is seeking greater independence in its funding, so it is approaching venture capital funding from the university side rather than the venture capital industry side. It has set itself the target of raising more than $100 million over five years to invest in university ideas and has raised $50 million so far from the Canada Foundation for Innovation (CFI), industry and the Provincial Government (Canadian Uni1).

The view of the late Professor Doriot of Harvard Business School was that, if you have a good idea, money will find you (Gupta, 2004). The experience of venture capital in the United States is that:

"... we mostly search ideas out. It is rare for a good idea to come over the transom. [In my experience] we saw a huge number of investment opportunities come across the transom, but we never invested in a single one. The good ideas always came to us from another venture capitalist who we know and had dealt with, or as a referral from a company we were already invested in. We are all well connected, we each have our [trusted sources]... and we all know each other, so we learn about a new idea quickly".

Entrepreneurship and Culture

Of all the factors that encourage entrepreneurial behavior among North American faculty members, the freedom to pursue personal commercial interests through consulting is one of the strongest. This is evident from the experience of U.S. Private University:

"I think that [giving professors a day a week to pursue business interests such as consulting and spin-off companies] is a critical part of the U.S. success, and I only really learnt that when I had someone come and teach in my class about federal labs. Now the federal labs don't give their employees a day a week. And if their employees want to take a year off to start a company, they won't let them come back. [So] I think letting faculty take leave of absence, [start a company], and then come back [is critical to success]". Every member of faculty gets a day a week to pursue commercial opportunities (U.S. Private University).

The situation is somewhat similar in Canada, where universities allow academic staff time to undertake private commercial activity such as consulting, but it is not generally as structured as the one-day-per-week available in the United States, but can amount to up to 20% of the teaching time (Canadian Uni2).

North American universities with an exceptional record in TT&C, particularly in respect of the numbers of spin-off companies, appear to be the ones that grant the greatest degree of freedom to academic faculty members to engage in commercial activities. It is not simply a matter of undertaking great research at Ivy League universities, even though they may achieve remarkable discoveries, for that is not the same as commercialization. As a United States venture capitalist explained:

"Princeton [University] is a place where pretty much everyone is living his or her dream. So, there's no way you're going to spin anybody out of Princeton, it's as simple as that. [And that is true of the Ivy League generally]. The exact opposite of that is MIT... What you could get from Princeton or the Ivy League generally... [is] great technology but, since you cannot spin a person out, what do you do? Because generally it takes at least one person, maybe more than one, to embody the creativity of the technology and to keep it moving forward in the early stages". What is different between these universities? "It's a matter of freedom. If you take a Princeton or most schools like Princeton, they have a pretty rigid set of rules". Faculty members have only one day a week to do what [they] want – consulting, forming a company, or whatever. "But MIT? Who knows where anybody is? MIT is totally free in that regard. There are guys on the faculty at MIT who are running companies... It's more related to how much freedom there is than anything else... The two [most free] places are MIT and Stanford...".

Entrepreneurship is presently enhanced in some universities by cross-appointment of staff. For example, at U.S. Private University some TTO staff have been appointed as academic faculty to take classes in entrepreneurship, while at Canadian Uni2, and many other universities, cross-appointments between clinical areas such as hospitals and the university are quite common.

U.S. Private University TTO, in conjunction with the School of Management, has established a technology entrepreneurship institute at the university which offers two main features. The Institute offers courses in entrepreneurship and, upon request, assists university faculty members in establishing new companies. Possibly partly as a result of this, U.S. Private University has observed a decline in numbers attending commercialization boot-camps which may reflect either a peaking in demand or demand being met through the other activities of the Institute (U.S. Private University).

Encouraging entrepreneurship is an area where individual universities have exhibited significant innovation. Canadian Uni1 created the position of entrepreneur-in-residence at the university in 2005 with a serial entrepreneur in the position. The entrepreneur's role was introduced as part of a broader program to promote the establishment of new ventures; his: "...mission [was] to identify and facilitate the creation of healthy companies from [the university] and its Affiliated Hospitals, and to provide them with the environment and resources to achieve their maximum potential". It has proven so successful that the university intends to expand the program to three part-time entrepreneurs able to provide assistance across a range of disciplines (Canadian Uni1).

Innovative universities are often trialing new ideas in entrepreneurship such as Canadian Uni1 with its program that integrates and includes:

"...workshops, courses, networking events, company in a box (which was just access to common legal documents and those sorts of things), mentorship (sitting on boards), early-stage access to capital, and we actually tried our hand at a bit of an accelerator, which was a just a shared office, and it didn't really work too well in our environment... And that's been expanded now to what we hope will be a [university] New Venture Program that will bring in the introduction of two funds: an idea fund and a seed fund as well as internships for students at both the undergraduate and graduate level" (Canadian Uni1).

The Role of Intermediaries

Intermediaries have played a role in university TT&C in the United States for many years. Reamer, *et al.* (2003, p.xviii) identified three types of intermediary:

- **Intermediaries working with technologies from all sources**;
- **Federal technology transfer intermediaries:** focus on transferring technology from federal laboratories; and
- **Federal technology contract intermediaries:** focus on assisting businesses in obtaining Small Business Innovation Research and Small Business Technology Transfer contracts

The role of intermediaries is recognized in the development of a model for linkages by scholars such as Mohannak (1999), Howells (2006) and Pollard (2006).

Probably the largest and best-known of the intermediaries in the United States is the Battelle Memorial Institute[31] (BMI) which, since 1929, has been involved in research and development including working with universities and other publicly-funded research entities to commercialize ideas. Among its many activities, BMI provides services to enhance small business including identifying technologies, co-bidding to add strength to proposals, and providing mentors and protégés. A more recent development in this regard is the Robert C Byrd National Technology Transfer Center[32] at Wheeling Jesuit University in West Virginia which claims its core capabilities are technology evaluation and market assessment, partnership development, computer information services and strategic technical services to bridge the gap between university or government laboratory and a commercial operation.

The strength and power of United States universities, as well as their willingness to take the initiative, is illustrated by the fact that U.S. Private University has opened a European-style manufacturing intermediary, as mentioned earlier. In Canada the lack of intermediaries to make the link between research and industry has been noted as a major impediment, being done only by government laboratories, and then on too restricted a scale (Canada Foundation for Innovation, 2002, p.12).

The United States has a history of not-for-profit organizations possessing huge resources that are prepared to act for the common good. Increasingly, these entities are prepared to fund the great expense involved in medical applications in such a way that permits patent problems to be largely avoided. As argued by a United States strategic analyst, the most successful technology transfer operations in the United States are becoming the:

"... new foundations: Case, Milken, and the like. These guys are coming and they are playing different types of intermediaries. They are saying: 'we know there are different types of viral infections in the United States which, if we can get these vaccines designed, manufactured and delivered, we can actually do something about it. But the pharmaceuticals don't see enough money in it. Investors don't see enough upside to it, no one wants to take the risk, but we know there's a problem'. Therefore, unlike government, which used to be the neutral risk-taker, [but which] can no longer take the risk because [Congress is no longer prepared to take such risks]. The real funding in this country for risk-taking, and for a whole set of new technology transfer is these foundations". People are not prepared to wait patiently for solutions to the grand challenges. So "what you have now are impatient forms of capital that are basically [addressing these grand challenges]. So that's part of these new intermediaries... " Foundations and the like represent "new forms of collaboration that did not exist in this country, or around the world. You have some people who are thinking of the higher good... ". In essence, any company that wants to exploit the intellectual property may

do so, "but what we're not going to do is [fight] over the research and the research agenda and who [wins] along the way". Disputes over the ownership of intellectual property will not be permitted by the foundations. "For the moment, foundations are trusted relationships... You are creating new forms of collaboration that do not exist in this country, or around the world".

There is evidence, also, of a change among some university TTOs taking on a multi-function role including technology transfer, commercialization and economic development. This can be seen at universities such as North Carolina and Arizona State. Universities like these recognize that there are several pathways to exploit the results of the research enterprise.

Venture capitalists have been supplying not only capital but advice and mentoring to emerging companies for many years and represent another form of outsourcing relevant to commercialization. Venture capital has existed for so long in the United States and has evolved to a degree of sophistication that it should rank as another important form of intermediary.

The growing scope for intermediaries like those identified here illustrates a number of points. First, because it changes the dynamics of the way certain innovation occurs, it highlights the need for any university TTO structure to be continually reviewed to ensure that it keeps pace with the rapidly-changing environment in which it exists. Second, the existence of so many loose and amorphous arrangements portends some of the difficulties facing universities that work on the principle of patenting and patent exploitation. Noting the difficulties that patent-holders are experiencing in the United States because of delays and aggressive patenting practices, the way ahead for university commercialization may be increasingly a mélange of practices with the emphasis on working for the public benefit continuing to be the strongest motivation.

CHALLENGES FACING NORTH AMERICAN UNIVERSITIES

Key Factors that Improve TT&C Performance

It would be convenient to conclude that there are a number of common threads that run through all successful TTOs and, to an extent, this is true. But there are also a number of threads that should, at first glance, be important in TT&C success but, after analysis, appear less important. Many analysts and researchers have, after weighty analysis produced lists of qualities and characteristics that universities and TTOs should possess, or of resources that governments should provide if TT&C success is to be achieved. But lists are at best guides and provide little insight into what makes any particular TTO a success, and whether and how such success could be replicated in other institutions (Rousseau, 2006). In particular, such an approach makes little allowance for that most esoteric and individual characteristic of any organization: its culture, and universities are no exception to this rule.

There are many factors that influence a university's success in TT&C. Of the three groups of factors identified by Tornatzky, *et al.* (2002), namely *mechanisms and facilitators*, *institutional enablers*, and *boundary-spanning*, universities may have some incidental influence on the first and third of these (*mechanisms and facilitators* and *boundary-spanning*), but they are substantially constructed by forces outside the control of individual universities. This external environment, or *common infrastructure*[33], to use the term of Gans and Stern (2003, p.13), is effectively the same for all universities within any one country. However, each university normally has control over the second group of factors, the *institutional enablers*. If the common infrastructure offers a workable national regime, then success or lack of it in TT&C by any individual university must be determined largely through the agency of suitable *institutional enablers*[34]. However, it is important to acknowledge that suitable institutional enablers will not necessarily be exactly the same for each university. The evidence above makes it clear that each university has its own individual character, culture, history and environment around which institutional enablers have been constructed.

In one sense, this conclusion is largely self-evident. If the common infrastructure around which university TT&C is built is unworkable, it would be reasonable to expect that all the universities in such a nation would produce no substantial TT&C results. Yet in every country there are examples of superior performance among some universities, and just as clear evidence of poor TT&C performance among others. Since MIT operates with the same common infrastructure as other universities in the United States, why is its performance in TT&C so evidently superior to most others?

Based on extensive interviews, Siegel, *et al.* (1999, 2004) concluded that there are three organizational factors that are most critical in determining TTO efficiency, "... reward systems for faculty, TTO staffing and compensation practices, and actions taken by administrators to extirpate informational and cultural barriers between universities and firms". The data from our case studies substantially support and expand on this conclusion of Siegel, *et al.* On top of these requirements for TTO efficiency, the remaining key factor in TT&C success in a university is senior management support. The findings here are consistent with the view expressed by Rousseau (2006) on the role of evidence-based practice (of which, it is suggested, the case studies are an outstanding example), that (p.267): "Evidence-based practice is not one size-fits-all; it's the best current evidence coupled with informed expert judgment". It is this role of informed expert judgment in applying evidence-based practice that is the vital factor in the success of better-performing TTOs.

The factors that are consistent in every instance of successful university commercialization performance evident from the case studies are two-fold. First, the university President and other executives concerned in commercialization have to believe in it and make a genuine commitment to its success.

The second factor is that the TTO has to be led by an individual that possesses an unusual combination of at least four attributes: the ability and willingness to work in the arcane structure of the university; the ability to be both an entrepreneur and a manager; the flexibility of mind and perceptiveness to see what is happening in a dynamic field as the environment evolves and matures; and a leader of people and business[35]. The capabilities of this individual allow him or her to select staff for the TTO that complement the business because, of course, quality staff are also essential[36]. Technical domain skills are not a significant skill required in the leader, although the ability to empathize with researchers is. The importance of the founder in an entrepreneurial entity – and a new university TTO, in particular, is an entrepreneurial entity – has been well established in the literature (Bruderl, *et al.*, 1992; Shane and Stuart, 2002). As a university TTO grows, other factors such as available capital and size and skills of the work-force begin to grow in importance, although it is suggested that the sense of entrepreneurialism, as well as the other attributes listed above, required of the leader of the TTO never cease to be relevant. The attributes mentioned here were recognized by Debackere and Veugelers (2005, p.340), when they said, in respect of researchers at their case study university who successfully commercialize: "Assistance and funding have helped in this process [exploiting the university's knowledge base], though they cannot act as a substitute for the ambition, the strategic thinking and the drive for implementation of the researchers themselves". It is suggested that these essential attributes are not limited to academic researchers alone but apply, also, to TTO staff generally, and the TTO Director in particular. Higher-performing TTOs identified among the case studies do not have a domain specialist as their Director; most successful Directors appear to have first degrees sometimes supplemented by a MBA. This notion of the importance of an entrepreneurial TTO led by a business-savvy director is consistent with the principal requirement usually identified by scholars as essential to TT&C success: the personal relationship. The essential pre-conditions to Silicon Valley were leadership and social networks (Rogers and Larsen, 1984), and the quality of the human infrastructure is just as relevant in a university TTO as it is in any other business undertaking. Thus, while the processes of the TTO are important, they are not sufficient in themselves to ensure success without the ability of the TTO, and its Director in particular, to develop the personal relationships and apply the principles of entrepreneurial business that are essential to TT&C success.

It is important to note that the conclusions drawn here on the characteristics possessed by universities successful in TT&C have been found by the authors to apply to universities in developed economies irrespective of size, prestige or location. Clearly, then, smaller and regional universities are subject to the same factors – in this matter they are not unique and should approach TT&C in the same way as universities generally, although it is acknowledged that they will still have particular issues resulting from their individual culture and profiles.

Successful TTOs have particular qualities of entrepreneurialism and commerciality, characteristics that are not part of university culture. As one TTO Director expressed it: "...we have a business culture, the university hasn't".

The data also suggest that success breeds success. There is evidence in reported research by the likes of Lee (2000) that university-industry collaboration is sustainable for the foreseeable future, while some are tempted to suggest, like Feller (1997, p.36), that "Technology transfer has the potential to become [a] self-propelling force within the university".

The second major result from this analysis is that the objective of university research has to be the greater welfare of the community. The relevant question becomes: does university emphasis on TT&C and the existence of a university TTO promote this primary objective of greater welfare of the community? Or is it neutral, or does it act as an inhibitor? And what are the metrics by which the question may be answered? It is reasonably self-evident that TT&C performed by universities is of value only if it serves the greater welfare of the community. But this issue is not presently capable of definitive answer because devising adequate metrics has proven to be all but intractable. Metrics represented by reasonable proxies go some way to providing an answer, but even some of the most thoughtful analyses struggle to make a convincing case using numerical results.

What constitutes the greater welfare of the community has many facets. If universities merely insert into the community the fruits of their work as quickly and economically as possible it may not improve welfare unless their work is of true value to the community. Therefore a university has to undertake research of value in order for society to benefit, but it can only do this if it has the best researchers supported by the best infrastructure. But the best researchers will, increasingly, only be attracted to universities where their work is being effectively disseminated into the community – that is, a good research university has to support TT&C[37]. Therefore, TT&C is as much a part of an effective research enterprise as any other part of the system. And the TT&C element of the research enterprise may be measured against such factors as those used above to identify the benchmark universities and then compare their practice to these benchmarks. This analysis suggests that there is a growing connection between TT&C and quality research.

It is important to recognize in undertaking this comparison that no two universities will be exactly the same. As a result there is no one single approach that can be adopted by them all, and producing lists of this or that provides little useful guidance. The important factor is to develop an environment in which each university can implement TT&C to the degree and in the way that best suits its circumstances, while encouraging each of them to use every avenue to continue to pursue the welfare of the community as their understanding and experience in different ways of undertaking TT&C develop (Litan, *et al.*, 2007).

Actions Available to Universities

While the two factors mentioned above are of primary importance to university TT&C success, they are not the only factors. The literature and our research have identified a number of other important factors, the *institutional enablers* (the factors within the control of the individual university), that are needed to encourage success, which are discussed below. But without the two key factors identified above in place any structure, no matter how good, is unlikely to produce acceptable results, while having the best institutional support, leading TTO management and world-class research can go a long way to overcoming even indifferent institutional processes.

First, the factors that may, at first glance, appear important. A common metric employed when measuring university commercialization is the amount spent on research. It is expected that the more a university spends on research the greater the quantum of research results and the greater the amount of commercialization. TTO Directors at successful universities say that if they had more TTO resources they could get even more commercialization results, and there is no reason to gainsay this. But there are examples of universities with quite small research budgets at the leading edge of commercialization. At the same time, many ivy league universities in the United States with large research budgets are not recognized as commercialization leaders. There are factors at work other than research expenditure by universities that correlate with superior commercialization performance. While volume of research may not be quite so important, quality of research is. One successful TTO Director commented that "While the university is small,... we just happen to be able to attract really good researchers, and we've got some fantastic world-class researchers here; so the quality of technology is high". The important factor is that a university must produce a sufficient volume of world-class research if it is to produce superior TT&C performance.

The proximity of universities to sources of venture capital could be expected to give them an edge when it comes to commercialization performance. But for every case where this could be true: Stanford and Berkeley in California, and MIT and Harvard in Massachusetts, there are contrary examples: Arizona State University and the University of North Carolina. There may be some advantage in proximity to major capital sources, but it is not compelling.

The ready availability of early-stage funding: pre-seed, seed, proof-of-principle and the like, could be an important issue in encouraging commercialization. This is a reasonable conclusion, although the amounts available for this are rarely large in any university, most commonly coming from retained earnings of the TTO, from the university itself, from government, from business and industry and, possibly, not-for-profit foundations. There is a case to be made that government should have no role in supplying such funding (although in smaller economies this may not be realistic). U.S. Public University has no internal funds for early-stage development and largely relies on industry taking an interest and funding potential technologies. But there are examples where universities with relatively young TTOs are able to use retained earnings to fund early-stage development. The use of internally-generated funds for research development enforces considerable discipline on the allocation of funding and may be a sound approach. But, in any event, the availability of funding that is supplied through government, industry and other sources, is not a differentiator between successful and other TTOs, because funding appears to be available to some extent for any good idea, irrespective of university background and size.

Access to an incubator can be important in some cases, more especially in life sciences where the cost of establishing wet laboratories can be prohibitive, but there is no need for a university to own or operate one. No case study university expressed any enthusiasm to own and operate an incubator, while all university TTO managements were opposed to the university being involved in their provision. To the extent that they are necessary or desirable, incubators should be provided outside the aegis of universities by government (less desirable) or private capital through foundations (more desirable).

Joint research centers can be an effective form of linkage between universities and industry, and may encourage a more commercial orientation on the part of universities. However, there are instances where they complicate issues such as intellectual property ownership vis-à-vis the university, and they generally do not benefit industry as a whole, only those companies that are part of the consortium. Where they supply a real industry need and their continued existence would be desirable, it is reasonable to expect that they would be sustained by private funding as occurs commonly in the United States.

Finally, there is evidence to suggest that clustering[38] of related businesses can form part of a virtuous cycle of reinforcing research, development, and commercialization (Porter, 1990, 1998). While they can have a profound effect on particular industries at particular times, they emerge spontaneously and appear incapable of being created artificially by government wish or fiat. Smaller economies are unlikely ever to have anything in the manner of a Silicon Valley, although valuable clusters of life sciences entities have been shown to evolve in some countries such as around the University of Cambridge in the U.K. and the University of Toronto in Canada.

Turning to the other *institutional enablers* that support successful TT&C, the two most important: institutional support and exceptional TTO leadership have been identified. The third element essential for TT&C success within the control of a university is a flow of sufficient world-class research. As discussed earlier it is the quality rather than the volume of world-class research that is important, but there must be a continuing flow available to sustain the commercialization enterprise.

Another *institutional enabler* important to commercialization success is the possession of a suitable intellectual property regime within universities. Collectively, universities (and other publicly-funded research entities) in Australia and the U.K. have developed sets of principles which are used to inform individual university intellectual property policies and practices. Neither the largely prescriptive approach used in the U.S. (the *Bayh-Dole Act*) nor the *laissez-faire* approach of Canada and New Zealand appear to be superior to that adopted in Australia and the U.K. United States-style prescription through the *Bayh-Dole Act* is unlikely to be necessary in other countries to overcome any existing problem[39], while analysis of the Canadian intellectual property environment shows that the significantly different policies and practices adopted by individual universities can sometimes make it more difficult to get ideas commercialized.

The importance of sound linkages between university and industry arises in almost every review of the environment as an essential component of effective TT&C. This research has shown that there are many modes of linkage; which among these many modes may suit an individual university will depend on a number of factors including the profile of the university, its culture, and its history and experience. It is impossible, for this reason, to offer a universal prescription as to which linkages should be adopted by any particular university. However, two key factors are important in ensuring that a university builds and maintains appropriate and strong industry links: institutional support for TT&C, and the leadership of the TTO, which are the same factors important in university TT&C success generally. In other words, by establishing the environment that fosters TT&C, the creation of the other important components necessary to make it work should necessarily follow-on. Through its structure and its academic faculty and staff a university will be able to identify and promote the linkages that will work best for it.

CONCLUSION

This chapter developed case studies of four North American universities with the intention of better understanding the factors that contribute to apparent success by universities in TT&C. The case study universities were examined according to fourteen factors which, from the literature, appeared to be sufficient to characterize universities' performance in TT&C.

Where the same common infrastructure exists among a group of universities, as it does for all United States universities, for example, then differences in performance between universities are largely a function of environmental factors, or *institutional enablers*, which are substantially within the control of individual universities. We found two common factors existed among universities that exhibited superior performance in TT&C. First, the university President and other executives concerned in commercialization have to believe in it and make a genuine commitment to its success. This must be evidenced both in rhetorical terms (support with words) and material terms (support with resources). Second, the TTO has to be led by an individual who possesses a number of attributes: the ability and willingness to work in the arcane structure of the university; the ability to be both an entrepreneur and a manager; the flexibility of mind and perceptiveness to see what is happening in a dynamic field as the environment evolves and matures; and be a leader of people and business.

Recognizing these two factors and the differences in culture among the disparate universities, it was concluded that university government had to encourage an environment in which each university can implement TT&C to the degree and in the way that best suits its circumstances. There is not one way to implement or operate TT&C that will work for every university, although the two key factors of executive support and TTO management capability remain of foremost importance in all environments.

The fourteen *institutional enablers* that we used to characterize the TT&C performance of our case study universities were:

- The regulatory environment;
- The character of the university;
- The TT&C structure employed by the university;
- The TT&C processes used by the university;
- The TT&C performance of the university;
- The TT&C performance of the TTO (as distinct from the university);
- Knowledge and awareness of TT&C among faculty members and researchers;
- The incentives offered by universities for faculty members and researchers to engage in TT&C;
- The nature and strength of the linkages between the university and industry;
- The impact technology parks and incubators had on a university's TT&C performance;
- The impact of intellectual property rules;
- The availability of financing for development of ideas – both at the early stage and later venture capital finance;
- The entrepreneurial culture of the university and its impact on TT&C; and
- The role played by intermediaries in promoting and supporting TT&C.

We found no universal prescription emerged from among these fourteen institutional enablers that could be applied to make universities superior performers. In each case allowance had to be made for each university's character, culture, history and environment. Of course, factors such as suitable organization structures and business processes, faculty education and incentives, industry linkages, intellectual property rules, and early-stage financing are important in TT&C performance. However, in any particular university these factors are rarely optimum, but become refined, improved and made workable by the two key factors identified above: genuine institutional support for TT&C; and a superior TTO Director.

REFERENCES

ACST - Prime Minister's Advisory Council on Science and Technology. (1999). *Public investments in university research: Reaping the benefits – Report of the expert panel on the commercialization of university research*. Industry Canada.

Adamson, W. (2004). *Finding the key elements of commercialisation – Process, skills and reward*, an independent White Paper prepared for the Australian Institute for Commercialisation, Retrieved October, 15, 2005, from: http://www.digitalinvestor.com.au/files/O2G65PO8WX/ADAMSON-Fixing-Key-Elements-Commercialisation.pdf.

Agrawal, A., & Henderson, R. (2002). Putting patents in context: Exploring knowledge transfer from MIT. *Management Science*, *48*(1), 44–60. doi:10.1287/mnsc.48.1.44.14279

Argyres, N., & Liebeskind, J. (1998). Privatizing the intellectual commons: Universities and the commercialization of biotechnology. *Journal of Economic Behavior & Organization*, *35*, 427–454. doi:10.1016/S0167-2681(98)00049-3

Audretsch, D. (2001). The role of small firms in U.S. biotechnology clusters. *Small Business Economics*, *17*(1-2), 3–15. doi:10.1023/A:1011140014334

Audretsch, D., & Lehmann, E. (2005). Do university policies make a difference? *Research Policy*, *34*, 343–347. doi:10.1016/j.respol.2005.01.006

AUTM - The Association of University Technology Managers. (2003). *AUTM licensing survey: FY 2003 licensing summary*. Deerfield, IL: The Association of University Technology Managers.

AUTM - The Association of University Technology Managers. (2007a). *The better world report part 2, technology transfer works: 100 innovations from academic research to real-world applications*. Deerfield, IL: The Association of University Technology Managers.

AUTM - The Association of University Technology Managers. (2007b). *AUTM U.S. licensing survey: FY 2005 licensing summary*. Deerfield, IL: The Association of University Technology Managers.

AUTM - The Association of University Technology Managers. (2007c). *AUTM Canadian licensing survey: FY 2005 licensing summary*. Deerfield, IL: The Association of University Technology Managers.

Bellone, C., & Goerl, G. (1992). Reconciling public entrepreneurship and democracy. *Public Administration Review*, *52*(2), 130–134. doi:10.2307/976466

Bok, D. (2003). *Universities in the marketplace: The commercialization of higher education*. Princeton, NJ: Princeton University Press.

Bozeman, B. (2000). Technology transfer and public policy: a review of research and theory. *Research Policy*, *29*, 627–655. doi:10.1016/S0048-7333(99)00093-1

Bruderl, J., Preisendorfer, P., & Ziegler, R. (1992). Survival chances of newly founded business organizations. *American Sociological Review*, *57*, 227–242. doi:10.2307/2096207

Canada Foundation for Innovation. (2002). *Conference on Innovation and Commercialization of University Research* (Vol. 2). Canada Foundation for Innovation.

Clark, B. (1998). *Creating entrepreneurial universities*. Oxford, UK: International Association of Universities and Elsevier Science.

Clayman, B., & Holbrook, J. (2003). *The survival of university spin-offs and their relevance to regional development*. Canada Foundation for Innovation.

Cohen, W., Nelson, R., & Walsh, J. (2002). Links and impacts: The influence of public research on industrial R&D. *Management Science*, *48*(1), 1–23. doi:10.1287/mnsc.48.1.1.14273

Collins, S., & Wakoh, H. (2000). Universities and technology transfer in Japan: Recent reforms in historical perspective. *The Journal of Technology Transfer*, *25*(2), 213–222. doi:10.1023/A:1007884925676

Colyvas, J., Crow, M., Gelijns, A., Mazzoleni, R., Nelson, R., Rosenberg, N., & Sampat, B. (2002). How do university inventions get into practice? *Management Science*, *48*(1), 61–72. doi:10.1287/mnsc.48.1.61.14272

Connell, D. (2006). *"Secrets" of the world's largest seed fund: How the United States Government uses its Small Business Innovation Research (SBIR) program and procurement budgets to support small technology firms*. Cambridge, UK: University of Cambridge, Centre for Business Research.

Debackere, K., & Veugelers, R. (2005). The role of academic technology transfer organizations in improving industry science links. *Research Policy*, *34*, 321–342. doi:10.1016/j.respol.2004.12.003

DEST Department of Education. Science and Training. (2003). Mapping Australian science and innovation: National and international linkages: Background paper. Canberra: Department of Education, Science and Training.

Domestic Policy Council - U.S. Domestic Policy Council Office of Science and Technology Policy. (2006). *American competitiveness initiative – Leading the world in innovation*. U.S. Domestic Policy Council Office of Science and Technology Policy.

Etzkowitz, H., & Leydesdorff, L. (2001). *Universities and the global knowledge economy*. London: Continuum.

Feldman, M., Feller, I., Bercovitz, J., & Burton, R. (2002). Equity and the technology transfer strategies of American research universities. *Management Science*, *48*(1), 105–121. doi:10.1287/mnsc.48.1.105.14276

Feller, I. (1997). Technology transfer from universities. In Smart, J. (Ed.), *Higher education: Handbook of theory and research* (*Vol. 12*, pp. 1–43). New York: Agathon Press.

Ferris, W. (2001). Australia chooses: Venture capital and a future Australia. *Australian Journal of Management*, *26*, 45–64. doi:10.1177/031289620102601S03

Fisher, D., & Atkinson-Grosjean, J. (2002). Brokers on the boundary: Academy-industry liaison in Canadian universities. *Higher Education*, *44*, 449–467. doi:10.1023/A:1019842322513

Gans, J., & Stern, S. (2003). *Assessing Australia's innovative capacity in the 21st century* (Working Paper 2003-16). Melbourne: University of Melbourne, Melbourne Business School.

GAO - U.S. General Accounting Office. (1990). *Case study evaluations (Transfer Paper 10.1.9)*. U.S. General Accounting Office.

Geiger, R., & Sá, C. (2008). *Tapping the riches of science; Universities and the promise of economic growth*. Cambridge, MA: Harvard University Press.

Goozner, M. (2004). *The $800 million pill: The truth behind the cost of new drugs*. California: University of California Press.

Greenberg, D. (2007). *Science for sale – The perils, rewards and delusions of campus capitalism*. Chicago, IL: The University of Chicago Press.

Gunasekara, C. (2005). The role of universities in shaping regional agglomeration: case studies in the Australian setting. *International Journal of Technology Transfer & Commercialisation*, *4*(4), 525–539.

Gupta, U. (2004). *The first venture capitalist – Georges Doriot on leadership, capital, & business organization*. Calgary, Canada: Gondolier.

Harman, G. (2002). Australian university-industry links: Researcher involvement, outputs, personal benefits and "withholding" behaviour. *Prometheus*, *20*(2), 143–158. doi:10.1080/08109020210137529

Hatakenaka, S. (2005). *Development of third stream activity: Lessons from international experience*. Higher Education Policy Institute. Retrieved September 9, 2007, from: http://www.hepi.ac.uk/downloads/DevelopmentofthirdstreamfundingSachiHatakenaka.pdf

Howard, J. (2005). *The emerging business of knowledge transfer, creating value from intellectual products and services (Report of a Study Commissioned by the Department of Education, Science, and Training)*. Canberra: Department of Education, Science and Training.

Howells, J. (2006). Intermediation and the role of intermediaries in innovation. *Research Policy*, *35*(5), 715–728. doi:10.1016/j.respol.2006.03.005

KCA - Knowledge Commercialisation Australia. (2006). *Big book of ideas 2006*. Knowledge Commercialisation Australia.

Kirp, D. (2003). *Shakespeare, Einstein, and the bottom line – The marketing of higher education*. Cambridge, MA: Harvard University Press.

Lambert, K. (2003). *Lambert review of business-university collaboration*. London: HMSO.

Lee, Y. (2000). The sustainability of university-industry research collaboration: An empirical assessment. *The Journal of Technology Transfer*, *25*(2), 111–133. doi:10.1023/A:1007895322042

Link, A., & Siegel, D. (2005). Generating science-based growth: An econometric analysis of the impact of organizational incentives on university–industry technology transfer. *European Journal of Finance*, *11*(3), 169–181. doi:10.1080/1351847042000254211

Litan, R., Mitchell, L., & Reedy, E. (2007). *Commercializing university innovations: Alternative approaches* (NBER Working Paper, 16 May 2007). Retrieved June 29, 2007, from: http://ssrn.com/abstract=976005.

Mansfield, E. (1991). Academic research and industrial innovation. *Research Policy*, *20*, 1–12. doi:10.1016/0048-7333(91)90080-A

Mansfield, E. (1998). Academic research and industrial innovation: An update of empirical findings. *Research Policy*, *26*, 773–776. doi:10.1016/S0048-7333(97)00043-7

Martin, M., Gruetzmacher, R., Lanham, R., & Brady, J. (2004). Assessing technology transfer and business development potential: Technology cluster analysis. *Economic Development Quarterly*, *18*(2), 168–173. doi:10.1177/0891242403261088

Mohannak, K. (1999). A national linkage program for technological innovation. *Prometheus*, *17*(3), 323–336. doi:10.1080/08109029908632135

Mowery, D., Nelson, R., Sampat, B., & Ziedonis, A. (2001). The growth of patenting and licensing by U.S. universities: an assessment of the effects of the Bayh-Dole Act of 1980. *Research Policy, 30*, 99–119. doi:10.1016/S0048-7333(99)00100-6

Mowery, D., & Shane, S. (2002). Introduction to the special issue on university entrepreneurship and technology transfer. *Management Science, 48*(1), v–ix. doi:10.1287/mnsc.48.1.0.14277

Narin, F., Hamilton, K., & Olivastro, D. (1997). The increasing linkage between U.S. technology and public science. *Research Policy, 26*, 317–330. doi:10.1016/S0048-7333(97)00013-9

National Academy of Sciences. National Academy of Engineering, & Institute of Medicine. (2006). Rising above the gathering storm – Energizing and employing America for a brighter economic future. The National Academies.

Nelson, R. (2001). Observations on the post-Bayh-Dole rise of patenting at American universities. *The Journal of Technology Transfer, 26*(1-2), 13–19. doi:10.1023/A:1007875910066

O'Shea, R., Allen, T., O'Gorman, C., & Roche, F. (2004). Universities and technology transfer: A review of academic entrepreneurship literature. *Irish Journal of Management, 25*(2), 11–29.

Owen-Smith, J., & Powell, W. (2001). To patent or not: Faculty decisions and institutional success at technology transfer. *The Journal of Technology Transfer, 26*(1-2), 99–114. doi:10.1023/A:1007892413701

Peters, L., Rice, M., & Sundararajan, M. (2004). The role of incubators in the entrepreneurial process. *The Journal of Technology Transfer, 29*(1), 83–91. doi:10.1023/B:JOTT.0000011182.82350.df

Pollard, D. (2006). Innovation and technology transfer intermediaries: A systemic international study. *Advances in Interdisciplinary Studies of Work Teams, 12*, 137–174. doi:10.1016/S1572-0977(06)12006-3

Porter, M. (1990). *The competitive advantage of nations*. New York: The Free Press.

Porter, M. (1998). Clusters and the new economics of competition. *Harvard Business Review, 76*, 77–90.

Press, E., & Washburn, J. (2000). The kept university. *Atlantic Monthly, 285*(3), 39–54.

Productivity Commission. (2007). *Public support for science and innovation (Productivity Commission Report)*. Commonwealth of Australia.

Reamer, A., Icerman, L., & Youtie, J. (2003). *Technology transfer and commercialization: Their role in economic development*. Washington, DC: U.S. Department of Commerce, Economic Development Administration.

Riddle, C. (2004). *Commercialization strategies of Canadian universities & colleges: Challenges at the university/college-industry interface, including intellectual property policies (A Study for the Advisory Council on Science and Technology)*. Government of Canada.

Rogers, E., & Larsen, J. (1984). *Silicon valley fever: Growth of high-technology culture*. New York: Basic Books.

Rousseau, D. (2006). Is there such a thing as "evidence-based management"? *Academy of Management Review, 31*(2), 256–269.

Shane, S. (2002). Selling university technology: Patterns from MIT. *Management Science, 48*(1), 122–137. doi:10.1287/mnsc.48.1.122.14281

Shane, S. (2004). Encouraging university entrepreneurship? The effect of the Bayh-Dole Act on university patenting in the United States. *Journal of Business Venturing, 19*, 127–151. doi:10.1016/S0883-9026(02)00114-3

Shane, S., & Stuart, T. (2002). Organizational endowments and the performance of university start-ups. *Management Science, 48*(1), 154–170. doi:10.1287/mnsc.48.1.154.14280

Shepherd, J. (2006, September). Transfers prove costly. [Higher Education Supplement]. *Times (London, England)*, 15.

Siegel, D., Waldman, D., Atwater, L., & Link, A. (2004). Toward a model of the effective transfer of scientific knowledge from academicians to practitioners: qualitative evidence from the commercialization of university technologies. *Journal of Engineering and Technology Management*, *21*(1-2), 115–142. doi:10.1016/j.jengtecman.2003.12.006

Siegel, D., Waldman, D., & Link, A. (1999). *Assessing the impact of organizational practices on the productivity of university technology transfer offices: An exploratory study* (NBER Working Paper 7256). National Bureau of Economic Research.

Siegel, D., Westhead, P., & Wright, M. (2003). Science parks and the performance of new technology-based firms: A review of recent U.K. evidence and an agenda for future research. *Small Business Economics*, *20*, 177–184. doi:10.1023/A:1022268100133

Sime, J. (2004). *The commercialisation of intellectual property from public sector research establishments: a discussion paper*. Unpublished paper based on a report to the Australian Institute for Commercialisation presented at the conclusion of a series of master classes conducted around Australia during 2003, Australian Institute for Commercialisation.

Slaughter, S., & Leslie, L. (1997). *Academic capitalism – Politics, policies, and the entrepreneurial university*. Baltimore, MD: The Johns Hopkins University Press.

Sobol, M., & Newell, K. (2003). Barriers to and measurements of the diffusion of technology from the university to industry. *Comparative Technology Transfer and Society*, *1*(3), 255–278. doi:10.1353/ctt.2003.0032

Stein, D. (Ed.). (2004). *Buying in or selling out? The commercialization of the American research university*. Piscataway, NJ: Rutgers University Press.

Thursby, J., & Thursby, M. (2002). Who is selling the ivory tower? Sources of growth in university licensing. *Management Science*, *48*(1), 90–104. doi:10.1287/mnsc.48.1.90.14271

Tornatzky, L., Waugaman, P., & Gray, D. (2002). *Innovation U.: New university roles in a knowledge economy*. Southern Growth Policies Board.

UVAPF - University of Virginia Patent Foundation. (2004). *Operating Manual.*

Vitale, M. (2004). *Commercialising Australian biotechnology*. Australian Business Foundation.

Washburn, J. (2005). *University Inc. – The corporate corruption of higher education*. New York: Basic Books.

Westhead, P. (1997). R&D "inputs" and "outputs" of technology-based firms located on and off Science Parks, R&. *Mana*, *27*(1), 45–62.

Yin (2003). *Applications of case study research* (2nd ed.). Thousand Oaks, CA: Sage.

Zemsky, R., Wegner, G., & Massy, W. (2006). *Remaking the American University – Market-smart and mission-centered*. Piscataway, NJ: Rutgers University Press.

Zhao, F. (2004). Commercialization of research: a case study of Australian universities. *Higher Education Research & Development*, *23*(2), 223–236. doi:10.1080/0729436042000206672

ENDNOTES

[1] The U.S. General Accounting Office has published a useful tome on the case study method applicable to these circumstances: (GAO, 1990).

[2] The industries and the number of firms surveyed by Mansfield (1991,1998) in each are: drugs and medical products (16); information processing (14); chemical (13); electrical (10); instruments (10); machinery (10): and metals (4)

[3] For example, Australian data published in the Big Book of Ideas 2006 by Knowledge Commercialization Australia (KCA 2006) disclosed 199 discoveries arising from Australian research in three categories: life sciences (108 discoveries or 54%), information and communications technology (31 discoveries or 16%), and advanced engineering (60 discoveries or 30%, of which 14 relate to medical advances).

[4] Gardasil is one of the trade names of a drug developed from research conducted at the University of Queensland, Australia, that inhibits the development of certain types of human pappilomavirus (HPV) known to lead to cervical cancer. It was patented in 1991 but only commercially available in 2006. Source: Commercialization Success Stories published by UniQuest (University of Queensland) 2007.

[5] While not an exception to the rule that ideas, as such, are almost never registrable, the form of expression of an idea may be protected by copyright in most cases, but copyright does not give any right to protection of the underlying idea (as distinct from its form of expression).

[6] But not just universities. In the United States there are a number of federal Government-sponsored laboratories, particularly in Defense, Health and Energy.

[7] Listed on the London Stock Exchange under the mnemonic: IVO. Further details from its website: www.imperialinnovations.co.uk

[8] Pub. L. 96-517, §6(a), Dec. 12, 1980, 94 Stat. 3018 (35 U.S.C. 200 *et seq.*) Also known as the *University and Small Business Patent Procedure Act* of 1980. This Act changed the prior position so that universities became entitled, subject to a number of specified conditions, to own and commercially exploit intellectual property arising from research conducted by them using federal funding.

[9] There is now a substantial body of books on this topic, such as Slaughter and Leslie (1997), Kirp (2003), Stein (2004), Washburn (2005), Zemsky, *et al.* (2006), Greenberg (2007), and Geiger and Sá (2008)

[10] Through so-called institutional patent agreements.

[11] Interviews were conducted at universities described in the analysis below as: U.S. Public University, and U.S. Private University in the U.S. ; and Canadian Uni1, and Canadian Uni2 in Canada. Both U.S. case study universities have had controversy in their past about TT&C.

[12] The U.S. also suffered an earlier crisis of concern about its loss of technological pre-eminence during the 1970s and early 1980s, leading to a raft of initiatives, the *Bayh-Dole Act* among them, that are usefully summarized in the article by Bozeman (2000).

[13] http://www.carnegiefoundation.org/index.asp

[14] The U.S. environment is also complicated by the existence of university *systems*, where a number of universities are, for some purposes, part of a system while, for other purposes, are counted as separate universities. The University of California system is probably the best known system, but many States have multi-university systems.

[15] These figures include universities and colleges affiliated with AUCC, the Association of Universities and Colleges of Canada: http://www.aucc.ca/can_uni/our_universities/index_e.html

[16] The numbers in this paragraph are calculated by the author based on AUTM member affiliation data published by AUTM in February 2007.

[17] The sixteen universities are: University of Wisconsin, MIT, University of Minnesota, University of California, University of Rochester, University of Washington, University of Florida, Wake Forest University, Emory University, New York University, University of Massachusetts, University of Iowa, Harvard University, University of Michigan, University of Colorado and East Virginia Medical School.

[18] The U.S. Private University Interview 2007 giving rise to this view is of particular authority because the interviewee for some years directed the AUTM data collection on which the observation is based.

[19] The University of British Columbia and the Université de Sherbrooke.

[20] Both programs authorized and managed under the following legislation: Small Business Reauthorization Act of 2000, P.L. 106-554 (15 U.S.C. 631 and 638); Title II of the Rehabilitation Act of 1973, as amended, P.L. 105-220 (29 U.S.C. 760-764); Title VI, Sec. 605 of the Higher Education Act, as amended (20 U.S.C. 1125); The Carl D. Perkins Vocational and Technical Education Act of 1998, P.L. 105-332 (20 U.S.C. 2301 et seq.); Education Sciences Reform Act of 2002, Title I-B (20 U.S.C. 9531-9534)

[21] Number from U.S. Private University interview

[22] These were known as Institutional Patent Agreements

[23] Quoted from par. 1(c) of the Small Business Innovation Research Program Policy Directive issued by the SBA, a copy of which may be located at: http://www.sba.gov/sbir/SBIR-PolicyDirective.pdf

[24] This description is derived from Sobol and Newell (2003, pp.269-270).

[25] Derived from the U.S. Department of Education website: http://www.ed.gov/programs/sbir/legislation.html

[26] The two universities that scored best as being "actively and successfully participating in, or linked to, state and local economic development" were not members of the AAU: Georgia Tech and NC State. Two other universities of the twelve identified in that study also were not members of the AAU: Virginia Tech and the University of Utah.

[27] These numbers are full-time equivalent, or FTE staff.

[28] For example, the issue of the efficacy of a central office TTO versus hub-and-spoke was the subject of much debate over several days in February 2007 amongst members of the North American TTO community in the large on-line network: Techno-L. (24 and 25 February 2007 – records are available)

[29] However this assumption may have been overturned to a large degree in Australia and other common law jurisdictions by virtue of a decision of the Federal Court of Australia in University of Western Australia -v- Gray (No 20) [2008] FCA 498 delivered on 17 April 2008. This decision is being appealed but, whatever the outcome, it serves to demonstrate a degree of uncertainty under which universities and their TTOs operate as it concerns intellectual property ownership rights.

[30] Congress also enacted a number of complementary Acts around this time including the *Stevenson-Wydler Technology Innovation Act 1980* (dealing with technology transfer from government laboratories) and the *National Cooperative Research Act 1984* (limiting the effect of anti-trust rules on certain co-operative conduct between researchers and industry), however they are not relevant to the present argument. Both Acts have been amended in material terms since their enactment.

[31] Information about BMI is located at: http://www.battelle.org/

[32] For more details, see: http://www.nttc.edu/

[33] Common infrastructure includes such factors as: investment in basic research; tax policies affecting corporate R&D and investment spending; supply of risk capital; aggregate level of education in the population; pool of talent in science and technology; information and communication infrastructure; protection of intellectual property; openness to international trade and investment; and overall sophistication of demand.

[34] Institutional enablers for the purpose of TT&C include university policies such as those covering intellectual property ownership, promotion of academic faculty, sharing of risks and rewards in TT&C, and the availability of funding for early-stage idea development.

[35] These characteristics may be compared and contrasted to those enunciated by Bellone and Goerl (1992) who said that private entrepreneurs possess four characteristics: autonomy, a personal vision of the future, secrecy, and risk-taking propensity. Certainly a person prepared to lead a university TTO probably needs a slightly different set of capabilities to those described by Bellone and Goerl by virtue of the significantly different environment in which they work.

[36] And the shortage of this class of person is also emphasized by Mowery and Shane (2002, p.viii) when they say "… management by universities of technology licensing activities requires a set of skills that are extremely rare within universities and in short supply more generally". This fact was identified in the data in this research, as well.

[37] It is not intended to confuse TT&C at this stage with any concept associated with income from TT&C and resulting financial consequences.

[38] Clustering may be defined as the grouping of a number of competing, collaborating and interdependent businesses working in a common industry concentrated in a geographic region.

[39] Although the potential problems that may arise from the 2008 case of University of Western Australia -v- Gray (No 20) [2008] FCA 498 may require this conclusion to be re-visited once appeals have been exhausted002E

Chapter 2
The Evolution of ICT Institutions in Thailand and Malaysia

Nicholas Maynard
RAND Corporation, USA

ABSTRACT

A country's national technology strategies can be an important contributor to economic development through its support of technology adoption and by advancing the national technology capacity. The development of a domestic information and communications technology (ICT) sector within a developing country requires the creation of specialized institutions that carefully coordinate their initiatives with the private sector. This case study research of Thai and Malaysian science and technology (S&T) institutions shows that this institutional and policy reform process is directly influenced by regional activities, as countries seek to match their regional peers for technology development. This effort to support ICT utilization requires governments to rapidly alter their policy goals and initiatives in response to shifts in technologies, global market demand, international investment, and local workforce capabilities.

INTRODUCTION

National public support for increased technology innovation and utilization can take many forms, including government-supported technology training; aggregating demand and serving as an anchor tenant; fostering e-government, e-health, and other services; universal service funds; and governmental safeguards for services such as e-commerce (Frieden, 2005). The communications technologies also need to be adapted to the needs of the local economic, political, and cultural environment, particularly if these services are originally introduced by an international entity. To meet local requirements, these national efforts require public-private-university coordination to successfully adapt information and communications technology (ICT) technologies transferred internationally and to enhance services created indigenously (Feinson, 2003; Balaji & Keniston, 2005).

The chapter offers a detailed look at both Thai and Malaysian ministries of ICT and S&T, includ-

DOI: 10.4018/978-1-61520-609-4.ch002

ing a discussion of the evolution of these institutions, the cross-border influence between the two countries, and the organizational challenges facing these agencies as they seek to implement their national technology strategies.

LITERATURE REVIEW

To understand the ICT policy choices of national governments, it is first important to note three major trends identified by the literature within the telecommunications sector. The first is the development of mobile and Internet technologies in addition to fixed line telephony (Baliamoune-Lutz, 2003). The second shift is the global trend away from monopoly operators to competitive carriers across these fixed, mobile, and Internet technologies (Wilson & Wong, 2003). The third shift under way is from governmental control to private ownership, or a mix of public and private with independent regulatory agencies (Levy & Spiller, 1994). Steinmuller suggests that ICTs, which can lower transaction costs, may be able to offer developing countries a conduit for avoiding stages that require high levels of capital and fixed asset concentration, as defined by Rostow's "stages of development" (Rostow, 1960), and moving directly to a knowledge-based economy (Steinmueller, 2001). As a result, many developing countries now view these technologies as an important conduit to fostering both productivity gains (McGuckin & Stiroh, 1998; Baumol & Solow, 1998) and economic development (Saunders, 1994).

Developing countries have accelerated their efforts to deliver affordable ICT access and improved utilization rates among their residents through a range of ICT policy initiatives (Graham, 2000). The two goals of increased access and utilization are important in enhancing a developing country's ability to compete globally for jobs and investment. Although these goals are touted frequently, they are not always tailored for a given country (Cohen-Blankshtain & Nijkamp, 2003). Policymakers must ask themselves how they define affordable access and improved utilization within the geographic, competitive, and political environment of the country (Javary & Mansell, 2002). Once these goals are defined, a set of policies can be implemented and a decision on the optimal ICT solutions can be made.

ICT infrastructure and applications will be adopted by a developing country in stages, with policymakers shifting their goals from supporting increased access, to developing a robust private sector, and finally to creating a globally competitive ICT industry (Grubesic & Murray, 2004). Although these goals are not mutually exclusive, there is a progression in policy and technological complexity as countries move away from directly supporting access infrastructure through a state-owned enterprise, to directing market competition through a regulatory agency, and then to indirectly supporting access through a ministry of ICT. In countries that support a domestic ICT industry, the government's role shifts to becoming a coordinator and advisor to the private sector. To overcome these changing priorities and governmental roles, countries are forced to reevaluate their goals on a regular basis, adjusting their policies and technology choices accordingly. As a result, ICT goals within a developing country will not be static; in fact, they must be flexible enough to adapt to the changing technological and economic conditions to achieve an optimum outcome (Strover & Berquist, 1999).

Institutional structure is defined as the level of institutional, legal, and regulatory structures put in place to support the creation of a national ICT strategy. Depending on the focus of the national policy, there can be a wide range of institutional structures across developing countries. For policies focused on indirect support to the market, such as financial subsidies to the private ICT sector, there are few institutional or regulatory hurdles for sustaining the effort. However, to fund and operate an ICT research park and the requisite infrastructure, similar to Malaysia's Multimedia Super Corridor (MSC), a governmental institution or other organization may need to be created that has the necessary authority to fund and coordinate these policies (Shari, 2003).

This institutional structure requires significant governance capabilities on the part of the new organization as well as a sustained financial commitment. Institutional structure is also linked to market and infrastructure requirements. For ICT strategies that are truly national in scope, rather than a regional effort, this may add a layer of bureaucracy for the ICT institution to coordinate. An increase in the complexity of the infrastructure for the strategy – for a national network, for example – will increase the institutional and regulatory hurdles for the ICT policy.

RESEARCH METHODOLOGY

Current ICT research focuses heavily on economic and market factors behind ICT adoption and utilization, and its impact on economic growth. Many of these studies have found significant benefits to developed countries of the Organization of Economic Co-operation and Development (OECD) and Newly Industrialized Economies (NIEs), but have not been able to conclusively demonstrate a correlation between ICT investment and economic growth in developing countries (Bassanini, 2002; Storm & Naastepad, 2005; Wilson & Wong, 2003). Rather than working to quantify the impact ICTs have on macroeconomic conditions, this research focuses on the process through which policies to support ICT adoption are successfully implemented across a range of emerging economies.

This case study approach outlines the economic, political, and technological environments of Thailand and Malaysia, highlighting the key policy and institutional choices that have resulted in their respective technological utilization rates. This case study analysis focuses on the growing gap between the two countries, where Thailand has not been able to keep pace with Malaysia, which has accelerated its ICT utilization rate and sector development. Malaysia has achieved this by successfully completing a series of ICT institutional and policy reforms that have created a competitive ICT market while directly supporting the local ICT industry. This research was completed using primary sources within the Thai and Malaysian governments, including the ministries of Finance, science and technology policy, and the national regulator. Interviews also include private sector service providers as well as non-profit and think tank organizations. This ICT institution study is driven by the following research questions:

- Are institutional reforms necessary to sustain the growth in the local ICT sector?
- How is development of a domestic ICT sector influenced by regional market and institutional factors?
- What lessons for other developing countries can be drawn from the challenges faced by Malaysia and Thailand in their institutional reform process?

INTERNATIONAL INVESTMENT AND TECHNOLOGY TRANSFER IN THAILAND AND MALAYSIA

The section below discusses foreign direct investment (FDI), multinational corporations (MNCs), and the role of technology transfer because these issues are vitally important to the creation and expansion of a developing county's ICT sector.

FDI in Thailand

An important component to the development of local industries and the acceleration of technology adoption is the foreign investment and technology transfer from global corporations located in Thailand (Jansen, 1995; Blomstrom &

Kokko, 1998). Like other developing countries, Thailand's high-tech industries in particular must rely on international investment and technology transfers (Kohpaiboon, 2006). Through technology and knowledge transfer from these MNCs to local branches and suppliers, a developing country can rapidly expand the capabilities of local industry to adopt and utilize technologies. Thailand has attracted significant amounts of FDI into its economy, reaching over 6 percent of gross domestic product (GDP) before the financial crisis in 1997. However, the country has not been able to expand this investment since the technology collapse in 2001 (World Bank, 2008). In addition, successful diffusion requires government intervention as well as coordination with the private sector to ensure that these technologies diffuse throughout the economy rather than concentrate within a single industry or region (David, 1997). Thailand has not been able to address this issue. In addition, the government has failed to directly link its approach to foreign direct investment with its efforts to enhance local technological capabilities. This is in contrast with Malaysia and Singapore, which were able to utilize FDI to expand local technology capabilities and accelerate adoption (Wong, 1999).

According to interviews that I completed with the Ministries of ICT, Finance, and others, there are increasing concerns within Thailand about losing investment from multinational corporations to neighboring countries. Although FDI incentive efforts are run by the Office of the Board of Investment, Thailand does not have the high levels of governmental coordination incentives, or marketing that other Southeast Asian economies use to increase their foreign investment. Several government officials each commented that Thailand is currently competing with low-cost countries, such as China and Vietnam, with which it cannot compete on price, and countries such as Malaysia and Singapore that have a much higher level of available talent for ICT and S&T. This situation makes it particularly difficult for Thailand to attract investment in technology sectors, which are deemed strategically important to the long-term growth of the economy.

A major component to enhancing local ICT industry development is the creation of a venture capital industry that invests in startups that are either commercializing local research or adapting international technology to the local market (Ramasamy, Chakrabarty, & Cheah, 2004). The local Thai venture capital (VC) community organized itself into the Thai Venture Capital Association in 1994, with members from the financial, accounting, legal, and advisory services industries. The VC community focuses its investments on firms that have reached their expansion stage. As a result, Thailand faces significant hurdles supporting early round startups with adequate financial backing due to the high levels of risk – a common problem for many developing countries. The Thai government has supported the expansion of the small venture capital community through tax incentives and a number of government-backed funds (Intarakumnerd, 2004). These VC funds target both small business development and areas affected by the financial crisis, seeking to fill in gaps left by the private sector.

FDI in Malaysia

Malaysia has long been successful at attracting FDI and MNCs in the manufacturing and IT sectors, reaching a level few developing countries can match (Jomo et al., 1997). Hundreds of MNCs and billions of dollars in FDI have flowed into the country, particularly into the free trade zones (FTZ) and technology parks set up near the regional manufacturing and technology centers. This includes over 1,700 technology companies – both domestic and international – in Malaysia's Multimedia Super Corridor (MSC), according to the MSC's online reporting. Government officials interviewed for this research suggest that there is close coordination across government ministries as well as between the private and public sectors to promote targeted ICT sectors. The public and private sectors also develop FDI incentive programs, creating packages tailored to specific firms and sectors.

Fueled by these foreign corporations, Malaysia has become an IT equipment exporter, boosting economic growth and local incomes. The government has been able to attract a wide range of software and IT MNCs, and meet its internal targets for relocating firms within the technology parks, even after the global technology downturn in the early part of the decade. This resilience despite downturns at home and abroad is a testament to the coordinated efforts of the public and private sectors to attract technology transfer, investment, and employment opportunities.

Few Association of Southeast Asian Nations (ASEAN) countries approach the high rate of FDI that Malaysia has maintained during the past 10 years, reaching 9 percent of GDP. In contrast, both Korea and Japan averaged around 2 percent of FDI during their development phases, preferring instead to use licensing agreements with MNCs to foster technology transfer rather than direct investments (Ismail & Yussof, 2003). This policy choice within Korea and Japan proved to be very beneficial in developing indigenous innovation capacity, where Malaysian policy is now focused (Hsiao & Hsaio, 2003). Jomo et al. (1997) argues that one of the main political reasons for allowing such a high influx of FDI into the country was to balance the economic power between the ethnic Malays and the minority ethnic Chinese, which have a disproportionate influence within the economy. The FDI was encouraged by the Malaysian government as a way to lower the Chinese minority's economic influence while supporting the growth of Malay-owned firms (Jomo, 2003). This has allowed the country to rapidly expand its global presence within the ICT sectors. However, it has also increased the country's dependency on MNCs and FDI. In addition, several sectors such as the electronics industry have not significantly expanded their roles and are limited to providing low-skill assembling within the global supply chain (Wah & Narayanan, 1999).

This predominance of foreign investment has created a gap in funding available to start-up technology firms in Malaysia and has led the government to develop entrepreneurial financing options for local firms. This has prompted the Malaysian Industry-Government Group for High Technology (MIGHT) to support the development of a financing corporation that coordinates with private sector financial institutions to lower the barriers for investment, according to officials within the organization. The finance corporation's activities include completing due diligence on prospective firms for investment while helping to foster relationships between start-up firms and private sector investment houses. These activities are vital to jump-starting local technology start-ups since the venture capital industry in the country is very small and relatively new. This government entity also identifies global technology niches – such as biotechnology, photonics, and advanced manufacturing – where Malaysian firms will have a competitive advantage due to the large base of local technology expertise.

TECHNOLOGY INSTITUTIONS

The efforts of the Malaysian and Thai technology institutions can be divided into three basic categories. The first includes network infrastructure deployment plans that have been based on South Korea's very successful government initiatives to support ICT network expansion (Frieden, 2005). These plans focus on increasing access, deploying fiber networks, and expanding universal service obligations. Some of these efforts have been successful, depending on the demand for ICT services that already exists within each country.

The second key area includes workforce skill development to assist in ICT knowledge transfers from MNCs while increasing local technology innovation capabilities. Thailand has moved to support training and skills development through its Ministry of Education and Ministry of Industry. However, Malaysia has made it a primary focus for its national development strategy.

The third area, IT industry development, was the primary area of focus for the private sector and the two national ICT strategies. In 2003, Thailand created Software Park and the Software Industry Promotion Agency to bolster its fledgling industry. In Malaysia, the IT sector has grown rapidly through the involvement of foreign capital and MNCs, both of which are coordinated through the Multimedia Super Corridor. Through the MSC, the government was able to attract a range of software and IT MNCs. The country is now focused on increasing the technology transfer and training from these global IT firms into its local ICT industries.

This section provides a background on the ICT and technology institutions of Thailand and Malaysia; examining the institutional development of science and technology Ministries in both countries as well as a discussion of the hurdles both countries must overcome to improve the design of their ICT institutions.

EVOLUTION OF THAILAND'S S&T INSTITUTIONS

Thailand watches its neighbors very closely to benchmark its own ICT policies and institutional structure. Thailand was implementing its first structural reform when Malaysia was undertaking a similar effort. As a result, Thailand mirrored its restructuring after its neighbor. Now that Thailand is working on a second reform, it has looked to Malaysia again to see how that country has evolved its ICT structure. Both countries' science ministries were previously the permanent secretary to their respective National IT Councils. In addition to Malaysia, Thailand's ICT agencies have looked to other countries in Asia to develop its ICT policies and institutions, including Singapore, Japan, Korea, and other ASEAN countries (Koh, 2006). Singapore has been widely viewed has having a successful ICT sector as well as an institutional and regulatory reform process (Painter & Wong, 2007).

The administrative organization charged with coordinating the design and implementation of Thailand's national ICT policies and initiatives is the National IT Committee (NITC). When NITC was created in the early 1990s, its efforts were directly supported by NECTEC, which worked with the committee to draft a national ICT policy. The Committee is officially headed by the prime minister and includes members of IT, telecommunications, banking, and other industry trade groups. A former official with the NITC stated that a deputy PM is typically assigned to run the Committee, in addition to other duties and obligations. As a result, the Committee's ability to develop and implement ICT policies would vary widely depending on the personal interests of the deputy assigned. According to government officials interviewed for this case study, those deputy ministers with a strong interest would inject their vision into the Committee, where it would cascade through the other ministries, accelerating efforts and improving initiative outcomes. In contrast, ICT policy design and implementation would suffer a significant slowdown under deputies who were not as interested in this NITC leadership position.

Thailand's National Electronics and Computer Technology Center

The national innovation agency, the National Electronics and Computer Technology Center (NECTEC), was created through the National Science and Technology Development Act by Parliament as an independent agency to focus on research and development (R&D) efforts within the electronics, information technology, and communications technology industries (Intarakumnerd, 2004; Hobday & Howard, 2007).

NECTEC was placed under National Science and Technology Development Agency (NSTDA), which is part of the Ministry of Science and Technology and was created as part of the same act. Officials interviewed for this research stated that NSTDA is focused on developing and supporting S&T policy within the Thai industry through four key divisions: nanotechnology, biotechnology, material sciences, and NECTEC.

The development agency works with major universities to develop and commercialize technology innovations and has recently set up a Technology Licensing Office to work with universities to license and commercialize their research. The Ministry of Science and Technology completed its own master plan for science and technology policy (NECTEC, 2003). This plan identified several industrial sectors to target for support by NSTDA and the other agencies within the ministry, including food processing, automotive, garment, and others. According to NSTDA officials, prior to this plan, university and government researchers had applied for funding based on their own disparate interests, which left R&D efforts uncoordinated within these sectors. To receive funding under the science master plan, researchers and ministries must demonstrate how their funding and policy support the government's S&T goals.

Officials at NECTEC stated that the agency focuses much of its work on applied R&D efforts, commercialization efforts, and technology transfer. Its agenda is tailored to ensure that there is commercial demand for the R&D activities currently being funded by public and private sources. In addition, the agency runs its own labs, staffed by researchers that complete university training through NECTEC scholarships both in Thailand and abroad. These researchers are contracted to the government for twice the length of their schooling. However, NECTEC loses many non-scholarship researchers to the private sector due to uncompetitive compensation packages. Similar to the National Science Foundation in the US, NECTEC provides grant funding to university researchers in addition to its own labs, particularly in areas where the agency is lacking in expert personnel (Liefner & Schiller, 2008).

The national innovation policies aimed at developing the ICT sector are designed by NECTEC and the Ministry of ICT (MICT) for implementation across the other ministries. Interviews with officials at NECTEC and the MICT confirmed that once NECTEC develops the policy, the Ministry of Science and Technology presents it to the cabinet for approval. If the policy is approved, it is set as a national policy for all of the ministries and departments to follow. Although NECTEC and MICT are responsible for the ICT policies within the country, both of these government organizations are short on funding and manpower. This is due to the lack of large-scale projects, which tend to receive the most funding and political attention. NECTEC, the MICT, and the Ministry of Science and Technology have very limited authority, political influence, and budget while the Ministries of Finance, Commerce, and Transport all have high levels of political power due to their budgeting authority for large infrastructure projects.

Software Industry Promotion Agency of Thailand

Formerly a part of the MICT, the Software Industry Promotion Agency (SIPA) is a public organization similar to NECTEC, with flexible regulations and employment contracts. According to discussions with a former executive of SIPA, the agency has focused on deploying several pilot programs aimed at improving government services. One pilot was completed using local software developers to build an online interface for the immigration service. Another example included the development of an online portal to promote Thai regions for business investment and job creation. This is similar to the efforts of local US chambers of commerce, aggregating key economic, demographic, and quality of life indicators for prospective investors and businesses. Although this is an easy step for a US region, in Thailand the data is spread across many disparate agencies and databases, making integration difficult. SIPA also tracks and provides information on available ICT experts and upcoming graduates to attract investment in the software sector.

The Thai government has set a target of creating an indigenous IT sector, but even with this public sector support, the private sector has made little headway. The government has been working on a series of IT bills to support the sector, but in the last nine years has only passed two. As a result, there is no master plan or policy strategy for developing this IT sector – although it is outlined in IT2010. There have been some successes by the Software Industry Promotion Agency and within the Software Park, a national technology park devoted to the industry. However, the Software Park has done little outside of training individuals and SMEs (Mephokee & Rvengsrighaiya, 2005; Gray & Sanzogni, 2004).

In addition, a Software Park was developed by the NSTDA as part of an industry support policy designed for the National IT Committee. The Software Park has the same independent agency status as NECTEC and offers office space and shared computer facilities to small software entrepreneurs. The park also offers a firm matching program, linking small software companies to larger Thai and international partners. Both SIPA and the Software Park have similar goals and there is overlap between the two efforts due to a lack of institutional coordination.

EVOLUTION OF MALAYSIA'S S&T INSTITUTIONS

Malaysia's Ministry of Science, Technology and Innovation (MOSTI) identifies target sectors for national technology policies for the next 10 years, focusing on areas where the Malaysian government can support private sector competitive advantages. Its activities include risk assurance for private sector development, and monitoring and evaluation of policy implementation. MOSTI was previously the permanent secretary for the National Information Technology Council (NITC), which is similar to Thailand's NITC; is also chaired by the prime minister; and is coordinated across the private sector and several ministries for ICT policy implementation in Malaysia, including education and energy, water, and communications (Lall, 1999).

In interviews, ministry officials described additional initiatives including technology development and innovation, as well as best practice development for R&D within the local S&T industries including the ICT sectors. MOSTI also supports improved coordination across private and public research centers through strong leadership directed by the prime minister's office. MOSTI views its role as an ICT enabler through increased investment, expanded R&D activities, and enhanced innovation capacity. MOSTI also is responsible for regulating and enforcing the Communications and Multimedia Act. This regulation is in coordination with the national regulatory commission.

Through the NITC, MOSTI conducted an assessment of national ICT needs in 2006 to revise the national ICT policies and initiatives. The Malaysian government is currently in its 2006-2010 economic plan and ministerial budgets have already been set. But MOSTI will utilize the 2007 budget to launch its initiatives and then move to revise the 2008 budgets during the mid-plan review. This flexibility of funding is key to Malaysia's ability to respond to the rapidly changing landscape within the ICT sectors. From the NITC planned report on ICT needs, focus areas on the ICT road map include sector development, infrastructure deployment, R&D, and community development.

Malaysian Industry-Government Group for High Technology

Malaysian Industry-Government Group for High Technology (MIGHT) is a non-profit, quasi-governmental organization that operates under the oversight of MOSTI, Malaysia's innovation and science agency. The organization is charged with supporting and operationalizing "Malaysia Incorporated," which is modeled after Japan and Singapore's public, private, and academic coordination to innovate and compete in the global marketplace. This coordinated initiative, called the "triple helix," to build innovation capacity is driven by the stakeholders in these three sectors across the country. MIGHT is made up of 15 to 20 business, government, and academic leaders and is co-chaired by the PM's science adviser. All policy proposals, once accepted by MIGHT, flow through the science adviser to the PM's office and then to the cabinet for approval and funding. MIGHT has successfully bridged the gaps across these three sectors and coordinated with competitive firms within the private sector to support globally competitive industry development.

In the early 1990s, there was little coordination within Malaysia's private sector let alone across the public or academic sectors, according to interview officials. In 1993, the organization was under the prime minister's office when it was moved to MOSTI (Kam, 1999). The decision was made to place the organization under the prime minister to bolster its ability to draw political support from the PM as well as the other executive agencies. Aside from these political considerations, this move was also necessary due to MIGHT's focus on specific privatized industries that cut across multiple agencies and as a result had previously not received the necessary governmental support to bolster their development. One example outside the ICT sector is the aerospace industry, which had previously fallen under the communications, transport, and finance ministries. With no single ministry leading the effort to support this industry within Malaysia, and no single ministry able to coordinate across the overlapping efforts, there was a lack of support and coordination. Under MIGHT's direction, a national aerospace blueprint was developed through 12 months of policy meetings by a technical committee that included the airlines and manufacturers, and a steering committee that was chaired by the prime minister's science adviser. Once presented to the prime minister, the government funded a multi-year effort that was coordinated by a purpose-built policy vehicle chaired by the prime minister – the Malaysian Aerospace Council – that evaluates the progress of the national aerospace blueprint every six months.

According to interviewed officials, MIGHT's process of governmental support is to use successful interventions, whether they are within the aerospace or ICT industries, as pilot programs for launching efforts across a range of technology sectors. For MIGHT to target a given sector, it must fall within the portfolio of several ministries, have already undergone the privatization process, and cannot have a private sector organization already operating to support sector development. This last requirement prevents the government from duplicating private sector efforts or reducing the incentive within the private sector to establish these coordinating bodies on their own. MIGHT also looks to see where Malaysia can find global technology niches, where it is feasible for the country's existing technical expertise to compete internationally while driving national exports and innovation. It also looks to leverage the existing industrial and technological base within the country. The other areas of focus for MIGHT include entrepreneurial development financing, pharmaceuticals, and radio frequency identification (RFID) tags.

Malaysia's National RFID Program

There is a clearly defined process for evaluating the technology niches that will receive support through the ICT road map. This includes an assessment of the market demand for a given information technology, the viability of this technology, Malaysia's ability to compete in this market, and the global competition within this sector. Once this technology for global export is chosen, a road map will be created for governmental support that will include key performance indicators (KPIs) developed and monitored by MOSTI.

According to interviewed officials at MIGHT, the goal of the national RFID program is to develop a cluster of firms capable of supporting the development, innovation, and marketing of this emerging technology. RFID tags are inexpensive chips that can be used to track inventory by a range of industries, most notably retail and logistics. The program's larger goal is to support the revival of the national semiconductor industry, which has seen its market share and margins shrink due to global competition and the global technology slowdown in the early 2000s (Rasiah, 2003). Although Malaysia is third in the world for semiconductor manufacturing, the country's design houses for hardware, software, and middleware had begun to whither and lose their attractiveness to top engineering talent during this downturn. Malaysian wafer facilities, built with government funding, were very competitive but had lost out to global companies with combined chip design and application capabilities. In response to this situation, the government decided that it could not rely on manufacturing capabilities alone and began working to support the development of chip applications along with the revival of the country's chip design houses. The government support for the revival of the national semiconductor industry included MIGHT negotiating a technology license from Japan for an RFID chip and then leasing this technology to a newly created private company, Senstech.

To build a sustainable level of supply and demand, the government has created a multistage plan for development and deployment. The first phase includes using the government as a test bed and to have the government act as a coordinator across the key private sector players to begin the transition away from bar codes. The government is offering these government agencies the right to use these RFID chips if they are suitable to the program. The government has also agreed to act as the anchor tenant for the technology once it is produced, purchasing the chips for governmental applications to jump-start the industry. These governmental purchases will be used for government programs as well as by state-owned enterprises in sectors such as cattle tagging, logging, and auto manufacturing. However, if agency or SOE testing demonstrates that these chips are not suitable, end users can transition to other available RFID sources. The second phase includes extending the program to wholesale and retail logistical support within Malaysia. To serve the needs of these sectors, the program will expand production to benefit from the economies of scale that are essential in lowering the costs of individual RFID chips. The final step will focus on the international markets and competing against global firms for exports. This phase's approach is key to developing a financially viable sector while positioning Malaysia's semiconductor at the forefront of this growing ICT sector.

Malaysia's Multimedia Super Corridor

The Multimedia Super Corridor is a large technology park devoted to ICT sector development and is the cornerstone of the government's efforts for supporting the development of ICT industries within Malaysia (Bunnell, 2002). The MSC is run by the Multimedia Development Corporation (MDeC), which has representation within Malaysia's National IT Council. Another key institution supporting the MSC is the MSC Implementation Council, which focuses on technology sector investment and development. The MSC Implementation Council is chaired by the prime minister and the MDeC acts as the permanent secretary. Other ministries provide oversight for these efforts, with MOSTI approving MDeC policies and programs, and the Ministry of Finance approving budgets and five-year development plans. MDeC's focus on the MSC is complementary to MIGHT, which has a broader portfolio of expanded technology industries nationwide.

The MDeC is also working to bolster the development of these SMEs through a program that targets these firms at each stage of development (Ramasamy, Chakrabarty, & Cheah, 2004). Starting with support for idea creation, seed capital, first stage, and all the way up to IPO, there is a tailored program within the MSC to support the investment, training, supplies, and industry connections of these firms (Jomo, 2003). There is also significant funding and expertise available to support R&D efforts. The MDeC is also working to improve commercialization efforts of university research by matching faculty and students with MSC companies as well as providing incubators for university-based start-ups.

The MDeC is currently implementing plans to expand the MSC to additional population centers around Malaysia (Bunnell, 2002). This effort is championed by each state within Malaysia, with the MDeC coordinating the development of the regional technology parks. To avoid duplicated efforts and state-level competition, each of the new MSCs must differentiate their economic and innovation goals to win approval to develop a park. One example is the MSC in Penang, which is focused on the electronics industry and leverages the state's robust semi-conductor industry (Wah & Narayanan, 1999). As of 2006, four additional MSCs had begun development with the support of government tax breaks, no labor restrictions, no hurdles for fund transfers, and government guarantees on communications services.

An important point, noted by MDeC officials interviewed, is that the MSC effort is driven by KPIs, which are benchmarked annually with goals for exports, employment by function, and investment, among others. To help ensure these targets are met, there is regular communication between MOSTI and MDeC as well as monthly dialogue between MDeC and its MSC tenants. MDeC also reviews the operations and tax status of each business and has revoked the licenses of both MNCs and local SMEs for breach of contract. A majority of the companies with revoked licenses have been local; these SMEs have difficulty maintaining their efforts to compete within ICT sectors.

CHALLENGES FACING NATIONAL ICT INSTITUTIONS

Institutional Challenges in Thailand

Within Thailand, leadership issues and conflicts of interest have both been major impediments to policy implementation. The country's National Information Technology Committee is officially headed by the prime minister, but this position was typically delegated to a deputy minister. As a result, the ability of the Committee to develop and implement ICT policies would vary widely depending on the personal interests of the deputy assigned. Policy implementation would suffer a slowdown under deputies who were not interested, according to participants who were later interviewed. More recently, the prime minister headed the NITC himself, but was also found to have a conflict of interest – potentially steering the Committee to benefit his personal business interests.

Another major hurdle for ICT institutions and agencies in Thailand is the lack of program evaluation. ICT programs are developed by ministerial chief information officers (CIOs). They are evaluated only at the project level and this limited monitoring is completed in a vacuum. The focus of the evaluation work is on gathering quantitative data on hardware deployment, not on qualitative data on ICT utilization and the economic benefit of the ICT initiative. Most important, there is no overall benchmarking or evaluation effort by the government against its own master plan – an essential component to ICT strategy success (Docktor, 2004).

Although there is no overarching benchmarking or evaluation effort by the government, one avenue for monitoring and evaluation is the submission of ministry-level ICT Master Plans to the MICT and the Budget Bureau (Banerjee & Chau, 2004). These Master Plans are developed by the ministries and mapped back to the national ICT plan for justification. However, there are no set criteria for doing this. Plans may be rejected due to political considerations or may be rubberstamped to avoid bureaucratic confrontation. When plans are reviewed, they are only completed at the project level with little connection to the government's larger efforts.

Another key component missing from the MICT's efforts is the establishment of public-private partnerships (PPPs) aimed at harnessing the expertise of the private sector and the financial support of the government to spur ICT sector development. This has been a key focus of discussion and is recognized as crucial within the ministry – particularly because neighbors such as Malaysia have successfully implemented public-private partnerships across several technology sectors (Shapira, et al., 2005). However, there is little activity or implementation around fostering partnerships, particularly in ICT applications. Currently, when the government implements a partnership for its e-governance programs, there is a tendency to become overly reliant on the vendor for customization or changes downstream, according to officials. This leaves government ministries vulnerable to high additional costs for these programs as their IT requirements change over time.

Institutional Challenges in Malaysia

By housing ICT infrastructure programs within the Ministry of Energy, Communications and Multimedia (KTAK), Malaysia can emphasize ICT infrastructure deployments while concentrating the necessary budget and authority on implementing those policies within a single agency (MCMC, 2006). However, officials with KTAK and the ministry both reported that this new institutional structure does not have the critical mass of ICT expertise that was within the original Ministry of ICT and is now spread out among separate agencies.

Although there have been many successful efforts to launch government technology initiatives, this has not been the case across all potential sectors (Wee, 2001). Interviewed officials with MIGHT suggested that in areas falling under the purview of a traditional sector ministry, such as agriculture, there have been fewer new initiatives to support private sector development and little institutional reform to be more responsive to market changes. MOSTI works in conjunction with the Economic Planning Unit, a civil service agency, to develop institutional restructuring plans for each of the other ministries. There is a review and restructuring process for reducing agency overlap, which has restructured and dismantled several agencies, most recently in 2004, to make the government more responsive to market changes. These changes were deemed necessary and supported by the prime minister to enhance governmental efforts to bolster national competitiveness. However, there is still a significant tendency for traditional sector agencies to strongly resist any governmental restructuring and revitalization effort.

CONCLUSION

ICTs can help accelerate the development process by reducing transaction costs, increasing transparency, and enhancing links across sectors and with neighboring countries. These national efforts then bolster research and knowledge transfer, which accelerates productivity growth and helps to diffuse economic benefits. To capture these benefits of ICT adoption, countries like Malaysia and

Thailand need a strong state intervention that supports and regulates the private sector in delivering affordable access across the country.

In addition, national governments must also focus on spreading the economic benefits beyond capital while ensuring that technology and expertise from MNCs are transferred to local firms across the economy – as Malaysia is currently striving to achieve. Governments must carefully craft incentives for local and international businesses to invest in local ICT industries while demonstrating their willingness to withdraw these incentives from local and international firms that no longer meet the standards for support.

REFERENCES

Balaji, P., & Keniston, K. (2005, July). Tentative conclusions. Information and communications technologies for development: A comparative analysis of impacts and costs. Department of Information Technology, Government of India. Retrieved from http://www.iiitb.ac.in/Complete_report.pdf

aliamoune-Lutz, M. (2003). An analysis of the determinants and effects of ICT diffusion in developing countries. Information Technology for Development, 10, 151–169.

Banerjee, P., & Chau, P. Y. K. (2004). An evaluative framework for analyzing e-government convergence capability in developing countries. *Electronic Government, 1*, 29–48. doi:10.1504/EG.2004.004135

Bassanini, A. (2002). Growth, technology change, and ICT diffusion: Recent evidence from OECD countries. *Oxford Review of Economic Policy, 18*(3), 324–344. doi:10.1093/oxrep/18.3.324

Baumol, W. J., & Solow, R. (1998, Fall). Comments. *Issues in Science and Technology, 15*(1), 8–10.

Blomstrom, M., & Kokko, A. (1998). In G. B. Navaretti Foreign investment as a vehicle for international technology transfer. In G. Barba Navaretti, P. Dasgupta, K-G. Maler, & D. Siniscalco (Eds.), Creation and Transfer of Knowledge: Institutions and Incentives. New York: Springer Verlag.

Bunnell, T. (2002, March). Multimedia utopia? A geographical critique of hgh-tech development in Malaysia's multimedia super corridor. *Antipode, 34*(2), 265. doi:10.1111/1467-8330.00238

Cohen-Blankshtain, G., & Nijkamp, P. (2003, August). Still not there, but on our way: thinking of urban ICT policies in European cities. *Tijdschrift voor Economische en Sociale Geografie, 94*(3), 390–400. doi:10.1111/1467-9663.00265

David, P. A. (1997). Rethinking technology transfers: Incentives, institutions and knowledge-based industrial development. In C. Feinstein & C. Howe (Eds.), Chinese Technology Transfer in the 1990s: Current Experience, Historical Problems and International Perspectives. Cheltenham, UK: Elgar.

Docktor, R. (2004). Successful global ICT initiatives: Measuring results through an analysis of achieved goals, planning and readiness efforts, and stakeholder involvement. Presentation to the Council for Excellence in Government.

Feinson, S. (2003, June). National innovation systems overview and country cases. Knowledge Flows and Knowledge Collectives: Understanding The Role of Science and Technology Policies in Development. In B. Bozeman, et al. (Eds.), Synthesis Report on the Findings of a Project for the Global Inclusion Program of the Rockefeller Foundation.

Frieden, R. (2005). Lessons from broadband development in Canada, Japan, Korea and the United States. *Telecommunications Policy, 29*, 595–613. doi:10.1016/j.telpol.2005.06.002

Graham, S. (2000, March). Symposium on cities and infrastructure networks: Constructing premium network spaces: reflections on infrastructure networks and contemporary urban development. *International Journal of Urban and Regional Research, 24*(1), 183. doi:10.1111/1468-2427.00242

Gray, H., & Sanzogni, L. (2004). Technology leapfrogging in Thailand: Issues for the support of ecommerce infrastructure. *Electronic Journal on Information Systems in Developing Countries, 16*(3), 1–26.

Grubesic, T. H., & Murray, A. T. (2004, Spring). Waiting for broadband: Local competition and the spatial distribution of advanced telecommunication services in the United States. *Growth and Change, 35*(2), 139–165. doi:10.1111/j.0017-4815.2004.00243.x

Hobday, M., & Howard, R. (2007). Upgrading the technological capabilities of foreign transnational subsidiaries in developing countries: The case of electronics in Thailand. *Research Policy, 36*(9), 1335–1356. doi:10.1016/j.respol.2007.05.004

Hsiao, F. S. T., & Hsiao, M.-Ch. W. (2003, February). Miracle growth in the twentieth century –International comparisons of East Asian development. *World Development*, 227–257. doi:10.1016/S0305-750X(02)00188-2

Intarakumnerd, P. (2004, April). Thailand's national innovation system in transition. First Asialics International Conference on Innovation Systems and Clusters in Asia: Challenges and Regional Integration. National Science and Technology Development Agency, Bangkok Thailand.

Ismail, R., & Yussof, I. (2003). Labour market competitiveness and foreign direct investment: The case of Malaysia, Thailand and the Philippines. *Papers in Regional Science, 82*, 389–402. doi:10.1007/s10110-003-0170-2

Jansen, K. (1995). The macroeconomic effects of direct foreign investment: The case of Thailand. *World Development, 23*(2), 193–210. doi:10.1016/0305-750X(94)00125-I

Javary, M., & Mansell, R. (2002). Emerging internet oligopolies: A political economy analysis. In Miller, E. S., & Samuels, W. J. (Eds.), *An Institutionalist Approach to Public Utilities Regulation* (pp. 162–201). East Lansing, MI: Michigan State University Press.

Jomo, K. S. (2003). Growth and vulnerability before and after the Asian crisis: The fallacy of the universal model. In Martin, A., & Gunnarsson, C. (Eds.), *Development and structural change in Asia-Pacific: globalising miracles or end of a model?* (pp. 171–197). London: RoutledgeCurzon.

Jomo K. S. (Author) › Visit Amazon's Jomo K. S. Page Find all the books, read about the author, and more. See search results for this author Are you an author? Learn about Author Central

Jomo, K. S., Chung, C. Y., Folk, B. C., Ul-Haque, I., Phongpaichit, P., Simatupang, B., & Tateishi, M. (1997). *Southeast Asia's Misunderstood Miracle: Industrial Policy and Economic Development in Thailand, Malaysia, and Indonesia*. Boulder, CO: Westview.

Jomo, K. S., Rasiah, R., Alavi, R., & Gopal, J. (2003). Industrial policy and the emergence of internationally competitive manufacturing firms in Malaysia. In Jomo, K. S. (Ed.), *Manufacturing Competitiveness in Asia: How International Competitive National Firms and Industries Developed in East Asia* (pp. 106–172). London: RoutledgeCurzon.

Kam, W. P. (1999). Technological capability development by firms from East Asian NIEs: Possible lessons for Malaysia. In Jomo, K. S., & Felker, G. (Eds.), *Technology, Competitiveness, and the State: Malaysia's Industrial Technology Policies* (pp. 53–64). London: Routledge.

Koh, W. T. H. (2006). Singapore's transition to innovation-based economic growth: infrastructure, institutions and government's role. *R & D Management*, *36*(2), 143–160. doi:10.1111/j.1467-9310.2006.00422.x

Kohpaiboon, A. (2006). Foreign direct investment and technology spillover: A cross-industry analysis of Thai manufacturing. *World Development*, *34*(3), 541–556. doi:10.1016/j.worlddev.2005.08.006

Lall, S. (1999). Technology policy and competitiveness in Malaysia. In Jomo, K. S., & Felker, G. (Eds.), *Technology, Competitiveness, and the State: Malaysia's Industrial Technology Policies* (pp. 148–179). London: Routledge.

Levy, B., & Spiller, P. T. (1994). The institutional foundations of regulatory commitment: A comparative analysis of telecommunications regulation. *Journal of Law Economics and Organization*, *10*(2), 201–246.

Liefner, I., & Schiller, D. (2008). Academic capabilities in developing countries – A conceptual framework with empirical illustrations from Thailand. *Research Policy*, *37*, 276–293. doi:10.1016/j.respol.2007.08.007

Malaysian Communications and Multimedia Commission and Ministry of Energy, Water and Communications. (2006). The national broadband plan: Enabling high speed broadband under MyICMS 886. Cyberjaya, Malaysia: Malaysian Communications and Multimedia Commission. Retrieved from http://www.mcmc.gov.my

McGuckin, R., & Stiroh, K. (1998, Summer). Computers can accelerate productivity growth. *Issues in Science and Technology*, *14*(4), 41–48.

Mephokee, C., & Ruengsrichaiya, K. (2005, December). Information and communication technology (ICT) for development of small and medium-sized exporters in East Asia: Thailand. United Nations Publication, Comisión Económica para América Latina y el Caribe (CEPAL), Project Document.

NECTEC. (2003). *Thailand: Information and communication technology master plan (2002-2006)*. Bangkok, Malaysia: National Electronics and Computer Technology Center.

Painter, M., & Wong, S.-F. (2007). The telecommunications regulatory regimes in Hong Kong and Singapore: When direct state intervention meets indirect policy instruments. *The Pacific Review*, *20*(2), 173–195. doi:10.1080/09512740701306832

Ramasamy, B., Chakrabarty, A., & Cheah, M. (2004). Malaysia's leap into the future: an evaluation of the multimedia super corridor. *Technovation*, *24*, 871–883. doi:10.1016/S0166-4972(03)00049-X

Rasiah, R. (2003). Foreign ownership, technology and electronics exports from Malaysia and Thailand. *Journal of Asian Economics*, *14*, 785–811. doi:10.1016/j.asieco.2003.10.006

Rostow, W. W. (1960). *The stages of economic growth*. Cambridge, UK: Cambridge University Press.

Saunders, R. J. Jeremy J. Warford, and Björn Wellenius. (1994). Telecommunications and economic development (2nd ed.). Baltimore: Published for the World Bank by the Johns Hopkins University Press.

Shapira, P., Youtie, J., Yogeesvaran, K., & Zakiah, J. (2005, May). Knowledge economy measurement: Methods, results and insights from the Malaysian knowledge content study. Triple Helix 5 Conference - Panel Session on New Indicators for the Knowledge Economy, Turin, Italy.

Shari, I. (2003). Economic growth and social development in Malaysia, 1971-98: Does the state still matter in an era of economic globalisation? In Andersson, M., & Gunnarsson, C. (Eds.), *Development and structural change in Asia-Pacific: globalising miracles or end of a model?* (pp. 109–124). London: RoutledgeCurzon.

Steinmueller, W. E. (2001). ICTs and the possibilities for leapfrogging by developing countries. *International Labour Review*, *120*(2), 193–210. doi:10.1111/j.1564-913X.2001.tb00220.x

Storm, S., & Naastepad, C. W. M. (2005). Strategic factors in economic development: East Asian industrialization 1950–2003. *Development and Change*, *36*(6), 1059–1094. doi:10.1111/j.0012-155X.2005.00450.x

Strover, S., & Berquist, L. (1999, November 22-24). Telecommunications infrastructure development: The evolving state and city role in the United States. Cities in the Global Information Society Conference. Newcastle upon Tyne.

Wah, L. Y., & Narayanan, S. (1999). Technology utilization level and choice: The electronics and electrical sector in Penang, Malaysia. In Jomo, K. S., Felker, G., & Rasiah, R. (Eds.), *Industrial Technology Development in Malaysia* (pp. 107–124). London, New York: Routledge.

Wee, V. (2001, June). Imperatives for the k-economy: Challenges ahead. InfoSoc Malaysia Conference, Penang, Malaysia.

Wilson, E. J. III, & Wong, K. (2003). African information revolution: A balance sheet. *Telecommunications Policy*, *27*, 155–177. doi:10.1016/S0308-5961(02)00097-6

Wong, P. (1999). National innovation systems for rapid technological catch-up: An analytical framework and a comparative analysis of Korea, Taiwan, and Singapore. Paper presented at the DRUID's Summer Conference, Rebild, Denmark.

World Bank. (2008). *World Bank development indicators data base 2006*. Washington, DC: World Bank.

Chapter 3
A Case on University and Community Collaboration:
The Sci-Tech Entrepreneurial Training Services (ETS) Program

S. Ann Becker
Florida Institute of Technology, USA

Robert Keimer
Florida Institute of Technology, USA

Tim Muth
Florida Institute of Technology, USA

ABSTRACT

Small businesses are viewed as the backbone of America and integral in the recovery of any economic downturn. Creative approaches to university and community collaboration are being explored to achieve high rates of success in launching, sustaining, and growing small businesses. One such approach, the Entrepreneurial Training Services (ETS) program, is being studied by Sci-Tech University as a means of technology innovation and regional economic development. The ETS Program has several unique features including: the entrenchment of a large number of adults in the program, an intensive training approach that is implemented in a short time frame, personalized mentoring offered to each entrepreneur in the program, and the leveraging of resources with a large, diverse group of community partners. The case profiles the region using Strengths, Weaknesses, Opportunities, and Threats (SWOT) Analysis, identifies an ETS framework on which the program is based and explains the process of implementation. The case concludes with challenges facing the university and local community in offering the ETS Program to a large and diverse group of entrepreneurs. It also summarizes benefits and successes from initial implementation efforts.

DOI: 10.4018/978-1-61520-609-4.ch003

BACKGROUND

"The entrepreneurial mystique? It's not magic, it's not mysterious, and it has nothing to do with the genes. It's a discipline. And, like any discipline, it can be learned" (Drucker, 1985; cited in Kuratko, 2005).

The United States views small businesses as the backbone of America and integral in the recovery of any economic downturn. According to Minniti and Bygrave (2004), the U.S. has achieved its highest economic performance during the last 10 years by fostering and promoting entrepreneurial activity (Minniti & Bygrave, 2004). New business incorporations have averaged 600,000 per year creating millions of jobs. In 1995, 807,000 new small businesses were established becoming an all time high (Kuratko, 2005). Fifteen percent of the fastest-growing new firms ("gazelles") accounted for 94% of the net job creation (Kuratko, 2005).

Entrepreneurship is a driving force in the U.S. economy through technology innovation and creation. In fact, Reynolds, Hay, and Camp (1999) state that about two-thirds of all new inventions are created by smaller companies. A simple illustration is recent innovation in business startups taking advantage of Web 2.0 technologies. Some well-known business startups include Facebook, Google, YouTube, MySpace, and LinkedIn. Many of these social media businesses have both university and community connections related to technology transfer, creation, and commercialization.

Entrepreneurs make two significant contributions in domestic and international market economies (Kuratko, 2005). The first contribution is a renewal process that pervades and defines market economies. Entrepreneurs play a crucial role in innovations that lead to technological change and productivity growth. Entrepreneurs are about change and competition because they change the market structure (2005; p. 578). Market economies are dynamic, organic entities rather than established ones. As such, entrepreneurial companies focus on future opportunities; they are not concerned about inheriting the past (Kuratko & Hodgetts, 2004).

A second contribution is that entrepreneurial companies provide a means by which millions enter the economic mainstream. Entrepreneurial companies enable women, minorities, those with disabilities, veterans, and immigrants the opportunity for economic success. Kuratko (2005) quotes from the US Small Business Administration (SBA, 1998), "Entrepreneurship plays the crucial and indispensable role of providing the "social glue" that binds together both high-tech and "Main Street" activities".

University and Community Collaborations

Policy makers in the private and public sectors have realized the importance of universities in regional economic development (Chakrabarti & Lester, 2002). Many universities have initiated programs in partnering with entrepreneurs and both small and large businesses for technology transfer and innovation. Several high-tech areas in the U.S. with strong collaborative ties to local and regional communities include: Boston (Massachusetts Institute of Technology), Silicon Valley (Stanford University), Austin (University of Texas-Austin), San Diego (San Diego State University) and Research Triangle (North Carolina at Chapel Hill), among others. Technology innovations in these areas spur economic growth in communications, energy, sustainability, healthcare, technology, and information assurance, among others.

Universities benefit from community collaborations particularly in the areas of technology transfer and innovation. Kim and Marschke (2007) point out that technology transfer is typically not achieved through the private sector reading scholarly output about innovations and inventions. It is achieved through sustained, close interactions between businesses and researchers in transferring knowledge. This insight points out the need for collaborative environments whereby researchers can share research outcomes with the private sector. A result is the establishment of business and university partnerships with the potential for product or service commercialization.

Another benefit is related to knowledge creation. A university setting promotes innovation because faculty and student researchers are often not constrained in their creative activities. What is perceived as a technology "failure" in a business setting may be viewed as a "success" in a university research setting as an outcome is knowledge creation. In a university setting, technology innovation is often the result of iterations of knowledge creation for which intellectual property is generated. This freedom to be creative allows local entrepreneurs to bring research ideas to the university to collaborate, innovate, and transfer new technologies to the private sector.

Research indicates that teaming researchers from universities and engineers from private industry may help overcome the problem of the "Valley of Death". This is a critical portion of the innovation cycle where technological advances are established, but never put to productive use or brought to market (Li, 2003; cited in D'Cruz & Ports, 2006). University entrepreneurship and business incubation programs demonstrate that entrepreneurship skills can be taught and small businesses supported through collaboration and networking (Tornatzky, et al., 2002; Palmintera, et al., 2000).

The current availability of funding through the 2009 Recovery Act has made it possible for universities to study innovative approaches to advancing technologies through community collaboration efforts. Through state and federal funding initiatives, such as the U.S. Department of Commerce University Center for Economic Development, there are unprecedented opportunities for university and community collaborations that promote local economic development. It has also made it possible for the advancement of technologies addressing domestic and global issues of energy, climate change, security, communications, healthcare, housing, and education.

Case Overview

This case is about a university's initiative to develop an Entrepreneurial Training Services (ETS) program to serve the community, build bridges between researchers and small businesses, and promote economic growth and job creation. Several key aspects are introduced in terms of the innovative nature of this pilot program. The Setting the Stage section provides background information about the university in existing entrepreneurial outreach programs. It also describes briefly the demographics and economic conditions of the local region. The Case Description section presents a SWOT analysis of the local region demonstrating the need for innovative programs in entrepreneurship. This section also describes unique program features, the targeted clientele, program components, and replication opportunities. The case concludes with both challenges and lessons learned.

Past researchers suggest that in order to create a more receptive environment for entrepreneurship, a number of fundamental societal changes must occur (D'Cruz & Ports, 2006). One such change is that universities become actively involved in economic development initiatives through education and training programs. Kirzner (1984) points out local communities are the breeding ground for entrepreneurship capable of creating an environment favorable to it. The ETS Program, as described in the following sections, is in pursuit of this initiative.

SETTING THE STAGE

Sci-Tech University's[1] goal for its Entrepreneurial Training Services Program: Provide technical

Figure 1. Example technology transfer university initiatives

The University of Florida (UF) has been very successful at technology transfer initiatives. Below, are examples of technology innovation by UF researchers (www.uf.edu).

•*Beta Biomed Improves Patient Monitoring* - Beta Biomed's unique sensor consolidates a number of devices commonly used to monitor patients, thus decreasing patient discomfort and hospital costs.
•*Sinmat Gentle Slurry Raises the Bar for Microcircuitry Surfaces* - Sinmat's slurries polish copper and low dielectric constant materials in a uniquely soft and gentle manner, reducing the industry-standard process into a single step.
•*Gatorade Gives Athletes Competitive Edge* - Originally developed by Dr. Robert Cade for the Gator football team, Gatorade contains compounds which can be used to reduce or prevent adverse physiological effects of physical exercise or environmental exposure.
•*Healthier Peanuts* - UF peanut breeder Dan Gorbet developed a new peanut variety that is richer in heart-healthy oleic fatty acids, yields more peanuts per acre, and has a longer shelf life.

assistance in entrepreneurship to the region's distressed workforce in collaboration with nonprofit, for-profit, economic development and government entities capitalizing on Sci-Tech's science and technology expertise for high-tech business startup, growth, and sustainability.

Sci-Tech University, located in the Southeastern part of the U.S., over the years has maintained a strong connection with regional partners particularly in the areas of business development and small business research initiatives. With recent funding support, Sci-Tech has expanded its community outreach by offering technical assistance services to include an intensive, Entrepreneurial Training Services (ETS) program and virtually-accessible resources (*i*Place for Entrepreneurs™) bringing together researchers, community partners, and entrepreneurs to collaborate, innovate, and transfer technology to the commercial sector.

A region benefits economically from a university's entrepreneurial outreach initiatives through business growth and job creation. A university benefits from technology transfer by creating long-term sources of revenue. Many universities have been successful at generating a significant revenue stream through technology transfer to the private sector. In illustration of this, Figure 1 shows several University of Florida technology innovations posted on its Office of Technology Licensing Web page (www.uf.edu).

Sci-Tech recognizes the need for establishing mechanisms for university and entrepreneurs to collaborate. Technology transfer and commercialization initiatives can start in a university setting. They can be fostered through relationships built between local entrepreneurs and university faculty members. They can also be supported through community resources provided by nonprofit, profit, and government entities.

Sci-Tech University, through its ETS Program, will target displaced workers associated with reductions in federal funding in defense-related industries. It is estimated that 10,000 to 15,000 regional workers will be impacted by impending workforce reductions. Many of these workers have technology-related expertise as engineers, software developers, and technicians. Many others have specializations in a range of science and technology fields. With a current 10% regional unemployment rate, the economic impact of workforce reductions in defense-related industries will be devastating unless technical assistance programs are put in place.

Sci-Tech will build upon its existing programs that offer technical assistance in business development. These include its Women's Business Center and Center for Small Business Support. These entities have been very successful at leveraging resources with non-profit, for-profit, and government entities for maximizing service offerings to nascent entrepreneurs and small businesses in a three county region. Partnering entities include the County Workforce Development Board, Chambers of Commerce, Economic Development Centers, the U.S. Small Business Administration (SBA) Small Business Development Center, Business Incubation Centers, and other regional organizations.

Sci-Tech University proposes to be innovative in its technical assistance services for business development through the use of Web 2.0 technologies (e.g., Web conferencing, social media, learning management systems). An objective is to make available technical assistance services to entrepreneurs 24/7, thus eliminating barriers associated with time constraints and geographic boundaries. These services are focused on personalization such that entrepreneurs have access to much needed resources in business function areas or technical areas of expertise.

The Region

Over a half million people live in the region, as estimated by the U.S. Census Bureau. Of those over 25 years, almost 90% have high school or higher education, 28% hold a bachelor's degree or higher, and 11% earned graduate degrees. Median household income is close to $49,000 (U.S.Census Bureau, 2009).

The regions current unemployment rate is 10%, which is higher than the current national rate of 9.6% (LaborMarketInfo.com, 2009). The unemployment rate is expected to increase with impending changes in the defense-related industries. Job losses are predicted to be around 10,000 to 15,000 within a 12 to 24 month time span. Corporate contraction is affecting regional unemployment with recent layoffs by high-tech companies.

The region has a number of large technology companies that directly support defense-related industries. The region has ocean and coastal research and technology facilities (e.g., State Sea Research Institute), and the State Solar Energy Center is the primary renewable energy and energy efficiency research institute. The region is also home to Sci-Tech University, widely recognized as a technology leader in education and research.

Sci-Tech University Support Systems

Sci-Tech University has the commitment, experience, infrastructure, and qualifications for providing regional technical assistance to nascent entrepreneurs and small businesses. The College of Business at Sci-Tech has a history of supporting small business development, as evidenced by faculty research, course offerings, technology transfer, and community outreach efforts. Graduate students are required to complete a business plan, and undergraduate students may complete a business plan through course electives. The College of Business promotes student involvement with local businesses through class projects and its internship program.

Sci-Tech has initiated efforts to bring researchers and small businesses together for technology transfer, innovation, and commercialization. These collaborations are primarily in high priority areas of sustainability, green energy, information security, and emerging technologies. Table 1 illustrates faculty researchers partnering with small businesses through a state-funded program. It is anticipated that the ETS Program would increase the number of collaborations through its training program and online support system.

Professors Susan Smith and Harold Thomas, initiators of the ETS Program, are College of Business professors. Professor Smith has directed several other projects with outcomes that met or exceeded funding agency expectations. Smith is the director of both the Women's Business Center and the Center for Small Business Support. Some of the measurable outcomes from both projects are shown below.

Table 1. Sample partnerships

Area	Priority Area	University Researcher
Clean Water Filters	Mechanical	Dr. Mary Hills
Energy Efficiency Sensors	Environment	Dr. Harold Chan
Chemical Reactions and Sustainability	Chemical	Dr. Sharon Addison
Heat Reduction Materials	Thermodynamics	Dr. Paul Hsu

Table 2. Measurable outcomes reported (approximately four year period)

Services Outcomes	
Site Training	28,209 student class seats.
Web Training	13,883 student views.
Small Business Use	1,014 small businesses voluntarily reported employees taking classes.
Networking	744 attendees at local monthly meetings.

Table 3. Measurable outcomes reported (one year period)

Services Outcomes	
Training	1,092 student class seats.
Mentoring	24 unique individuals.
Networking	364 attendees at monthly meetings.

Measurable outcomes for the Center for Small Business Support are shown in Table 2. This program offered training in business tools and government contracting. Though funded support ended in 2005, Sci-Tech views the project as highly successful with significant contributions to the community at large.

The Women's Business Center is currently funded by a federal grant. The WBC provides technical assistance to women entrepreneurs and women-owned small businesses in several counties. Table 3 shows the most recent outcomes achieved by the Center.

These centers show the capability of Sci-Tech to develop effective community outreach programs with the potential for significant economic benefit to the local area. They point out the lack of existing programs that target the growing number of local displaced workers with high-tech skills. They also point out the need for an innovative program whereby potential entrepreneurs may partner with researchers on technology innovation and gain skills for technology commercialization.

CASE DESCRIPTION

Professors Smith and Thomas received federal funding to build upon Sci-Tech's infrastructure of providing technical assistance services to entrepreneurs and small businesses. They developed the concept of the ETS Program as a physical and virtual means of providing business development services while promoting collaboration among researchers and small businesses in the region's high priority areas.

They started by identifying creative aspects of the ETS Program in achieving measurable outcomes reported to the funding agency. These include the following:

Intensive, Hands-On Approach. The ETS Program is innovative in its intensive, hands-on approach to business development offered in a short time frame. Each client will be provided in-step training, mentoring, and counseling services to support the successful completion of a business plan. The overall objective is to offer personalized support starting with a business opportunity and ending with launch-ready status. Though other universities offer similar programs, the Sci-Tech version proposes to entrench an entrepreneur in a multi-dimensional support system offered onsite, in community-sponsored facilities, and through the use of Web 2.0 technologies.

Integration of Knowledge, Expertise, and Experience. The ETS Program is innovative in its integration of academic, industry, and small business expertise. Clients will be matched with faculty mentors by area of expertise and with industry and small business mentors by industry sector or area of expertise. Faculty experts provide guidance on marketing, financials, operations, human resources, and technology. Industry experts provide insights on starting, sustaining, and growing a business within an industry sector. They also provide expertise in finance, capital, legal, taxation, and licensing issues. Small business volunteers are used extensively in workshop and training seminars to share lessons learned and best practices. The overall objective is to offer a wide range of resources to support small business development.

Support Network with Web 2.0 Technologies. The proposed ETS Program is innovative in its use of Web 2.0 technologies to promote peer-to-peer networking, remote counseling and mentoring, and 24/7 access to training resources. Web 2.0 technologies, integrated into the program, will include social networking (e.g.,LinkedIn.com, FaceBook.com), blogging (e.g., ETS blog, Twitter.com microblog), and social media (e.g., RSS feeds, news alerts). Web conferencing (e.g., AdobeConnect) supports training and mentoring services for clients on different work shifts, having personal obligations, or with long commutes. The overall objective is to offer virtual support that transcends geographic, personal, and time boundaries. Each client will be introduced to Web 2.0 technologies for staying informed, distance learning, networking with peers, and promoting his or her small business.

Leveraged Resources with Community Partners. The ETS Program is innovative in its leveraging of resources with community partners in mentoring, counseling, and training services provided. Some examples include: the county's Workforce Development Board assists in worker recruitment and job placement, the Small Business Administration (SBA) provides talks on small business loans and SCORE volunteers provide additional mentoring support; the Women's Business Center offers resources that target women entrepreneurs and women-owned small businesses; the local Women's Center provides microloans to qualified entrepreneurs; and several local Chambers of Commerce offer facilities for hosting networking events and small business fairs.

The professors identified the comprehensive scope of the program as being a challenge in achieving measurable objectives. Unlike past entrepreneurial programs launched by the university, the ETS Program involved a large number of constituents for which personalized services were to be provided. The ETS Program also involved a diverse group of participants involved at various stages in the program. These would include: faculty researchers, administrators, experts in the field, small businesses, nonprofit and profit entities, and government agencies.

SWOT Analysis

Professors Smith and Thomas initiated efforts to develop the ETS Program such that it could begin with a short time frame. Professors Smith and Thomas set out to study the strengths, weaknesses, opportunities, and threats (SWOT) as foundational to developing and sustaining the ETS Program. They reviewed the Regional Planning Council's *Comprehensive Economic Development Strategy* (CEDS, 2009) (www.ecfrpc.org/). Pertinent information, related to the ETS Program, is shown in Table 4.

They quickly realized that the economic impact of workforce reductions in defense-related industries would be significant for the region. Projected job losses, compounded with a higher than the national average unemployment rate, threatens the local county in becoming a "distressed county" as defined by the federal government. The federal definition of "economically distressed," is based on a county having either: (1) Per capita income of 80 percent or less than the national average; or (2) unemployment rate one percent greater than the national average for the past 24 months.

Detailed SWOT Analysis

The professors conducted research to further understand each component of the SWOT analysis and its impact on the ETS Program. Their findings are briefly summarized.

Table 4. SWOT analysis for the region (ecfrpc.org, 2009)

Economic Development Strengths	
Knowledge Structure	Universities and community colleges in the region offer a broad range of education services in high-tech, environmental, business, and related areas.
Emerging Technologies	Our region is part of the High Tech Corridor, which promotes high-tech economic development in national priority areas.
High-Tech Industries	The region is home to high-tech industries in military, security, bio-medical, energy, technology, and education.
Economic Development Weaknesses	
Lack of Growth Infrastructure	The region lacks an infrastructure for population growth in protecting environmental lands.
Workforce Transition	The region lacks a comprehensive plan to support those impacted by labor reductions in defense-related industries.
Education System	The region lacks adequate funding for K-12 education system impeding the development of a high-skilled workforce.
Economic Development Opportunities	
High-tech Infrastructure	Regional organizations form a "community cluster" in providing a high-tech infrastructure for entrepreneurs.
Workforce Transition	High-tech small businesses are created by transitioning workers selecting entrepreneurship as a career option.
Education Improvements	High-skilled workforce is increased through K-12 science, technology, engineering, and mathematics (STEM) education.
Economic Development Threats	
Fragmented Leadership	Fragmented decisions on land use, transportation, and economic development are counterproductive to growth opportunities.
High-Tech Contraction	Corporate contraction due to the economic stagnation and cut backs in federally funded programs results in attrition of high-skilled workers.
High-Tech Job Loss	Unemployment and economic stagnation results from job losses associated with reductions in defense-related industries.

Weaknesses and Threats

Regional Job Loss. Government predicts direct reductions in workforce to be at least 3,500 jobs at a local federal facility. Some industry officials say the number could be as high as 15,000. The best-case scenario would result in the loss of about 10,000 other jobs in the surrounding community; the worst-case number is close to 30,000. It is estimated that job losses will be equivalent to 3% of the county's entire nonagricultural employment base.

Corporate Contraction. Employees face unemployment due to corporate contraction in preparation for significant cut backs in government spending in defense-related products and services. The recent 140 job losses at High-Tech Corporation[2], and 50 at Universal Technology, Inc., have been attributed to reductions in defense-related programs. Corporate contraction will result in fewer high-tech positions necessary to absorb high-tech job layoffs. Side effects include increasing difficulty in attracting high-skilled workers to the region, and graduating students in high-tech fields leaving the region for external job opportunities.

Loss of High-Skilled Workers. Highly skilled workers are accepting employment in other geographic locations (e.g., Alabama, Colorado, New Mexico, Texas), as worker layoff dates approach. These losses in human capital have a long-term economic impact on the region. Many forecast this "brain drain" as being a significant to the economic well-being of both the region and the State.

Small Business Government Contracts. "We will have a gap in the contractor workforce at the federal facility that will not be filled," said a high level administrator in response to significant reductions in federal funding. DefenseR-yx Corporation, one of the largest contractors on defense-related projects, estimates the loss of its multi-billion dollar service contract as having a trickle-down, economic impact. In the past, DefenseR-yx awarded subcontracts to numerous small businesses in meeting its government contract deliverables. In just one year, DefenseR-yx Corporation awarded small business contracts close to US $50 million. A portion of these subcontracts are "set aside" for small business types such as: veteran-owned, small disadvantaged, and service-disabled veteran-owned. A reduction of government contracting set asides for small businesses, such as those provided by DefenseR-yx, will have an economic impact on the region (Appendix B provides definitions for various small business types that would be impacted.)

Research and Innovation Government Contracts. In 2008, the federal agency awarded across its sites almost 400 Small Business Innovation Research (SBIR) and Small Business Technology Transfer (STTR) contracts. These awards have become foundational to technology innovation in the region's economic development priority areas of: defense, energy, technology, security, and the environment. The reduced federally-funded program will most likely have a significant impact on SBIR/STTR awards given out to small businesses in the region.

Table 5 has definitions for each type of small business grant. The STTR grant is of particular interest to Sci-Tech given it provides funding for small businesses and researchers to collaborate and innovate on established priority areas. Sci-Tech is concerned about the reduction of these types

Table 5. Definitions of SBIR/STTR grants (www.sbir.gov)

SBIR	Small Business Innovation Research (SBIR) grants are awarded to small businesses by federal agencies to provide funding for early-stage innovation of products or services that are often too risky for private investors (e.g., Venture Capitalists).
STTR	Small Business Technology Transfer Program (STTR) uses a similar approach to the SBIR program to expand public/private sector partnerships between small businesses and nonprofit U.S. research institutions.

of grants given they provide a revenue stream in support of a range of cutting-edge research activities.

Strengths and Opportunities

Knowledge Structure. Sci-Tech, as a premier science and technology institution of higher education in the region, offers a broad range of education, training, and community resources. Faculty researchers are studying technological advances in sustainability, information security, green energy, emerging technologies, and communications.

Entrepreneurs in High-Tech Industries. The region is part of a High Tech Corridor and home to many small businesses started by entrepreneurs previously employed in defense-related industries. By tapping into high-tech entrepreneurship talent, there is the potential for economic growth through business startup and technology innovation.

Career Option for Transitioning Workers. With the impending workforce reduction, entrepreneurship may be an appealing career option for many high-skilled workers. Many engineers, instrumental in technology innovation, may consider starting a business. Keeping them local would promote economic expansion in the region while reducing the loss of intellectual power when displaced workers move to other geographic regions.

Collaboration with Research Experts. Sci-Tech has research expertise associated with its centers and laboratories including the Energy Research Center, Information Security Center, Biomedical Sciences Center, and High Resolution Imaging Research Program. Sci-Tech offers entrepreneurs and small businesses a high-tech knowledge-base from which a range of collaborative opportunities exist. Small business and faculty collaborations have been established through a regional nonprofit outreach program in pursuit of SBIR/STTR federal grants.

The professors concluded that economic conditions mandated an intensive, hands-on training program for adults and displaced workers in the region to be implemented as quickly as possible. The next step was to further profile clientele to be served by the ETS Program.

Clientele

Professors Smith and Thomas identified four targeted client groups to be served by the ETS Program. Each is briefly described.

High-Skilled Workers at the Federal Facility. High-skilled workers at the facility are directly impacted by federal cut backs in defense-related funding. Many are engineers or other high-skilled workers employed by the federal government or supporting companies. The State is counting on these high-skilled workers staying in the region to attract new high-tech businesses. However, job opportunities in other regions of the United States could result in significant worker attrition and economic loss for the region. Sci-Tech intends on promoting its ETS Program as an alternative to moving out of the State. Many clients will view this favorably, as they have established roots in the region.

High-Skilled Workers in the Region. Many high-skilled workers will be indirectly impacted by federal cut backs in defense-related funding; as well as, the economic downturn in the region. (The State recently recorded an unemployment rate of close to 11%). These workers are employed by both small and large companies supporting defense, energy, communications, and information security. The loss of these high-skilled workers, in transitioning to other locations, would further devastate the regional economy.

Small Businesses in the Region. Many regional small businesses, particularly those geographically close to the federal facility, will be impacted by workforce reductions. Many need assistance in revising their business plans to identify new product and service opportunities and target new markets. The ETS Program will offer technical assistance such that small businesses can

Figure 2.

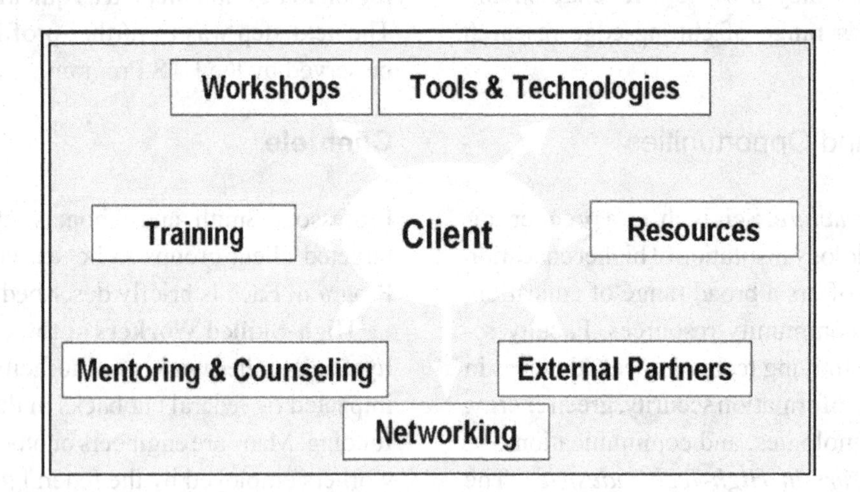

implement business practices that will promote sustainability and growth.

Researchers and Innovators. Sci-Tech's faculty members, primarily in science, technology, and engineering fields, need assistance in identifying collaborative opportunities. Professors Smith and Thomas have initiated discussions with faculty members working with small businesses for technology innovation and commercialization. Several faculty members, holding patents on intellectual property, welcome the opportunity to transfer their technology to the private sector. Sci-Tech's administration is excited about the possibility of revenue generation through technology transfer and commercialization efforts. Several other universities have had great success with revenue generation as a result of technology transfer initiatives (refer to Figure 1 for examples).

The Entrepreneurial Training Services (ETS) Program

Professors Smith and Thomas put together an ETS framework that encompasses technical assistance support with the end goal of having each entrepreneur or small business in the program complete a business plan to be reviewed by a panel of experts.

The ETS framework, shown in Figure 2, has entrepreneurial workshops, training seminars, and mentoring and counseling services all of which guide a client through the business planning process. Networking events bring together entrepreneurs, business experts, mentors, and small businesses for collaboration and information sharing. External partners, profit and nonprofit, help minimize the duplication of local and regional technical assistance services. Community, state, and national resources support an entrepreneur in business startup, sustainability, and growth. Tools and technologies are used by clients to build support networks and for lifelong learning. Table 6 briefly identifies each element in the ETS Framework.

Project Timeline

Professors Smith and Thomas put together an aggressive project timeline in serving the clientele previously identified. Figure 3 shows the timeline of program activities over a nine month time period for which three (3) technical assistance service tracks will be offered targeting nascent entrepreneurs and small businesses in the region. It is anticipated that after the ETS Program tracks are concluded, a small business fair will

Table 6. ETS framework elements

Workshops	Client explores a career roadmap to entrepreneurship.
Training	Client learns basics of business development; completes a business plan; and has it reviewed by a panel of experts.
Counseling	Client meets with counselor for expert guidance in business function area.
Mentoring	Client meets with mentor for business development guidance.
Resources	Client has access to resources for business startup, sustainability, and growth.
Technology	Client uses Web 2.0 technologies for marketing, networking, and communication.
Networking	Client participates in networking events as support systems and lifelong learning.

Figure 3. ETS program timeline shown for November to July

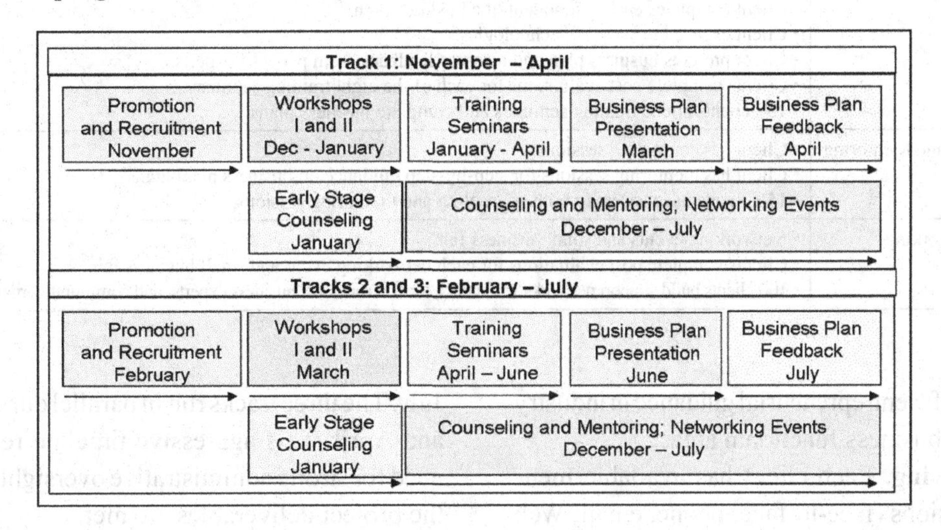

be co-hosted by Sci-Tech University, profit and nonprofit organizations, and local, regional, and federal agencies.

Each element of the ETS framework is briefly described in terms of the projected program timeline.

Recruitment. The first month of each ETS Program track focuses on recruitment activities. Information sessions are held onsite at the federal facility to recruit clients directly impacted by reductions in funded programs. They also take place at community events located throughout the region. Multiple media sources (newspaper, radio, business magazines) are used to recruit nascent entrepreneurs to enroll in the ETS Program.

Workshops and Early Stage Counseling. Each workshop is held 5 pm to 8 pm on a weekday in a geographically convenient location. These workshops are intended for exploring entrepreneurship as a career path. Early stage counseling immediately follows the workshops as a means of answering questions about the program.

Business Skills Training and Business Plan Development. Each training seminar is held on a weekday in a geographically convenient location to the client. The training seminars offer an intensive, hands-on approach to completing a business plan. The training program ends with business plan feedback provided by a panel of experts.

Counseling. Each client has available counseling sessions (face-to-face, phone, email, Web conferencing) to be used anytime after attending the workshop sessions. These technical assistance

Table 7. Project deliverables

Technical Assistance	Deliverables
Recruitment	• ETS Web site with registration capability. • Promotional materials, press releases, brochures. • 200 plus clients registered for Workshops I and II.
Workshops	• Workshop materials and sessions. • Client exposure to entrepreneurial business environment. • Lessons learned and best practices from local entrepreneurs. • 200 clients attend Workshops I and II; complete post evaluations.
Early Stage Counseling	• Client has session with business area or industry sector expert for needs assessment. • 170 client contacts for counseling guidance. • 150 clients sign up for training seminars.
Training	• Training materials and seminars. • Client completes each component of a business plan. • Client exposed to Web 2.0 technologies. • Client presents business plan and receives feedback from panel of experts. • Clients complete post evaluations for each of the eight training seminars. • 150 clients attend training seminars and complete business plans.
Counseling and Mentoring	• Client has counseling sessions. • Client has mentoring sessions for step-by-step guidance on business planning. • 150 clients have access to both counseling and mentoring sessions.
Networking	• Networking events and small business fair. • Clients complete post evaluations for each networking event and small business fair. • 100 clients build support networks with peers, local businesses, business experts, and community organizations.

services offer entrepreneurial guidance in industry sectors or business functional areas.

Mentoring. Each client has available mentoring sessions (face-to-face, phone, email, Web conferencing) to be used during specified training breaks or as needed during the training time period. Mentoring technical assistance services support each activity in business planning.

Networking. Each networking event is held monthly in a geographically convenient location to the client. A one day small business fair is held in July after business plan activities are completed. Networking services play an integral role in building entrepreneurial support systems for business startup, sustainability, and growth.

A detailed project timeline in Appendix A shows the progression of technical assistance services in each track in support of project objectives. The timeline shows the first track starting in November with a holiday break in December and then continuing through April. The second and third tracks start in February and continue through July. The three tracks run in parallel during March and April. This aggressive timeline reflects the need for strong administrative oversight to ensure the project deliverables are met.

Project Deliverables

The ETS Program deliverables, for each technical assistance element, are identified in Table 7. These include workshop and training materials; workshops sessions, training seminars, counseling and mentoring sessions, and networking events; sufficient number of clients recruited and completing the ETS Program; and panel reviews of each business plan. Other deliverables include client evaluations; performance tracking; and monthly and final reports.

Table 7 shows the intensive, hands-on approach being taken to promote business development. Though details of each component are not presented, it is noted that the mentoring and training services focus on the successful completion of each

Figure 4. Organizational structure of the ETS program

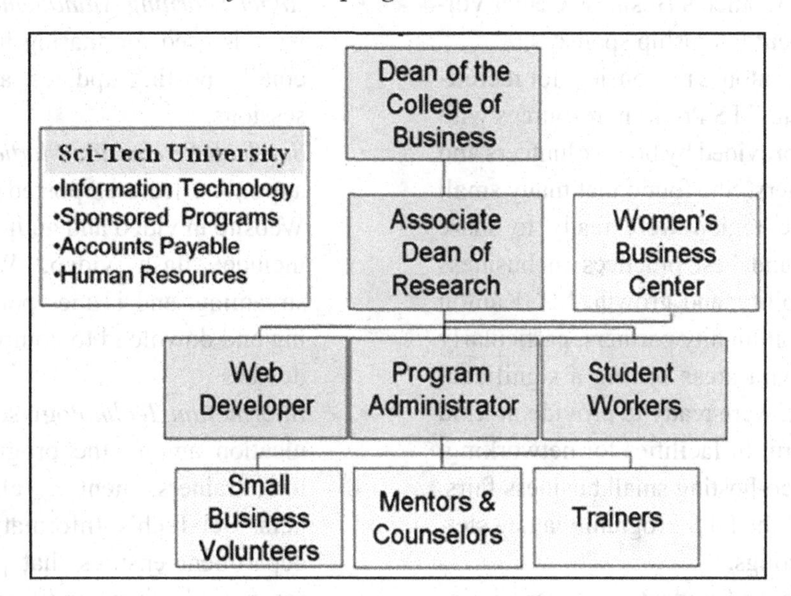

component of a business plan. The end goal is a business plan that may be presented to a panel of experts inclusive of Angel Investors and Venture Capitalists. Community experts, small business volunteers, and faculty researchers were made available to entrepreneurs for guidance in technology transfer, intellectual property, marketing, or financing, among other areas.

Project Oversight

Professor Smith has management oversight of the ETS Program. Professor Smith has extensive experience in contract administration, information and communication technology, and business development. She is the Associate Dean of Research and is responsible for ensuring project deliverables are met within the specified timeline.

The Program Administrator, Ms. Rachelle Kanton, is responsible for daily operations in implementing the project plan. The Program Administrator has operational oversight of instructors, counselors, and mentors. The Program Administrator is the community liaison in leveraging resources with local profit and nonprofit entities. She is responsible for tracking and reporting project performance to Sci-Tech administration.

Sci-Tech provides operational support for the ETS Program through its internal organizational components. The roles of these organizational components are:

- **Sponsored Programs**. Ensures university and grantor policies are used in contract administration.
- **Human Resources**. Enforces hiring practices and equal employment opportunity procedures.
- **Accounting**. Processes payroll, invoices, and check requisitions.
- **Information Technology**. Provides technology support and maintains Web and email systems.

Figure 4 shows the Associate Dean, Program Administrator, and ETS Program staff as part of the College of Business organization structure. Small business volunteers, mentors, counselors, and trainers are actively involved in the College of Business; and many serve as board members, business faculty, Women's Business Center volunteers, and student internship sponsors.

Ms. Rachell Kanton is responsible for recruitment in leveraging ETS Program resources with in-kind services provided by both volunteers and community partners. She found that many small businesses in the region were ready to share lessons learned and best practices in business startup, sustainability, and growth. Ms. Kanton also found that community partners, particularly those affiliated with areas having a significant economic impact, were ready to provide in-kind support in the form of facilities for networking, staff support for co-hosting small business fairs, and promotion of the ETS Program via newsletters and Web postings.

Several faculty and student volunteers, in information technology and computer science fields, offered their services in the prototyping of the Web 2.0 technologies to be used creatively in connecting entrepreneurs in a network of support.

Sci-Tech Software Tools and Technologies

Sci-Tech University will provide Web 2.0 technologies in support of the ETS Program. This is an important element of the ETS Program for several reasons. Entrepreneurs need skills in business technologies for long term adaptability, survival and growth. Adeptness with on-line collaboration and technologies requires training and hands-on experience. Faculty participants, in the form of trainers or mentors, may find offsite travel difficult due to onsite teaching schedules. Web 2.0 technologies provide a means of connecting a faculty mentor with a client removing time and distance barriers.

The Web 2.0 technologies include:

- *Web conferencing software*, Adobe Acrobat Connect, is used for distance counseling and mentoring of individuals and groups.
- *Angel Learning Management System* software is used for sharing files, exchanging emails, posting updates, and hosting chat sessions.
- *Sci-Tech Production Studio* is used to create online lectures posted on the program Website in video and audio formats. These include: flash video, Windows media streaming, and iTunes podcasts for viewing and download to computer and mobile devices.
- *Information Technology* supports communication among the program administrators, trainers, mentors, clients, and partners. Sci-Tech's Information Technology department ensures that program data is secure and private and there is reliable access to email, web services, and networking systems.

*i*Place for Entrepreneurs™ Web Portal

Professors Smith and Thomas will focus on the use of Web 2.0 technologies, as a reflection of the innovative spirit of the ETS Program. Professor Smith developed the *i*Place for Entrepreneurs™ Web portal concept to bring researchers and entrepreneurs together to innovate and invent technologies for commercialization. Faculty and student volunteers will explore the use of emerging technologies in the development of an intensive, personalized support system offering real-time technical assistance services.

Building Virtual Bridges. Users of the *i*Place for Entrepreneurs™ Web portal are provided access to a data repository of researcher profiles by area of expertise and entrepreneur profiles by high-tech areas. Profiles will be searchable offering automated support for matching researchers and entrepreneurs with common interests. Entrepreneurs will post requests for collaboration; and researchers will post requests for consulting and research. Figure 5 illustrates capabilities of

Figure 5. iPlace for Entrepreneurs™ web portal illustration of capabilities

the Web portal in building virtual bridges among ETS staff, researchers, entrepreneurs, community partners, and small business owners.

Research Collaboration. The *i*Place for Entrepreneurs™ Web portal offers a means of promoting collaboration in the pursuit of SBIR, STTR, and other federal grant opportunities. The Web portal will have links to grants.gov, fedbizzopps.gov, sbir.gov, sba.gov, and each federal agency. The overall goal is to promote research particularly in regional economic priority areas of: energy, sustainability, information security, communications and technology.

Connections Anytime and Anywhere. Many high-tech industries in the region rely on shift work. Training schedules may impose barriers to workers on various shifts (e.g., noon to 8 pm). The *i*Place for Entrepreneurs™ Web portal will post training seminars supported by a variety of electronic media including video streaming, podcasts, and RSS feeds. It will also bring together through virtual means researchers, community partners, small businesses, and entrepreneurs for networking and information sharing.

Training on Technology for Competitive Advantage. Technical assistance will focus on the use of information technology to ensure small business adaptability, survival, and growth in a global marketplace. Sci-Tech's production studio will produce videos of small businesses accessible via the *i*Place for Entrepreneurs™ Web portal. Entrepreneurs will be trained on a range of technologies such as: web conferencing, electronic press releases and newsletters, blogs; RSS, podcasting; and streaming media.

Replication and Sustainability Initiatives

Professors Smith and Thomas will focus on both replication and sustainability initiatives given the short time frame of funding support for the ETS Program. Their vision is that the ETS Program can be readily replicated throughout the United States for small business development and job creation. The key to replication is having community partners work together, in implementing technical assistance service components. They identified the following steps in replicating the ETS Program.

- **Implement the ETS Framework.** The ETS framework provides a foundation for implementing an entrepreneurial program that provides step-by-step support to each worker enrolled in the program. It is important to offer a sufficient number of technical services within a specified time frame in order for a client to be successful in completing a business plan.
- **Build Community Partnerships.** The ETS Program mandates effective partnering with a range of community organizations. Each partner has resources to be leveraged with the ETS Program in maximizing technical assistance provided to displaced workers in the region. The Workforce Development Board has recruitment capability for reaching workers across a region. The Business Incubation Center has facilities to support prototyping and manufacturing of high-tech products. The Small Business Administration and local banks offer resources on financing options. The local Chamber of Commerce offers at no cost meeting facilities for networking events.
- **Involve Local Small Businesses.** Each community has small businesses with inspirational stories of personal passion, drive, failure, and success. Most welcome an opportunity as volunteers to share their stories so workers can assess the entrepreneurial career path. They also are a knowledge base of "do's and don'ts" that can be shared with entrepreneurs for repeating what works and avoiding mistakes.
- **Offer Each Client a Safety Net of Mentors, Counselors, and Trainers.** Existing community resources may only "scratch the surface" of business plan development. As such, nascent entrepreneurs may become discouraged when trying to complete a business plan. By having a support system in place, each client has available a set of resources in offering support and answering questions related to business sector or functional area.
- **Provide Step-by-Step Guidance in Completing a Business Plan.** It is important to provide practical yet comprehensive guidance in completing each component of a business plan. This includes step-by-step instructions, business plan examples, individual exercises, and small group discussions. This learning-rich environment promotes confidence and commitment in completing a business plan.
- **Give Clients Feedback from Experts in the Field.** A client will gain invaluable feedback from experts on the viability of a business plan in terms of starting, sustaining, and growing a business. This integral component of the ETS Program offers a resource that otherwise may not be available.
- **Build Physical and Virtual Support Networks.** Peer-to-peer networking plays an important role in small business success. Networking offers an opportunity for entrepreneurs to promote their business, make connections for bootstrapping, and share lessons learned and best practices. Today's technologies offer networking support that transcends time and location boundaries.

The professors have implemented a process loop whereby each replication step is monitored during the implementation of the ETS Program. The overall objective is to identify areas for improvement for future implementations of the program.

CHALLENGES

There are several challenges facing Sci-Tech University in the implementation of the ETS Program. Each of these challenges is further discussed.

Knowledge-base. The ETS staff has to identify a sufficient number of mentors to guide each entrepreneur in the business planning process. Business faculty, industry experts, and small business volunteers have stepped forward to participate in mentoring activities. However, there may still be gaps in specific industry areas of expertise or business experience. Professor Thomas proposes that this challenge will be met through the implementation of the **i***Place* for Entrepreneurs™ Web portal where research experts external to the region can register online; thus, they become part of the mentoring process.

ETS Program Sustainability. Professors Smith and Thomas have already taken steps to pursue additional funding in support of the ETS Program. However, there is significant competition from local and regional community programs given current economic conditions. Smith and Thomas are developing a "program income" plan whereby clients would pay a nominal fee for enrolling in the training program. They also are exploring sponsorships with industry partners in hosting small business summits related to areas such as: technology innovation, subcontracting on federal grants, and training programs for high-tech workers.

Technology Transfer and Commercialization. Though resources are limited, the ETS staff has identified a need for small businesses training in the government contracting arena. This is especially important for small businesses to be competitive when applying for SBIR/STTR grants. Sci-Tech would like to strengthen its technology transfer initiatives such that funding for research and development increases through SBIR/STTR and other grant opportunities.

Small Business Access to Capital. Financial resources for entrepreneurs have become severely limited due to current economic conditions in the US. Angel investors and Venture Capitalists have become selective in funding new ventures. Local banks are more restrictive in providing bank loans and some no longer provide access to SBA loans. The ETS staff proposes to work with community partners on exploring financial options made available to local entrepreneurs.

Program Content. The clients enrolled in the ETS Program have a wide range of educational, experience, and expertise backgrounds. Some have technical backgrounds associated with the defense-related industries. Others are support personnel and administrators. Most have limited or no experience or education in business areas. Some clients were found to have minimal or no computer skills. All of these factors pose major challenges in developing and presenting content that meets the needs of adult learners with varying educational and professional backgrounds.

Program Schedule. The ETS Program schedule, as shown in Appendix A, requires significant oversight to ensure timely and sufficient technical assistance services are provided to each client. There are numerous scheduling activities that need to be performed and thereafter monitored including the use of facilities, presentations by small businesses and business experts, mentoring sessions appropriately timed, completion of business plan components, and business plan reviews by a panel of experts.

INITIAL SUCCESSES

To date, the ETS staff reported the following program successes that illustrate the potential for widespread use in economically-challenged communities.

1. Sci-Tech University was able to mobilize varied and numerous community resources in support of new business creation.
2. Approximately 45% of the original, introductory workshop attendees committed to and completed the ETS business training seminars and business plan development. (The workshops explained the opportunities and

challenges of pursuing the entrepreneurial career path.)
3. Public relations, promotional efforts, and the "cause" behind the ETS Program, motivated academic and professional experts to support the Program on a no-fee basis.
4. ETS client activity generated business for professional partners in the areas of patents, company formations, accounting and marketing.
5. ETS client word of mouth recommendations have generated new inquiries and a waiting list of people wishing to participate in a future ETS Program.
6. In addition to business creation outcome, participation in the ETS Program produced a more highly skilled, better trained group of workers.

Thus far, the community response has been overwhelmingly favorable with an increasing number of business experts and university personnel volunteering their services to the Program. As a result of the wide range of community support, the ETS staff and Sci-Tech University administrators are scheduling a Small Business Summit at the end of the first track of the ETS Program. They will invite community partners, faculty researchers, small business volunteers, and government entities to participate in roundtable discussions on the economic impact and long-term sustainability of the Program.

ETS PROGRAM BENEFITS

Community partners are confident that the ETS Program will have a positive impact on the local economy. The ETS staff and College of Business administration have an entrepreneurial history of implementing business centers and programs as a means of community support.

Sci-Tech is uniquely positioned to meet the entrepreneurial aspirations and needs of the County. Clients in the ETS Program enjoy numerous benefits that can not be found through other community resources. The ETS Program components, developed and delivered by one source (Sci-Tech), ensure continuity and consistency of entrepreneurial content and message.

Sci-Tech may be considered the "one-stop shop" for entrepreneurial content and resources. Clients do not need to search and seek out other community organizations to meet their needs. Clients save valuable time that can be devoted to building their business.

Sci-Tech, and specifically the ETS team, forms stronger, more supportive relationships with clients. By working with clients throughout the entire ETS Program and through all the ETS components, Sci-Tech comes to know the client better than any other organization. This knowledge is critical to increasing client entrepreneurial success. This deep client understanding in turn makes Sci-Tech the best facilitator for matching clients with community partners and resources.

The ETS Program serves as an outstanding model in preparing potential entrepreneurs for acceptance into the Regional Business Incubation Center. Sci-Tech is able to evaluate the progress of ETS clients and determine which clients have advanced sufficiently to be considered as candidates for the local business incubation program.

REFERENCES

CED. (2009). *East Central Florida regional planning council's comprehensive economic development strategy*. Retrieved September 4, 2009, from http://www.ecfrpc.org

Chakrabarti, A. K., & Lester, R. K. (2002, August 20). Regional economic development comparative case studies in the US and Finland. In *Proceedings IEEE Conference on Engineering Management*. Cambridge, UK.

D'Cruz, C., & Ports, K. (2006). Space Coast innovation and technology commercialization outreach. In *Proceedings of the United States Association of Small Business and Entrepreneurship (USASBE),* Tucson, AZ.

Drucker, P. F. (1985). *Innovation and entrepreneurship*. New York: Harper & Row.

Kim, J., & Marschke, G. R. (2007). *How much US technological innovation begins in universities? Economic Commentary* (pp. 1–3). Cleveland: Federal Reserve Bank.

Kirzner, I. (1984). The entrepreneurial process. In Kent, C. A. (Ed.), *The environment for entrepreneurship*. Lanham, MD: Lexington Books.

Kuratko, D. F. (2005). The emergence of entrepreneurship education: Developments, trends, and challenges. *Entrepreneurship Theory and Practice, 29*(5), 577–597. doi:10.1111/j.1540-6520.2005.00099.x

Kuratko, D. F., & Hodgetts, R. M. (2004). *Entrepreneurship: Theory, process, practice*. Mason, OH: South-Western College Publishers.

Labormarketinfo.com. (2009). State of Florida Agency for Workforce Innovation, Labor Market Statistics. Retrieved September 4, 2009, from http://www.labormarketinfo.com

Li, R. (2003, October 27). Alignment of funding mechanisms with scientific opportunities. *Regional Forum on Research Business Models — Workshop Summary*, Berkeley, CA, (OSTP/NSTC Committee on Science).

Minniti, M., & Bygrave, W. D. (2004). *Global entrepreneurship monitor*. Kansas City, MO: Kauffman Center for Entrepreneurial Leadership.

Palmintera, D., Bannon, J., Levin, M., & Pagan, A. (2000). Developing high technology communities: San Diego. Produced under contract to Office of Advocacy, U.S. Small Business Administration, by Innovation Associates, Inc., Reston, Virginia.

Reynolds, P. D., Hay, M., & Camp, S. M. (1999). *Global entrepreneurship monitor*. Kansas City, MO: Kauffman Center for Entrepreneurial Leadership.

SBA. (1998). *The new American revolution: The role and impact of small firms*. Washington, DC: Small Business Administration, Office of Economic Research.

Tornatzky, L., Waugaman, P. G., & Gray, D. (2002). *Innovation U: New university roles in a knowledge economy*. Southern Growth Policies Board.

U.S. Census Bureau. (2009). U.S. Census Bureau State and County Quick Facts. Retrieved September 4, 2009, from http://quickfacts.census.gov

APPENDIX A: PROJECT TIMELINE FOR THE ETS PROGRAM

<table>
<tr><th colspan="4">FIRST TRACK BEGINS</th></tr>
<tr><td colspan="4">November: Recruitment and Promotion</td></tr>
<tr><td>Wk 1</td><td>Recruitment and Promotion Activities</td><td colspan="2"></td></tr>
<tr><td>Wk 2</td><td>Information Session Onsite</td><td colspan="2"></td></tr>
<tr><td>Wk 3</td><td>Recruitment and Promotion Activities</td><td colspan="2"></td></tr>
<tr><td>Wk 4</td><td>Information Session Onsite, Networking</td><td colspan="2"></td></tr>
<tr><td colspan="4">December: Workshops I and II</td></tr>
<tr><td>Wk 1</td><td>Workshop Preparation</td><td colspan="2"></td></tr>
<tr><td>Wk 2</td><td>Workshop I</td><td colspan="2"></td></tr>
<tr><td>Wk 3</td><td>Workshop I & II</td><td colspan="2"></td></tr>
<tr><td>Wk 4</td><td>Holiday</td><td colspan="2"></td></tr>
<tr><td colspan="4">January: Workshops; Counseling, Training, Networking</td></tr>
<tr><td>Wk 1</td><td>Workshop II; Early Stage Counseling;</td><td colspan="2"></td></tr>
<tr><td>Wk 2</td><td>Early Stage Counseling; Networking</td><td colspan="2"></td></tr>
<tr><td>Wk 3</td><td>Building a Business Plan</td><td colspan="2"></td></tr>
<tr><td>Wk 4</td><td>Sales, Marketing, Customer Service;</td><td colspan="2"></td></tr>
<tr><td colspan="4" align="right">SECOND AND THIRD TRACKS BEGIN</td></tr>
<tr><td colspan="2">February: Training; Mentoring, Counseling</td><td colspan="2">February: Recruitment and Promotion</td></tr>
<tr><td>Wk 1</td><td>Build a Marketing Plan; Monthly Report</td><td>Wk 1</td><td>Recruitment and Promotion Activities</td></tr>
<tr><td>Wk 2</td><td>Mentoring Services; Networking</td><td>Wk 2</td><td>Information Session Onsite; Networking</td></tr>
<tr><td>Wk 3</td><td>Finance for Non-Financial Managers</td><td>Wk 3</td><td>Recruitment and Promotion Activities</td></tr>
<tr><td>Wk 4</td><td>Business Operations Overview</td><td>Wk 4</td><td>Information Session Onsite</td></tr>
<tr><td colspan="2">March: Training; Mentoring, Counseling</td><td colspan="2">March: Workshops I and II</td></tr>
<tr><td>Wk 1</td><td>Mentoring Services; Feb Monthly Report</td><td>Wk 1</td><td>Workshop Preparation; Monthly Report</td></tr>
<tr><td>Wk 2</td><td>Business Administration; Networking</td><td>Wk 2</td><td>Workshop I ; Networking</td></tr>
<tr><td>Wk 3</td><td>Business Plan Completion</td><td>Wk 3</td><td>Workshop I & II</td></tr>
<tr><td>Wk 4</td><td>Mentoring Services</td><td>Wk 4</td><td>Break</td></tr>
<tr><td>Wk 5</td><td>Business Plan Presentation</td><td>Wk 5</td><td>Workshop II; Early Stage Counseling</td></tr>
<tr><td colspan="2">April:Mentoring; Panel Feedback</td><td colspan="2">April Workshops; Counseling; Training</td></tr>
<tr><td>Wk 1</td><td>Panel Feedback</td><td>Wk 1</td><td>Early Stage Counseling; Monthly Report</td></tr>
<tr><td>Wk 2</td><td>Mar Monthly Report; Networking</td><td>Wk 2</td><td>Building a Business Plan; Networking</td></tr>
<tr><td>Wk 3</td><td></td><td>Wk 3</td><td>Sales, Marketing, Customer Service</td></tr>
<tr><td>Wk 4</td><td></td><td>Wk 4</td><td>Building a Winning Marketing Plan</td></tr>
<tr><td colspan="4" align="right">May: Training; Mentoring; Counseling..</td></tr>
<tr><td colspan="2"></td><td>Wk 1</td><td>Mentoring Services; Monthly Report</td></tr>
<tr><td colspan="2"></td><td>Wk 2</td><td>Finance for Non-Financial Managers</td></tr>
<tr><td colspan="2"></td><td>Wk 3</td><td>Business Operations Overview; Networking</td></tr>
<tr><td colspan="2"></td><td>Wk 4</td><td>Mentoring Services</td></tr>
<tr><td colspan="4" align="right">June: Training; Mentoring; Counseling..</td></tr>
<tr><td colspan="2"></td><td>Wk 1</td><td>Business Administration; Monthly Report</td></tr>
</table>

FIRST TRACK BEGINS		
	Wk 2	Business Plan Completion; Networking
	Wk 3	Mentoring Services
	Wk 4	Business Plan Presentation
		July: Mentoring; Panel Feedback ....
	Wk 1	Panel Feedback
	Wk 2	Networking; Jun Report
	Wk 3	Small Business Fair; Final Report

APPENDIX B: DEFINITIONS OF SMALL BUSINESSES (WWW.SBA.GOV; WWW.TRADE.GOV)

Veteran-owned Small Business	Small business that: (i) is at least 51% unconditionally owned by one or more veterans (as defined by <federal regulations>; or, (ii) in the case of any publicly owned business, at least 51% of the stock of which is unconditionally owned by one or more veterans; and whose management and daily business operations are controlled by one or more veterans.
Service-disabled Veteran-owned Small Business	Small business that: (i) is at least 51% unconditionally owned by one or more service-disabled veterans as defined by <federal regulations> with a disability that is service connected as defined by <federal regulations> or in the case of any publicly owned business, at least 51% of the stock of which is unconditionally owned by one or more service-disabled veterans; and (ii) whose management and daily business operations are controlled by one or more service-disabled veterans or, in the case of a veteran with permanent and severe disability, the spouse or permanent caregiver of such veteran.
Small Disadvantaged Business	Small business that: (i) has received certification as a small disadvantaged business consistent with <federal regulations>; (ii) no material change in disadvantaged ownership and control has occurred since its certification; (iii) where the business is owned by one or more individuals, the net worth of each individual upon whom the certification is based does not exceed $750,000, after taking into account the applicable exclusions in <federal regulations>; (refer to sba.gov for the full definition).
Woman-owned Small Business	Small business that: (i) is at least 51% unconditionally owned by one or more women; or in the case of any publicly owned business, at least 51% of the stock of which is unconditionally owned by one or more women; and (ii) whose management and daily business operations are controlled by one or more women.

Section 2
Organizations, People, Processes, and Paradigms in Technology Innovation

Chapter 4
Technology Innovation Adoption and Diffusion:
A Contrast of Perspectives

Michael Workman
Florida Institute of Technology, USA

ABSTRACT

The literature on technology innovation adoption and diffusion is vast. In this chapter, we organize and summarize some of the major perspectives from this body of literature, contrasting various theoretical perspectives on how innovations are adopted and shaped by organizational processes and structure. We first introduce the technology acceptance model, and innovation diffusion theory; and then we categorize viewpoints about organizational innovativeness. Drawing from this framework, for our case study background we introduce adaptive structuration theory, redefining some of its conceptual relationships in "structuration agency theory," putting primacy on the actions of agents and the means by which they operate through and around institutional structures. We then present a case study example of an expert decision support system, and we conclude with a discussion of implications for managers and entrepreneurs.

INTRODUCTION

An innovation is defined as the act of changing the established order, or introducing something new (Webster's Dictionary, 1978). Thus there are two sides to the innovation coin, the diffusion of them, and the production of them. One aspect these two sides of the coin share in common is some measure of risk-taking behavior, but the creation of innovations tends to be an individualized thought put into action, whereas the diffusion of them tends to be the result of a set of institutionalized and collective socialized actions, although sometimes initiated by a "champion" (Lake, 2009).

In this chapter, we will concentrate on the adoption and diffusion of technological innovations within organizations. We will first briefly introduce innovation research situated in the field of management science and briefly discuss how it has evolved relative to our topic. In the sections that follow, we

DOI: 10.4018/978-1-61520-609-4.ch004

will review two seminal streams of innovation adoption theory by introducing the technology acceptance model (Davis, 1989) and diffusion of innovations theory (Rogers, 1983). We will then categorize and summarize the major theoretical perspectives on how innovations develop from, or are shaped by, organizational processes and structures–referred to as organizational innovativeness (Chakravarthy, 1997). We will then present a brief case study to illustrate adoption and diffusion theory and structuration with an expert decision support system in an organization, and conclude with a discussion of implications.

The study of innovations has its basis in management science. Some of the most frequently cited examples are novel approaches or inventions such as Henry Ford's mass production of the automobile, or Eli Whitney's approach to the mass assembly of rifles, or his cotton gin, or about Thomas Edison and the invention the light bulb, or Alexander Graham Bell and the telephone; but the development of innovations occur with much less fanfare and more frequently than we might expect, with many failing to gain enough traction to be adopted or diffuse in organizations.

Management science has evolved from its early roots in the 19th-Century scientific management era to embrace the notion that an organization adapts and structures itself according to its business environment and needs, and that human actions emerge from, and business conditions grow out of, dynamic organizational systems (Sine, Mitsuhashi & Kirsch, 2006). This evolution in managerial thinking began in the 1970s and gave rise to "systems theory." As part of that renaissance, in the 1980s and early 90s, socio-technical systems theory (Trist, 1971) began to make its way into management practice in which both the technical and socio-cultural aspects of organizational systems were considered to be interdependent (Manz & Stewart, 1997). Socio-technical systems:

"...reflects the goal of integrating the social requirements of people doing the work with the technical requirements needed to keep the work systems viable with regard to their environments. These are considered interdependent because arrangements that are optimal for one may not be optimal for the other. Also tradeoffs are typical, and thus there is a need for both dual focus and joint optimization" (Fox, 1995, p. 92).

Although dual focus and joint optimization were imagined to be an ideal, socio-technical systems-driven practices initially ran counter to the entrenched mechanistic ones (Burns & Stalker, 1961) and was met with much resistance (Quinn, 1992). While process engineers, total quality management proponents, and members of process standards bodies fought to maintain their control over processes and developments throughout the 1980s and 1990s by means of certifications that tested conformance to a given standard with a presumed single correct solution, many organizations began concentrating instead on innovation through diversity of inputs in reaction. According to Hamel and Prahalad, (1993), there was a growing rejection of the notion that through benchmarking and standardization, everyone should have a race down identical paths where no one wins.

Consequently, creativity was touted as the means to achieve innovativeness. Creativity is by definition variation or deviation from the norm (Daft & Lengel, 1986); and since it often evolves out of crises (Kuhn, 1996), the phrase *necessity is the mother of invention* was an appropriate cliché. Thus newer organizational models came to operate less from a systematic perspective and more from a systemic one. In other words, while it was well recognized that standardization and striving toward some well-defined end using some well-defined means were crucial in certain areas and aspects within an organization, by and large, this forced convergence was too constraining to enable, let alone encourage, innovation (Stacy, 1992).

In that light, the drives toward innovation and differentiation became the overarching goals of many organizations (Porter, 1996), and organizational strategies were formulated and aligned to generate a level of variety that matched the level of flexibility required by the organization's purpose and environment (Denton & Wisdom, 1992). It was perceived that any innovation was strategic in nature and would need to combine flexible technologies, processes, and people to accommodate the spectrum of issues across multiple system boundaries (Mintzberg, 1994). This was an important concept because even "modest increases in variety usually produce dramatic increases in flexibility and in the capacity for adaptive response" (Pava, 1983, p. 55).

Nevertheless, the alignment and optimization of organizational systems meant that technological systems had to diffuse rapidly throughout organizations. Innovation theory (Rogers, 1983) postulates that the adoption of a technological innovation encompasses many facets spanning personal, social, and strategic aspects of organizations, and can derive from either an administrative core or from the "grass roots" level. Along with globalization and multiculturalism found in many organizations as the result of multinational conglomerations and outsourcing, it led to increased complexity, necessitating that managers and entrepreneurs attend not only to the technological aspects of variety but the socio-cultural aspects of variety as well (Semler, 1997).

By the time socio-technical systems theory had taken hold in organizational practice, it was firmly believed that inadequate attention to a particular technological environment or their impacts on social structures in organizations would result in problems such as a lack of *ownership* of, or lack of commitment to, organizational goals and innovations (Manz & Stewart, 1997). Consequently a marriage was made of the rational-technological elements of innovation and the socio-cultural and qualitative viewpoints about innovativeness such as those espoused by adaptive structuration theory (DeSanctis & Poole, 1994).

By emphasizing the role that social systems played in technology adoption, innovation, and organizational structuring, this ultimately led to an ideational integration about the technical, behavioral, and social systems in management science (Ngwenyama & Lee, 1997). This was considered to be important because dynamic interrelationships among people (their beliefs, values, intentions, and behavior) and their social and technological systems all collide in organizations and have direct bearing on an organization's innovativeness (Boxall, 1996; Porter, 1979).

THEORY FRAMEWORKS

Technology Acceptance Model

The study of how innovations are adopted and spread through organizations has been widely studied, primarily guided by two seminal sets of work, the technology adoption model (Davis, 1989) and the diffusion of innovations theory (Rogers, 1983), although other theories have also been utilized such as Ajzen's (1991) theory of planned behavior. The technology acceptance model (Davis, 1989), or TAM, describes how innovative technology impacts people at work. In his foundational study, Davis (1989) presented two variables as predictive constructs of technology adoption and implementation behavior, which were called "perceived usefulness" of technology and "perceived ease of use" of technology.

In his model, perceived usefulness and perceived ease of use predict attitudes about innovations that in turn predict people's intentions to adopt and use them. The perceived usefulness construct is defined as "the degree to which a person believes that using a particular system would enhance his or her job performance, whereas perceived ease of use refers to the degree to which a person believes that using a particular system would be free of effort" (p. 320). Davis (1989) also emphasized the subjectivity in innovation adoption since "a decision maker's choice of strategy is theorized to be based on subjective as opposed to objective accuracy and effort" (p. 321). Thus, TAM focused almost exclusively on individual acceptance of innovations in organizations.

Innovation Diffusion Theory

Besides the TAM, one of the most influential among the seminal works on innovation adoption and diffusion was the lineage of research conducted by Rogers (1983), known as the innovation diffusion theory. Unlike the TAM, rather than concentrating exclusively on individuals and their roles in adoption of innovations, innovation diffusion theory is generally studied as an organization-level construct, although individuals and their characteristics are still considered to be important components of this theory framework (Wolfe, 1994). To highlight this, an innovation in this sense is "an idea, practice, or object that is perceived as new by an individual or other unit of adoption" (Rogers, 1983, p. 11).

From innovation diffusion theory, the factors purported to influence diffusion of new technologies include adopter characteristics, the social network to which the adopters belong, innovation attributes, environmental characteristics, the process by which an innovation is communicated, and the characteristics of those who are promoting the innovation (Rogers & Kincaid, 1981). According to Wolfe (1994), innovation adoption falls under administrative or technological domains. Administrative innovations involve changes in the organization's structure or administrative processes, and tend to derive from management. Technological innovations derive from technical needs, and often from the "grass roots" level in the organization.

Innovation diffusion occurs in stages: initiation, adoption, implementation and institutionalization (Rogers, 1983). Adoption entails a decision making influence or force. Implementation involves the details of the innovation reflecting organizational contingencies. When an innovation becomes routine, it becomes an institutionalized structure (Chakravarthy, 1997). According to Taylor and Todd (1995), three characteristics of an innovation relate to technology adoption and diffusion (implementation) behavior. These characteristics are complexity, compatibility, and perceived relative advantage. Complexity in this framework is akin to the TAM perceived ease of use; and compatibility is akin to the TAM perceived usefulness.

The perceived relative advantage is a product of a cost-benefit assessment. "As the perceived relative advantages and compatibility of information technology usage increase, and as complexity decreases, attitude towards information systems usage should become more positive, and such an outcome would be consistent with the general diffusion of innovations literature and with specific results observed for information technology" (Taylor & Todd, 1995, p. 152). Elaborating on the concept of perceived relative advantage, other research on decision making about innovations (Gustafson & Reger, 1995) has included the maximization of benefits and the minimization of costs adjusted by or filtered through cognitive schematic frames.

According to this view, people do not simply adopt an innovation from rational aspects such as economics alone, but from perceived relative advantage influenced by their cognitive schema and previous experiences. These schematic frames represent preferences and affect, and are therefore not part of a formalized or systematic choice; consequently, the decisions about adopting new technological innovations may be subject to overconfidence errors such as optimism in response to perceived risks (c.f. Tversky & Kahneman, 1983). Thus, while (according to this theory) innovation adoption and diffusion decision making may be deliberate processes, they may not always be rational processes (Gustafson & Reger, 1995).

Finally, Rogers (1983) defined what has become commonly referred to as the "adoption curve" in which he articulates five categories of adopters. This is sometimes referred to as the S-curve adoption of an innovation because of its non-linear plotting over time. The categories Roger's defined were: innovators, early adopters, early majority, late majority, and laggards. Each has associated with it a category of risk tolerance, or risk homeostasis. That is to say, people differ in their assessments of risk (Wilde, 2001) and some people have a greater risk tolerance than others (Lauriola, Levin, & Hart; 2007), which can be plotted on the technology adoption curve –from innovators to laggards.

Innovation as a Socio-Cultural Phenomenon

Another important body of literature has considered the socio-cultural influences in technology innovation adoption and diffusion. For instance, Sheppard, et al. (1988) conducted a meta-analysis in which they revealed that when people perceive that they have little control over an outcome such as the adoption of a new technology, there is negative impact on intentions regarding the necessary risk-taking to adopt an innovation, and that they may make attempts to dissuade others from adoption of an innovation as well, either directly or through normative actions. The perception of control has socio-cultural components called "contextual adequacy" and "symbolic meaning."

Contextual adequacy relates to the sufficiency of an innovation to satisfy perceived needs (Fussell & Benimoff, 1995). The symbolic meaning aspect of an innovation may influence perception of constituent's control in a related but separate way from contextual adequacy (Ngwenyama & Lee, 1997). That is, the mental models people formulate about what the innovation may mean for a given adopter goes beyond the innovation's technological characteristics, such as how they "feel" about it, the implications for their jobs, or relates to their intrinsic resistance to change. To summarize the technology innovation adoption and diffusion theory, according to Webster (1998):

"Innovation characteristics theory assumes that individuals adopt innovative technologies to perform individual tasks and therefore technologies need to be designed with the correct characteristics of these individuals, while implementation processes theory assumes that groups perform interdependent tasks and therefore managerial effort needs to be focused on the process of implementation" (p. 258).

BACKGROUND

From the preceding section on innovation adoption and diffusion theory, we suggested that the research bases on innovation diffusion have been influenced largely by concentrating on individuals acting in their organizational structural and social environments. In this section, we will examine and contrast several perspectives on how technological innovations "come about" and become institutionalized in organizations, referring to this as organizational innovativeness. The purpose of this section is to provide the background for our case study.

We will begin with the view that is consistent with the individual-focused approach, where an individual or a small group of actors are presumed to drive or shape innovation adoption, and then progress into a more systemic view of how collective social action may do this. We conclude this section by considering the possibilities for how actors may affect innovation adoption and diffusion either directly or indirectly, but ultimately, we propose that technology innovation adoption and diffusion are the result of collective social activity governed in part by how actors utilize organizational processes and structures to influence adoption and diffusion of innovations.

Innovation as a Deliberative Strategic Plan

As might be inferred from the innovation adoption theory in the previous section, an innovation is widely assumed to be a strategic organizational concept (c.f. Jarzabkowski, 2008). Still, the concept of strategic innovative behavior has little consensus in the literature about what it means (Raisch & Birkinshaw, 2008). To some, it a process that *can be engineered* (Mintzberg, 1994) while to others it is the characteristic of *future visioning* (Denton & Wisdom, 1992) and for others it is exploiting *core competencies* (Hamel & Prahalad, 1993). Regardless, the body of research literature on strategic innovation divides along the lines in a continuum where at one polar end consists of processes that are scientific, planned or structured (e.g. Semler, 1997) and on the other polar end are forms of artistry, imagination, and intuition (e.g. Stacy, 1992). Thus there are different perspectives about strategic organizational innovativeness.

In a deliberative paradigm, a strategic innovation is a product development or service activity that (1) is used to gain competitive advantage (Porter, 1979, 1997), and (2) requires "long-term" commitment in time and resources to produce, provide, or implement (Simon & Houghton, 2003). Strategic innovations therefore differ from tactics and operations in that the former is a blend of various analyses, behavioral techniques, and the use of power and organizational politics to bring about broadly conceived outcomes (Quinn, 1992). Tactics are managerial actions that enact a strategic innovation (Greer, 1995), and operations are the daily routines that result from management actions such as planning and production and quality control (Synnott, 1987).

From this point of view, realizing a strategic innovation in organizations comes by way of a deliberative process through a "value chain" (Porter, 1979), either as a formalized and structured procedure or as an imaginative and creative visioning process. In either case, from this perspective, innovation adoption and diffusion is a focused and systematic method (Simon & Houghton, 2003).

Innovation as a Convergent Set of Processes

In the 1970s as part of the Tavistock studies that took place in the United Kingdom, a new view of organizations as parts of interconnected systems emerged (Trist, 1971). From this socio-technical systems theory perspective, strategic innovation results from organizational alignment of strategic and operational activities (Semler, 1997). In this sense, it shares the perspective of the strategic innovation planning lens in which innovativeness is a formalized and deliberative action. It differs however in the sense that this is not seen as a single deliberative process but rather as a convergence of many (and possibly disparate) processes (Greer, 1995).

In realizing an innovation from the convergence perspective, managers and entrepreneurs apply principles defined in socio-technical systems theory, first by performing a system scan (Fox, 1995) in which the mission of the organization is identified and stakeholders' interests are reconciled by what is viable with regard to the outside environment. Various techniques ranging from "Delphi" to stochastic modeling technology have been used for this (along with their cousin "action planning") and continue to be popular approaches employed by many managers and entrepreneurs.

The *systems scan* is followed by a technical analysis in which the technology inputs and outputs (rather than tools, processes or techniques) are defined. In an end-to-end process the systems are analyzed separately from jobs and people –from the input side of the system to the output side, and from this activity a unit-operations flow chart is produced. Variances are noted (other than those representing human error or breakdowns) and recorded and a key-variance matrix table is generated with the primary goal of eliminating deviations. This way, it can be determined what actions are necessary to control the variances. Finally, the social analysis is performed. This function (called focal-role analysis) determines the role-expectations and work-related interactions of those in positions most involved with the control of key variances (Fox, 1995).

Simultaneously with this effort to drive out variance and gain conformity to a given set of standards, the process seeks to establish sets of core competencies, and align the collective values and beliefs of the organizational members with organizational objectives, and the social relationships are noted for work-related interactions of "focal persons and four survival criteria: sound key-variance control, adaptation to the external environment, integration of in-system people activities, and long-term development" (Fox, 1995, p. 99). The product of this activity is called a "grid of social relations."

Once the system is defined in this manner, social-technical systems innovations could be implemented, which according to Pava (1983, p. 64) involve: (1) Guiding the explicit acknowledgement and designation of major deliberations and their corresponding discretionary coalitions. (2) Establishing responsibilities of each coalition for every deliberation. (3) Delineating human resource policies that support effective deliberations among discretionary coalitions. (4) Suggesting structural changes in the organization pertaining to responsibilities and coordination that enhances major deliberations and their associated coalitions, and, (5) proposing technical enhancements to assist discretionary coalitions engaged in major deliberations about innovations.

Innovation as an Emergent Phenomenon

The notion that innovativeness is an emergent phenomenon discards the concept that organizational activities are static enough to record and asserts that capturing or bracketing its fluid structures is impractical, if possible. The view of innovation through emergence is more akin to the strategic innovation perspective in-so-far as innovations are seen as the product of imagination and creativity, but it differs in the sense that these processes are not considered part of a formal deliberative or systematic method. Also in contrast to the socio-technical systems approach which suggests that innovations are driven through the alignment of organizational systems, from the emergent perspective, innovation evolves out of chaos and crises (Stacy, 1992).

According to Kuhn (1996), when a normative approach to solving problems (which Kuhn called a paradigm) becomes insoluble, a radical transformation occurs (which Kuhn coined a paradigm shift). During the latter 1990s, it became apparent in the literature on strategic innovations that it was not feasible in the manner of socio-technical systems theory to maintain a peaceful coexistence between those who believed that innovation could be designed and those who believed innovation was pure serendipity. To put it another way, the efforts to drive out variance and drive towards conformance to a standard regarding innovative development, versus generating the variety needed for continuous innovation do not peacefully coexist (Mintzberg, 1994).

As an example, it became apparent that trying to maintain a "grid of social relations," let alone make note of all the interrelated organizational objectives involving those social relationships was impractical if even possible (Stacy, 1992). In addition, new organizational structures had evolved including the notion of teamwork and extreme forms of organizational autonomy such as self-managed teams (Workman & Bommer, 2004). The emphasis on teamwork relies on the essential thought that deliberations are democratic and that ideational generation and decisions about them require consensus rather than compromises. In this way, all organizational behaviors were perceived as strategic, and that values and beliefs of participants would help to foster innovativeness through the collective efforts and activities, including conflicts (Mintzberg, 1994).

Thus from this vantage point, since organizations are comprised of people, ultimately, it is the collective capabilities of those people that orchestrate the achievements in organizations. Here, the development of capabilities results from dynamic and active processes that happen *ad hoc*. This contrasts with the convergent prescriptive, which was criticized by Antonacopoulou and FitzGerald (1996) because, they asserted, it is misleading to consider innovation as a narrowly defined and rigid process that can be formulated *a priori*. To them –innovativeness is an open-ended activity. In this way it is not be the capabilities of people that are developed and applied in building an innovation, but rather it constitutes the capabilities of people in action through continuous learning and experimentation that produces radical mutations we call innovation (Stacy, 1992).

In a more contemporary form, the view of innovativeness embodies a holistic "open systems" perspective that neither concentrates exclusively on converging mechanistic efficiencies nor on fostering creative effectiveness (Sine, et al., 2006). Instead it embraces self-organization and the informal and fluid processes that germinate spontaneous ideas, but it also recognizes that these seemingly "chaotic" organizations and reorganizations have structure to them. More particularly, the continuous establishment and disestablishment of social networks and the natural shifts in organizational power energizes action through both cooperation and conflicts that erupt and dissipate. And fundamental to this more organic view of innovation as an emerging phenomenon is the realization of the differences between problems that have a single correct solution and are programmable, and those that are equivocal and involve multiple subjective points of view about innovative solutions (Daft & Lengel, 1986).

Innovation through Structuration

Structuration theory helps to explain the role of human agency in the reciprocal relationship between human social systems and their social structures, which Giddens called the "duality of structure." It is important to note that structuration theory (Giddens, 1984, 1991) has taken some criticism in the "post-modern" management theory literature for being poorly operationalized (Gephart, Boje, & Thatchenkery, 1996; Ritzer, 2005; Stones, 2005), and as well in the information technology literature for having difficulty bracketing actions and transitions in structures, especially in light of technological advances that can affect structuration (De Sanctis & Poole, 1994; Orlikowski, 1996).

Consequently, adaptive structuration theory (De Sanctis & Poole, 1994) was developed from structuration theory to capture how people adapt to technology as well as adapt the technology to their own usage behaviors. For instance, studies (c.f. Maznevski & Chudoba, 2000) have shown that the social structures and power distributions affected how people utilize new technology. Other studies (De Sanctis & Poole, 1994; Orlikowski, 1996) have demonstrated how groups used the same decision support system differently for determining project priorities.

In these studies, depending on the social structures, the use of power, and the results of conflicts that arose within the groups, very different structuration took place. More specifically, some group members tended to rely more on influence and persuasion while others tended to use force. Therefore, from a socio-cultural perspective, adaptive structuration theory, or AST, (Desanctis & Poole, 1994) has provided "a model that describes the interplay among advanced information technologies, social structures, and human interaction in the adoption of innovations" (p. 125). The social context provides the rules and resources that serve as templates for a group's actions relative to how they might respond to changes, such as in the introduction of an innovation.

Note that according to AST, the social context of technology adoption and their use relies heavily on leader support and on task structures (Kahai, Sosik & Avolio, 1997). In this way, AST explains the use of information technology and its resulting outcomes (Desanctis & Poole, 1994). These structures are also important to technology innovations; for example, innovation theory generally suggests that innovations are typically derived from the administrative core where leader support is an essential (Wolfe, 1994). In terms of task structure, technology adoption would be impacted by the suitability of the technology to the worker's tasks, and the worker's group interactions. Leader support and task structure "pertain to the systems normative framework for appropriate behaviors" (Kahai, et. al., 1997, p. 123). But this theory alone has not fully explained how innovation becomes "embraced" by organizational members, as illustrated by Workman (2007) in the façade exposed concerning the use of an innovative expert system foisted upon network engineers by their management, as described in our case study that follows. It is important therefore to consider structuration from an agency (social-actor) perspective.

Structuration Agency Theory

Agency in a structuration theory sense is anyone who acts within the formalized social structure of the organization (Jarzabkowski, 2008). The structuration agency use of the term "agency" represents individual behaviors that operate within a broad network of socio-structural influences (Bjorklund, 1995; Chomsky, 1996), which can be within or outside the formally defined organizational structures. Bandura (2001) described this triadic phenomenon as "agentic transactions, where people are producers as well as products of social systems" (p. 1). As defined in this theory, agency exists on three levels: (1) direct personal agency (an individual's actions), (2) proxy agency, which relies on others to act on one's behalf to secure individually desired outcomes, and (3) collective agency, which is exercised through socially coordinative and interdependent effort (Chomsky, 1996; Bandura, 2001). The notion of agency from this perspective contrasts with non-deterministic (chaotic) and non-rational "natural" processes that generate the environments in which people operate either formally or informally (Beck, Giddens & Lash, 1994).

The reciprocal relationships between agentic action and social structures are referred to as the "duality of structure" by Giddens (1984). In these terms, structure is defined by the regularity of actions or patterns of behavior in collective social action, which become institutionalized, and agency then is the human ability to make rational choices and to affect others with consequential actions based on those choices that may coincide with or run counter to institutionalized structures.

Structuration on the other hand is a dynamic activity that emerges from social interaction (Giddens, 1984). Particularly, social action relies on social structures, and social structures are created by means of social action. Thus structures derive the rules and resources that enable form and substance in social life, but the structures themselves are neither form nor substance. Instead, they exist only in and through the activities of human agents. For example, people use language for communications with one-another, and language is defined by the rules and protocols that objectify the concepts that people convey to each other (Chomsky, 1996). The syntax structure of language is the arrangement of words in a sentence, and by their relationships one to another (e.g. subject-predicate noun-verb phrase) the sentence structure establishes a well-

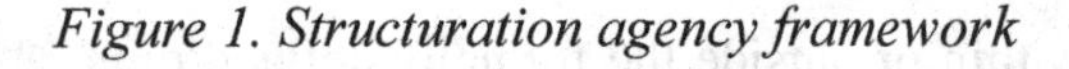

Figure 1. Structuration agency framework

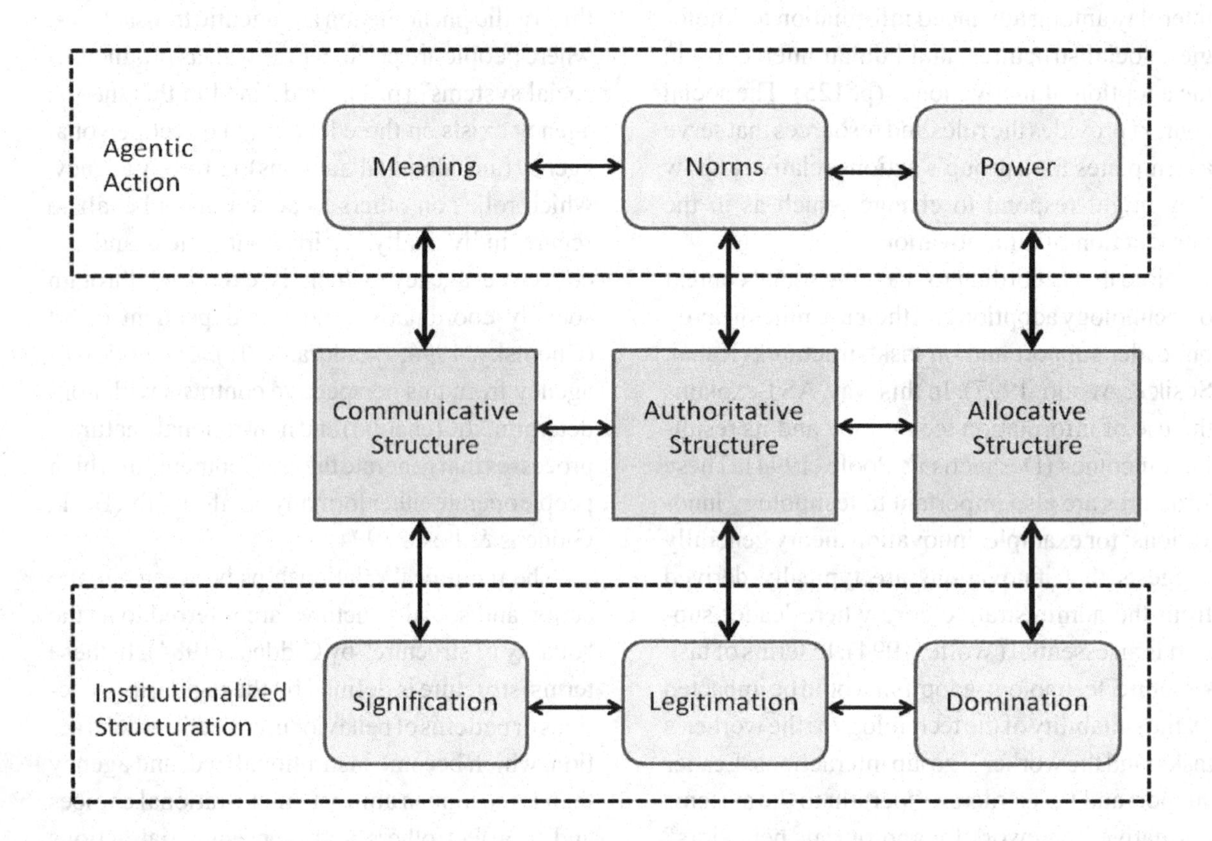

defined grammar that people use to communicate (Verspoor & Lowie, 2003).

However, language is also generative and productive and an inherently novel activity, allowing people to create sentences using the syntax rather than to simply memorize and repeat them (Chomsky, 1996). In similar fashion, institutionalized structures regulate agentic behavior, but agents may also disrupt institutionalized structures (Jarzabkowski, 2008). The defining features of structuration theory that explain how these processes work are: signification, legitimation and domination. Signification concerns how meaning is co-created and interpreted by agents, legitimation encompasses the norms and rules for acceptable behavior, and domination refers to power, influence, and control over resources (Giddens, 1991). Collectively, the signification, legitimation and domination constitute the institutionalized structuration processes. Agentic interaction with these processes creates the communicative structure, authoritative structure, and allocative structure, respectively. In the next section we present a case study to illustrate an example of agency and structuration related to technology innovation adoption and diffusion.

CASE STUDY

Globetel (a pseudonym) operates the world's second-largest computer network. Network engineers residing in various parts of the world often need to make changes to the network, but given the vast complexity of the network, a change made to a network segment in one geographical location might have a significant negative impact on another network segment in a different geographical location. In one particularly critical incident, network engineers made a change in the Dallas Texas network operations center (NOC) that led to a collapse of the entire Illinois network segment, causing that segment of the network to be out of operation for several days.

Because network outages such as these were extremely costly and disruptive to many organizations, executive management created a "skunk works" team, which was a term given by Kelly Johnson at Lockheed Martin to an advanced development team that bypassed traditional organizational processes and structures to shorten the development time for innovations. The skunk works team set out to build an expert decision support system (EDSS) that could make predictions about how a proposed network change might impact the overall network infrastructure and traffic.

An EDSS is similar in many ways to conventional decision support systems (DSS). Like DSS, EDSS usage consists of a series of analytical modeling steps that define the problem context and gather relevant information (Taylor & Karlin, 1993). Going beyond DSS, however, EDSS does not terminate with the gathering and rendering of information for human consumers. Instead, the EDSS operates upon the information with reasoning and inference to generate solutions to given problems in the form of stochastic models and prescribed steps and courses of action for the users to follow (Gregor & Benbasat, 1999).

The literature on innovation adoption and diffusion (e.g., Taylor & Todd, 1995) has frequently commented on the importance of innovative technology use in organizational outcomes. For example, technologies can be instrumental in facilitating decision making (DeSanctis & Poole, 1994), may improve communication (Kraut et al., 1998), and may enhance effectiveness, productivity and job performance (Goodhue & Thompson, 1995); yet because the social and communicative structures are often ignored in the introduction of innovations, even with a "champion" for an innovation, systems that are shown to be effective may not diffuse throughout an organization (Brown & Duguid, 2000).

When the EDSS technology was introduced by management into the engineering organization, many were skeptical that the technology could be more accurate than the experience and knowledge of the engineers, and there were outward expressions of distrust. Since the EDSS originated from the administrative core in Globetel, and because the EDSS dictated to the engineers what actions to take, there was initially a strong resistance to adopt the innovation. Some of the engineers expressed that they felt "left out" of the skunk works, and some worried about the meaning of the technology relative to their jobs, and some stating that management was "trying to shove the technology down their throats." There were also expressions that the technology was too rigid. Since the technology required well-defined progressive stages confined to a usage paradigm, the perception of rigidity negatively affected some attitudes towards the technology. In such cases, social communicative and normative influences may impinge on the adoption of the technology (Taylor & Karlin, 1993; Taylor & Todd, 1995).

Unlike pressure levied through an allocative structure that tends to produce resistance, using the communicative and authoritative structures, people sometimes strive more progressively to conform to normative pressures as social influences increase (Salanick & Pfeffer, 1978). Combined with the use of power and force to gain compliance, negative peer normative influence can create an amplification of the resistance. A sample of a dissuasive statement given by a case study participant was, "Can't you figure it out? Why do you have to use the tool for that?" Thus strong sentiment against adopting a technology by peers may increase dissuasive influences and inhibit diffusion. Yet in spite of this, management's response to try to overcome the resistance was to further increase the pressure through the allocative structures by threatening to levy punishments on those who neglected to use the technology.

As indicated in adoption and diffusion theory, human-technology and human-information usage represent different contexts (Gregor & Benbasat, 1999; Morris & Venkatesh, 2000; Tan & Hunter, 2002). For instance, people may utilize a technology because it is expected as part of standard organization practices (legitimation) having coming from an administrative (and authoritative structure), or acceded to by domination (through the use of power to reward or punish), but this forced compliance may lead to counterproductive behaviors in protest (Brown & Duguid, 2000).

When human-induced errors continued at high rates and network outages persisted, a study was initiated by management to find out if the technology was effective. As investigators studied the technology logs and compared them to courses of action engineers took, along with "trouble tickets" called into the technology support center, they determined that the technology was effective in reducing human-induced errors. However, they also learned that the more pressure managers placed on engineers through the use of force to utilize the technology there was a concomitant non-linear relationship between executing the EDSS and ignoring the technology recommendations. In addition, investigators found that negative attitudes toward the EDSS acted as a "social contagion" and spread throughout the engineering organization even among those who had not yet tried the technology, further inhibiting diffusion (Workman, 2005).

In summary therefore, in this case, managers had operated primarily through the use of power and force, relying on authoritative and allocative structures to require the use of the technology. As greater pressure was placed on engineers to use the EDSS, many engineers would simply use the technology to generate the models and procedures, but instead of following the directions given by the technology, they would often ignore the directions and perform their tasks according to their own judgments.

In retrospect, signification from the communicative structure was seen as an important missing focus in this case because technology use does not necessarily lead to the use of the information manipulated or produced by the technology (Plumlee, 2003). It has been found, as in this case, that social influences and individual attitudes may interact in such a way that when people have poor attitudes or perceptions about a technology and yet social forces strongly encourage its use, people may feign the use of technology but without commitment or without following its prescription (Terry, Hogg & White, 1999). Thus management realized the need for joint optimization by communicating, negotiating, and cooperating with the constituents in the organization for introduction of the innovation.

A task force was established to gather input from the engineers and to address their concerns. The engineers were incorporated into further developments of the EDSS technology, providing for example, input into the knowledgebase and user interface changes. Eventually expressions about the EDSS then began to change. A sample of a persuasive supporting statement given by a participant in the case study was, "Why do you do it the hard way? Just use the tool for that." Ultimately, the EDSS became widely utilized and recognized by engineers as an important facilitation in reducing errors.

DISCUSSION AND LESSONS LEARNED

Technology innovation adoption and diffusion have been actively researched; however, given the vast of body literature it can be difficult to determine under what circumstances innovations are adopted and diffused, and what factors may lead to resistance. In this chapter, we summarized and organized major streams of theory and perspectives on innovation and adoption and diffusion. We utilized structuration agency as a framework in our case study to show how actors and structures play out in the adoption and diffusion of an innovative technology.

In the case presented, we observed that if an innovative technology is not utilized, the potential benefits from the technology may be lost, or may contribute to the failure of the technologies to live up to their promised payback (Brown & Duguid, 2000). Innovations such as EDSS not only have ability to perform the more mundane tasks such as data gathering, they also enhance the ability to more effectively solve problems with the technology (Taylor & Todd, 1995). Nevertheless, because EDSS provides a structured approach to problem solving and decision-making within a particular usage paradigm, some users may develop negative attitudes towards the technology (e.g., DeSanctis & Poole, 1994).

When peers or groups of individuals have positive or negative perceptions of an innovation, they tend to use the communicative and authoritative structures to encourage or discourage diffusion by conveying meaning and normative expressions. This process may occur either directly or by proxy agency. Direct and indirect agency levies social influences in different ways, but in either case, they may encourage or discourage performance of a behavior to varying degrees that correspond to the amount of social force applied (Ajzen, 1991; Salanick & Pfeffer, 1978). Moderate approval or disapproval of performing an act may be viewed as encouragement or discouragement, whereas when expressed or normative social forces reach extremes, social influences become perceived as pressure to act or refrain from acting (Terry et al., 1999).

Finally, in addition to the normative influences on technology adoption, social influences are also exerted through observational models. For instance, watching the successful effort by others is likely to increase one's own effort towards that behavior, as well as the converse (Bandura, 1977).

Hence, in addition to individual assessments of what "important others" think about one performing a behavior, observation of these peoples' behavior likewise influences the performance of the behavior such as adopting and using an innovative technology (Ajzen, 1991). As a result, people who observe others using a system with positive results are encouraged to use the new system (Compeau & Higgins, 1995; Kraut et al., 1998).

REFERENCES

Ajzen, I. (1991). The theory of planned behavior. *Organizational Behavior and Human Decision Processes*, *50*, 179–211. doi:10.1016/0749-5978(91)90020-T

Antonacopoulou, E. P., & FitzGerald, L. (1996). Reframing competency in management development. *Human Resource Management Journal*, *6*, 27–46. doi:10.1111/j.1748-8583.1996.tb00395.x

Bandura, A. (2001). Social cognitive theory: An agentic perspective. *Annual Review of Psychology*, *52*, 1–26. doi:10.1146/annurev.psych.52.1.1

Beck, U., Giddens, A., & Lash, S. (1994). *Reflexive modernization. Politics, tradition and aesthetics in the modern social order*. Cambridge: Polity Press.

Bjorklund, D. F. (1995). *Information processing approaches: An introduction to cognitive development*. Washington, D.C.: Brooks-Cole.

Boxall, P. (1996). The strategic HRM debate and the resource-based view of the firm. *Human Resource Management Journal*, *6*, 59–70. doi:10.1111/j.1748-8583.1996.tb00412.x

Brown, J. S., & Duguid, P. (2000). *The social life of information*. Boston: Harvard Business School Press.

Burns, T., & Stalker, G. M. (1961). *The management of innovation*. London: Tavistock.

Chakravarthy, B. (1997). A new strategy framework for coping with turbulence. *Sloan Management Review*, (Winter): 69–82.

Chomsky, N. (1996). *Language and problems of knowledge*. Mendocino, CA: MIT Press.

Compeau, D. R., & Higgins, C. A. (1995). Computer self-efficacy: Development of a measure and initial test. *Management Information Systems Quarterly, 19*, 189–211. doi:10.2307/249688

Daft, R. L., & Lengel, R. H. (1986). Organizational information requirements, media richness and structural design. *Management Science, 32*, 554–571. doi:10.1287/mnsc.32.5.554

Davis, F. D. (1989). Perceived usefulness, perceived ease of use, and user acceptance of information technology. *Management Information Systems Quarterly*, (September): 319–338. doi:10.2307/249008

De Sanctis, G., & Poole, M. S. (1994). Capturing the complexity in advanced technology use: Adaptive structuration theory. *Organization Science, 5*, 121–147. doi:10.1287/orsc.5.2.121

Denton, D. K., & Wisdom, B. L. (1992). Shared vision. In Thompson, A. A. Jr, Fulmer, W. E., & Strickland, A. J. III, (Eds.), *Readings in strategic management* (pp. 52–56). Boston: Irwin.

Fox, W. M. (1995). Sociotechnical system principles and guidelines: past and present. *The Journal of Applied Behavioral Science, 31*, 91–105. doi:10.1177/0021886395311009

Fussell, S. R., & Benimoff, I. (1995). Social and cognitive processes in interpersonal communication: Implications for advanced telecommunications technologies. *Human Factors, 37*, 228–250. doi:10.1518/001872095779064546

Gephart, R. P., Boje, D. M., & Thatchenkery, T. J. (1996). Postmodern management and the coming crises of organizational analysis. In Gephart, R. P. (Eds.), *Postmodern Management and Organization Theory* (pp. 1–20). Thousand Oaks, CA: Sage.

Giddens, A. (1984). *The constitution of society: Outline of the theory of structuration*. Cambridge, UK: Polity Press.

Giddens, A. (1991). *Modernity and self-identity. Self and society in the late modern age*. Stanford, CA: Stanford University Press.

Goodhue, D. L., & Thompson, R. L. (1995). Task-technology fit and individual performance. *Management Information Systems Quarterly*, (June): 213–232. doi:10.2307/249689

Greer, C. R. (1995). *Strategy and human resources. A general managerial perspective*. Englewood Cliffs, NJ: Prentice-Hall.

Gregor, S., & Benbasat, I. (1999). Explanations from intelligent systems: Theoretical foundations and implications for practice. *Management Information Systems Quarterly, 23*, 497–527. doi:10.2307/249487

Gustafson, L. T., & Reger, R. K. (1995). Using organizational identity to achieve stability and change in high velocity environments. *Academy of Management Journal, Best Papers Proceedings,* 464-468.

Hamel, G., & Prahalad. (1993). Strategy as stretch and leverage. *Harvard Business Review*, (March-April): 75–85.

Jarzabkowski, P. (2008). Shaping strategy as a structuration process. *Academy of Management Journal, 51*, 621–650.

Kahai, S. S., Sosik, J. J., & Avolio, B. J. (1997). Effects of leadership style and problem structure on work group process and outcomes in an electronic meeting system environment. *Personnel Psychology, 50*, 121–146. doi:10.1111/j.1744-6570.1997.tb00903.x

Kraut, R. E., Rice, R. E., Cool, C., & Fish, R. S. (1998). Varieties of social influence: The role of utility and norms in the success of a new communication medium. *Organization Science*, *9*, 437–453. doi:10.1287/orsc.9.4.437

Kuhn, T. S. (1996). *The structure of scientific revolutions*. Chicago: University of Chicago Press.

Lake, M. (2009). The art of creation in science: A consonant paradox. *Market Times*, *19*, 278–197.

Lauriola, M., Levin, I. P., & Hart, S. S. (2007). Common and distinct factors in decision making under ambiguity and risk: A psychometric study of individual differences. *Organizational Behavior and Human Decision Processes*, *104*, 130–149. doi:10.1016/j.obhdp.2007.04.001

Lee, P. M., & O'Neill, H. M. (2003). Ownership structures and R&D investments of US and Japanese firms: Agency and stewardship perspectives. *Academy of Management Journal*, *46*, 195–211.

Manz, C. C., & Stewart, G. L. (1997). Attaining flexible stability by integrating total quality management and socio-technical systems theory. *Organization Science*, *8*, 59–70. doi:10.1287/orsc.8.1.59

Maznevski, M. L., & Chudoba, K. M. (2000). Bridging Space Over Time: Global Virtual Team Dynamics and Effectiveness. *Organization Science*, *11*, 473–492. doi:10.1287/orsc.11.5.473.15200

Mintzberg, H. (1994). The fall and rise of strategic planning. *Harvard Business Review*, (January-February): 107–114.

Morris, M. G., & Venkatesh, V. (2000). Age differences in technology adoption decisions: Implications for a changing work force. *Personnel Psychology*, *53*, 365–401. doi:10.1111/j.1744-6570.2000.tb00206.x

Ngwenyama, O. K., & Lee, A. S. (1997). Communication richness in electronic mail: Critical social theory and the contextuality of meaning. *Management Information Systems Quarterly*, (June): 145–166. doi:10.2307/249417

Orlikowski, W. J. (1996). Improvising organizational transformation over time: A situated change perspective. *Information Systems Research*, *7*, 63–92. doi:10.1287/isre.7.1.63

Pava, C. (1986). Redesigning sociotechnical systems design: Concepts and methods for the 1990s. *The Journal of Applied Behavioral Science*, *22*, 201–221. doi:10.1177/002188638602200303

Pfeffer, J., & Salancik, G. R. (1978). *The external control of organizations: A resource dependence perspective*. New York: Harper & Row.

Plumlee, M. A. (2003). The effect of information complexity on analysts' use of that information. *Accounting Review*, *78*, 275–296. doi:10.2308/accr.2003.78.1.275

Porter, M. E. (1979). *How competitive forces shape strategy*. New York: Free Press.

Porter, M. E. (1996). What is strategy? *Harvard Business Review*, (November-December): 61–78.

Quinn, J. B. (1992). Managing strategic change. In A. A. Thompson, Jr., W. E. Fulmer & A. J. Strickland III (Eds.), Readings in Strategic Management (4th ed., pp. 19-42). Boston: Irwin. (Reprinted from Sloan Management Review, 21, 3-20).

Raisch, S., & Birkinshaw, J. (2008). Organizational ambidexterity: Antecedents, outcomes, and moderators. *Journal of Management*, *34*, 375–409. doi:10.1177/0149206308316058

Ritzer, G. (2005). Structuration theory. *Contemporary Sociology: A Journal of Reviews, 36*, 84-85.

Rogers, E. M. (1983). *Diffusion of innovations*. New York: Free Press.

Rogers, E. M., & Kincaid, D. L. (1981). *Communication networks: Toward a new paradigm for research*. New York: Free Press.

Semler, S. W. (1997). Systematic agreement: a theory of organizational alignment. *Human Resource Development Quarterly*, *8*, 23–40. doi:10.1002/hrdq.3920080105

Sheppard, B. H., Harwick, J., & Warshaw, P. R. (1988). The theory or reasoned action: A meta-analysis of past research with recommendations for modifications and future research. *The Journal of Consumer Research*, *15*, 325–343. doi:10.1086/209170

Simon, M., & Houghton, S. M. (2003). The relationship between overconfidence and the introduction of risky products: Evidence from a field study. *Academy of Management Journal*, *46*, 139–149.

Sine, W. D., Mitsuhashi, H., & Kirsch, D. A. (2006). Revisiting Burns and Stalker: Formal structure and new venture performance in emerging economic sectors. *Academy of Management Journal*, *49*, 121–132.

Stacy, R. D. (1992). *Managing the unknowable. Strategic boundaries between order and chaos in organizations*. San Francisco: Jossey-Bass.

Stones, R. (2005). *Structuration theory*. New York: Palgrave-Macmillan.

Synnott, W. R. (1987). *The information weapon: Winning customers and markets with technology*. New York: John Wiley & Sons.

Tan, F. B., & Hunter, M. G. (2002). The repertory grid technique: A method for the study of cognition in information systems. *Management Information Systems Quarterly*, *26*, 39–57. doi:10.2307/4132340

Taylor, H., & Karlin, S. (1993). *Introduction to stochastic modeling*. London: Academic Press.

Taylor, S., & Todd, P. A. (1995). Understanding information technology usage: A test of competing models. *Information Systems Research*, *6*, 144–176. doi:10.1287/isre.6.2.144

Terry, D. J., Hogg, M. A., & White, K. M. (1999). The theory of planned behavior: Self-identity, social identity and group norms. *British Journal of Psychological Society*, *38*, 225–244. doi:10.1348/014466699164149

Tihanyi, L., Johnson, R. A., Hoskisson, R. E., & Hitt, M. A. (2003). Institutional ownership differences and international diversification: The effects of boards of directors and technological opportunity. *Academy of Management Journal*, *46*, 195–211.

Trist, E. (1971). New directions of hope. *Regional Studies*, *13*, 439–451. doi:10.1080/09595237900185381

Tversky, A., & Kahneman, D. (1983). Extensional versus intuitive reasoning: The conjunction fallacy in probability judgment. *Psychological Review*, *90*, 293–315. doi:10.1037/0033-295X.90.4.293

Verspoor, M., & Lowie, W. (2003). Making sense of polysemous words. *Journal of Language Learning*, *53*, 547–586.' doi:10.1111/1467-9922.00234

Webster, J. (1998). Desktop video teleconferencing: Experiences of complete users, wary users, and nonusers. *Management Information Systems Quarterly*, (September): 257–286. doi:10.2307/249666

Webster's Dictionary. (1978). *Webster's new 20th century dictionary*. New York: Harper-Collins.

Wilde, G. J. S. (2001). *Target risk*. Toronto: PDE Publications.

Wolfe, R. A. (1994). Organizational innovation: Review, critique and suggested research directions. *Journal of Management Studies*, *31*, 405–427. doi:10.1111/j.1467-6486.1994.tb00624.x

Workman, M. (2005). Expert decision support system use, disuse, and misuse: A study using the theory of planned behavior. *Journal of Computers in Human Behavior*, *21*, 211–231. doi:10.1016/j.chb.2004.03.011

Workman, M. (2007). Advancements in technology: New opportunities to investigate factors contributing to differential technology and information use. *Journal of Management and Decision Making*, *8*, 221–240.

Workman, M., & Bommer, W. (2004). Redesigning computer call center work: A longitudinal field experiment. *Journal of Organizational Behavior*, *25*, 317–337. doi:10.1002/job.247

Chapter 5
The Challenge of a Corporate Matchmaker

Francisco Chia Cua
University of Otago, New Zealand

ABSTRACT

The common structured procurement process of the Request for Information (RFI), Request for Proposal (RFP), and Business Case Development (BCD) is thought to establish ties with the right vendors and to strengthen relationships among other stakeholders. This single-case study gathered information through archival documents, observations, and in-depth interviews and examined whether RFI-RFP-BCP processes fostered favourable relationships with vendors. The study revealed certain disadvantages of the process.

"All change is not growth; as all movement is not forward." - *Ellen Glasgow*

BACKGROUND

Right after the University of Australasia (name disguised) determined its vision to be a "research-led University with an international reputation for excellence, it also made a commitment to invest in three things: capability, human resources, and information technology. Its current enterprise system, in particular, was in need of a major overhaul. The university's gatekeepers understood that the only way to achieve their vision would be to introduce new enterprise systems as soon as possible.

This case study examines the process by which this new system was introduced and finally adapted by the University, using Everett Rogers' Diffusion of Innovations (DOI) theory as the primary model.

In every social system, there are some individuals and groups which are quicker to adopt innovations than the others. The quickest and most venturesome are known as "Innovators," while the last to embrace change are called "Laggards." The University of Australasia fell into this latter adopter category.

DOI: 10.4018/978-1-61520-609-4.ch005

Table 1. Questions and evaluation criteria

		Trigger	Opinion leader	Matchmaking	Some concerns
1.	What drove the innovation?	✓			
2.	Who was the opinion leader? How was the need for change communicated?		✓		
3.	How did the executive sponsor explore the opportunity?			✓	
4.	How did the executive sponsor communicate his agenda?			✓	
5.	How did the vendors react to the communication?				✓
6.	Was the outcome of the matchmaking in line with the expectation?				✓

EVALUATION CRITERIA

In appraising the matchmaking stage, a small number of broad questions (Ellet, 2007) using Rogers' DOI theory (1962, 2003) will serve as the evaluation criteria (see Table 1).

THE MATCHMAKING STAGE IN THE EXPLORATION PHASE OF THE INNOVATION PROCESS

Criteria 1: The Trigger

The innovation process can be dated from 25 July 2006, when Craig Smith, the University's Director of Finance, submitted a business case document to the University board, recommending the replacement of the old financial management information systems (referred in this case study as enterprise systems). This document was a mere formality, however, as the Board, as well as the University's Chief Operating Officer (COO), had already agreed in principle to the change.

This COO, David Ramirez, happened to be the last Director of Finance before Craig took over. During his term, he had asked a third party organisation to appraise the University's current enterprise systems. At the time, however, the University had more urgent investment priorities than the information systems of the Finance Department.

Only when Craig assumed leadership of this division did the replacement of the old enterprise systems became Finance's number one priority. Nevertheless, David gave his full support and commitment to the new agenda (Figure 1).

This updating of the old systems was actually Craig's first priority as early as April 2005, when he was first hired by the University. As a new employee, he brought a very progressive attitude to the department. His biggest challenge when he assumed the reins, therefore, was dealing with a mindset more or less averse to change and development. He believed that he needed an experienced third-party change agent to help set his plans into motion.

Stanley Lim, the division's financial analyst since July 1997, brought Craig up to speed on the state of the old enterprise system, called BIZFIN. Table 2 lists the players. Though his first responsibility was to come up with a replacement for BIZFIN, the general lack of urgency towards the project had resulted in years of data gathering and little else. By the time Craig decided an innovation was in order, Stanley had collected about seven years of information.

Such a long "gestation period," so to speak, reveals something of the Finance Division's mindset. Along with the positive bias towards

Figure 1. The exploration and exploitation phases of the deployment of new enterprise systems

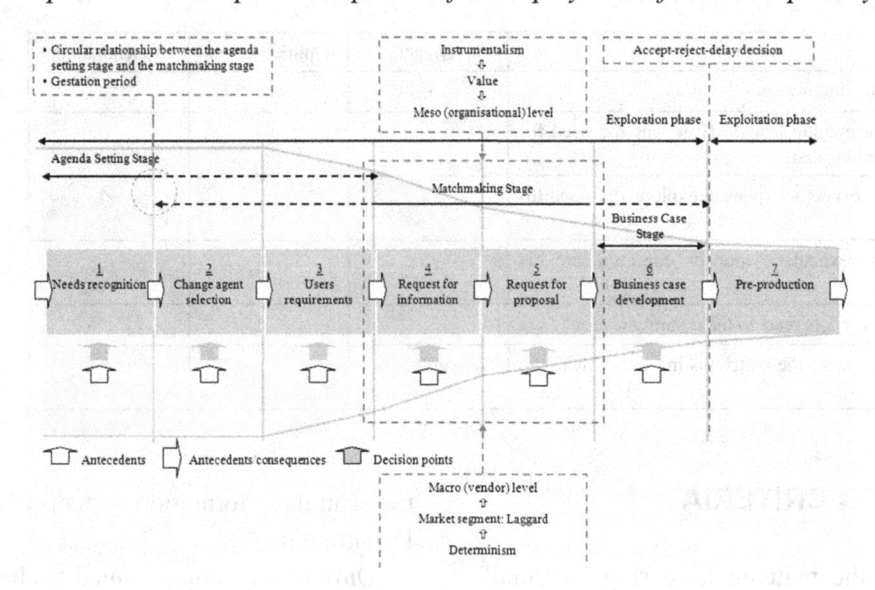

Table 2. List of Players

Craig Smith	Executive sponsor, Director of Finance of University of Australasia
David Ramirez	Chief Operating Officer, superior of Craig Smith, of University of Australasia
Stanley Lim	Opinion leader, Financial analyst of University of Australasia
Kevin Peters	Third party change agent of Providence Consulting

deploying a new enterprise system, there was a negative bias towards BIZFIN.

Craig's shorter tenure made him more pro-active (Greenalgh *et al.*, 2004) and willing to accept new ideas and interact with other people through dialogue. The DOI theory calls such a positive bias towards development "innovativeness" (Agarwal & Prasad, 1998; Auh & Mengus, 2005; Gatian *et al.*, 1995; Hirschman, 1981; Hult & Hurley, 1998; Hult *et al.*, 2004; Lefebvre & Lefebvre, 1992; Subramanian, 1996; Woodside, 2005). However, an innovative mindset alone is not enough to drive the deployment of a new system. In fact, all innovations are primarily driven by situational factors (Midgley & Dowling, 1978).

In the case of the UoA, perhaps the most significant situational factor was the increasing inadequacy of BIZFIN.

First of all, the widespread use of the "untamed" or "unmanageable" feral systems (Houghton & Kerr, 2004, 2006) throughout the University was clear evidence that BIZFIN wasn't doing its job. These systems were developed by individuals, departments, or divisions to get around the constraints of the old enterprise systems. (It could be said that the greatest advantage of BIZFIN was the degree to which it managed to enhance users' creativity enough for them to come up with original ways of getting around its limitations.) While the frustrated users had developed feral systems to supply any missing functions and information; they also ended up having to enter their transactions twice, first into the old system and again into the "untamed" feral system. This was hardly ideal.

Doubling the transactions meant doubling the risk of data-entry errors. The users frequently found themselves with two contradictory sets of data, adding to their frustration. The situation worsened as the managers became doubtful of any data and demanded that everything be double checked by the already overtasked employees.

Any proposed innovation ought to reduce, if not totally remove, such feral systems, in order to ensure the quality of information produced by the enterprise system. Two necessary conditions are: (a) the functionality required by the university users and (b) the seamless integration into other existing systems.

The feral systems were definitely one trigger to innovation, despite the general lack of urgency to overhaul the poorly functioning old system.

In addition, University users were becoming increasingly dissatisfied with BIZFIN, in the light of the rapid growth of new technologies. For one thing, BIZFIN was written in COBOL and used DOS operating systems, both of which were quickly becoming obsolete. Many other big organizations which had jumped on the BIZFIN bandwagon in the late 1980s and early 1990s were starting to notice its relatively steep learning curve and unfriendly user interface, compared to other systems which had since become available. This tied in with the second concern that the incumbent vendor's inability to develop new versions of the software, which resulted in the perceived risk of insufficient support for an already problematic system.

What was not immediately apparent to the users, but obvious to the employees maintaining the systems, was the increasing risk of relying on older models of hardware, operating systems, and software. Along with the increasing scarcity of COBOL experts (leading to the increasingly high cost of COBOL expertise) and the dwindling BIZFIN customer base, all users were facing an inevitable systems change.

This complex situational trigger reflects the deterministic viewpoint (eg, Apte & Karmarkar, 2007, p vii; Marx & Engels, 1973; McLuhan, 1964; Porat & Rubin, 1977; Toffler, 1970; Ellul, 1964; Orwell, 1949) of innovation. Commenting on these internal triggers in an interview, Stanley explained: "Certainly, BIZFIN did not meet the needs of the wider university. .. We use whatever reasons we could get in order to bolt the business case." Despite Craig's contrary opinion that BIZFIN was adequate for the needs of the Finance Division; the general consensus in the University was that there was an urgent need to deploy a new enterprise system.

Yet there were at least one very significant external trigger which also came into play. It was actually the Research department, rather than Finance, which tipped the balance towards replacing the old enterprise system. As previously mentioned, the University had a vision to be an international leader in research. To be faithful to this vision, it would consistently need to meet the government's research funding and compliance requirements—a responsibility for the Finance which its existing feral systems might not be able to live up to. Nor would the feral systems alone be enough to monitor a single study, if this study were spread out over several years and conducted by different groups of researchers.

In short, there was external pressure from the government as well as internal pressure from the Research department. Ultimately, however, the greatest trigger was the collective demands of all people in the university who relied on the enterprise system. The new system had to meet all their requirements, not just those of the government and the users in the Research department.

Clearly, an instrumentalist viewpoint (eg, Jasanoff *et al.*, 1995; Luftman & Koeller, 2003; Ogburn, 2007; Tönnies, 2001) can complement the more traditional deterministic way of looking at the case. While the software as a tool for their needs, the executive sponsor had a different way of appraising the system's use. The former focused on the system's functionality (or lack thereof) in day-to-day operations; the latter prioritised the new vision to make the University an international leader in research. Yet both viewpoints led to the same conclusion in favour of innovation.

Criteria 2: The Powerful Opinion Leader Who Created the Right Mindset for Change

Something else Stanley discussed in his in-depth interview was the importance of creating adequate awareness in order to facilitate the right attitude towards an innovation. "It is what our clients, the university users, need that is important. There were reasons why there was so much anxiety regarding the quality of information from the divisions. The reason for this is that we are not providing the services needed in order for them to produce the information that was wanted. It took quite a bit of time to bring that around."

Stanley would become the opinion leader who passed the same awareness to the executive sponsor, whose decision-making powers outweighed those of the rest of the users. His report emphasised the role of the Finance Division in facilitating the quality information received by the rest of the users all throughout the University. Only after he had brought the issue to Craig's attention did anyone on the executive level look into deploying a new enterprise system and launching a formal innovation process.

Criteria 3: The Matchmaking Stage

Figure 1 gives a visual representation of the Exploration Phase of new system deployment.

Setting the Agenda into Motion (1)

The executive sponsor's greatest challenge is setting the agenda into motion so that the process moves from the Agenda Setting stage to the Matchmaking stage (Figure 1). Craig's experience collaborated this principle: "All the implementations and all the project sub-processes have been relatively small when compared to [this initial step]." In fact, Craig could also have been referring to David's experience. Both executives had, while in the position of Director of Finance, the same agenda to evaluate and replace the current enterprise systems. While David failed to push the agenda, Craig succeeded. This can be attributed to his more innovative mindset.

Craig had a year and a half to study the situation and continuously evaluate his plans. Aside from Stanley's briefing, he consulted other internal departments and discussed the issue with his superior David (current Chief Operating Officer). He also canvassed the opinions of other Chief Financial Officers from similar organisations in the country, often engaging in site visits, both domestic and international, as well. This comprehensive investigation provided him with many insights, some of which he discarded (Van de Ven *et al.*, 1999) but most of which fostered his efforts to provide a reasoned action (Ajzen, 1985, 1991; Ajzen & Fishbein, 1980) of technology acceptance (Davis, 1986, 1989; Davis *et al.*, 1989; Mazis *et al.*, 1975; Venkatesh and Davis, 2000, p 186).

One can already see, at this point, the tight coupling (circular relationship in FIGURE 1) between the agenda-setting stage and the matchmaking stage. Both attempt to explore and reconcile the goals with the alternatives, and the ends with the means (Mintzberg *et al.*, 1976, p 265).

Change Agent Selection (2)

A formal matchmaking stage begins with a Request for Information and continues with a Request for Proposal. Determined to follow this model as strictly as possible, the University appointed Kevin Peters of Providence Consulting to be a third party change agent, putting him fully in charge of both RFI and RFP. According to Craig, he chose

Providence over a second prospect because the former adhered to a more structured approach to executing and managing the matchmaking stage. Such an approach was more compatible with the University's risk-averse mindset, which tied the expectation of quality outcome to a very formal procurement process.

User Requirements (3)

Kevin took charge of articulating the agenda via consultation sessions with over seventy individuals in all academic and service divisions. In his selection-evaluation report, dated March 2005, he documented thirty-seven of these sessions. His final analysis in favour of innovation was subsequently approved by the project steering committee. Only after this point did Kevin launch the Request for Information stage.

RFI (4)

Up to this point, the introduction of a new enterprise system was a purely internal matter. Having received the desired approval, Kevin was free to make contact with software package vendors. On 3 November 2005, the University uploaded its detailed requirements for a new system into a government electronic tendering facility and put an additional advertisement in a newspaper.

By 7 November, the document had been downloaded 150 times; and by the 17 November deadline, the University received twenty formal responses covering thirteen different products.

To evaluate each response, Stanley, Kevin and Craig considered vendor culture, proprietary lock-in, staff support ratio, and experiences in implementation. Kevin likely provided additional insider knowledge of the software industry in general and prospective vendors in particular. By the end of the RFI stage on 30 November, their team had narrowed the list down to four: Vendor F, Vendor N, Vendor O, and Vendor S.

At this point, one could critique the short timeline. Prospective vendors had, at the most, fifteen days to respond to the RFI. Such a time limit must have made it difficult for them to adequately pinpoint the needs of UoA. They could only "matchmake" the features of their software to the requirements specified in the RFI. What was a thorough report from the point of view of the University might not necessarily have given the vendors a comprehensive picture of the situation.

RFP and Selection (5)

The four short-listed vendors received a Request for Proposal (RFP) on 1 December 2005 and were asked to submit their responses by 9 January 2006. Only Vendor F and Vendor O continued to participate after this point.

As Craig remembers the time: "We looked at that at the time when [Vendor N] pulled out because they could not comply with the users' requirements. We looked at the other possible solutions and decided that it was not worth it. .. that we could not get any more value out of it. Especially on Christmas Eve. .. we were a little bit p***** off." He decided to focus on the strong candidates he saw in Vendor F and Vendor O. (Vendor O, in particular, had been telling the market that only they had the desired shared services module.)

However, the two short-listed vendors who did not participate in the RFP told a different story. Contrary to Craig's feedback that Vendor N had determined it could not meet the requirements and thus voluntarily pulled out of the race, the company actually had a very different reason. In an interview, Vendor N revealed that a certain account executive left the company around the time of the University's procurement process. The source added: "It is disappointing for me to learn that we were short-listed but did not complete the second stage of the process."

Vendor S pulled out for yet another reason. Firstly, it did not think there had been enough time to explore the University's unique situation and propose a configuration to suit its specific requirements. Though the firm did ask for an opportunity to do so, it was rebuffed. In a formal RFI-RFP process, all prospective vendors must receive only as much or as little information as their competitors.

The second reason was that Vendor S had determined the cost of sales to be too high to balance against the chance of winning. This was closely tied to the third reason, the expectation that it would probably not win the account, due to a certain situational factor. That factor, the involvement of Providence Consulting, was also the fourth reason.

As Vendor S revealed in an interview: "In the past, we've observed that [Vendor F] wins whenever [Providence] is running the process. In fact, they have won four out of five times. So we always look at who is running the processes before making a judgment.. . We look at things like [Providence's] office in [City X] being in the same building with [Vendor F's office]. Again, I am not saying they run a shanky or unfair process. It may have been completely fair. But we look at that and we say, well, let's look at our track record when [Providence] has been [in charge]. .. Let's look at our gut feel about our chance to succeed and what will come from it."

Ultimately, the company's "gut feeling" was that it would lose the account to Vendor F. This turned out to be an accurate prediction.

Meanwhile, on the University's side, a six-person team further evaluated the final two competitors. The steps included determining the weights, scoring the proposals, hearing vendor presentations, and referencing the site visits (during information-seeking stage), analyses and follow-up activities.

Those vendor presentations were attended by over thirty representative wider users. After hearing out both proposals, the users described System O as complex and "clunky," but liked the apparent user friendliness of System F. System F also had the advantage of the total cost of ownership, as the named user-licensing scheme of System O made its annual on-going cost more expensive.

In the end, though both vendors met the requirements (eg, functionality, product history, and vendor details), their proposed software packages elicited different reactions from the prospective users.. System F had a 93% fit and a score of 8.3, while System O had a 90% fit and a score of 8.0.

Business Case (6)

It was time for Craig to draft a business case document, giving upper management four options: (a) Preferred – System F; (b) Next – System O; (c) Shared Option – System O; and (d) Do Nothing.

This moved the ball into the University Council's court. The decision makers on the Council made the decision to replace BIZFIN and accept Option A.

Criteria 4: Some Concerns

Contract

One thing Craig said he would do differently if he had the first phase to go through all over again was ask for Kevin's help much earlier, rather than letting the contract drag on for too long. He cited Kevin's experience as "very valuable" to the University even before the procurement process began.

Worst Fear

He also shared his worry about "backfilling" the positions during the pre-production stage of the exploitation phase (Figure 1). "The people [from Finance] we put into the [Project] team are very good. [Their absence] will weaken the [Finance] Division in the short term."

The University had already learned its lesson about backfilling from a previous project on Human Resource systems. When the positions of the staff involved in that project were not been backfilled, the University ended up having to re-implement the HR systems! It was an undesirable consequence which could have been very easily avoided.

This was Craig's concern as the new project's executive sponsor. He could not limit his governance to a vacuum with only the project team and the new enterprise systems (Sohal & Fitzpatrick, 2002; Weill & Ross, 2004; Willcocks et al., 2006), but also had to consider the needs of the entire University.

Preference for Discovery

The misalignment of expectations is also worth noting. Vendor S declined to participate in the RFP because it did not think it was on the same page as the University. From its point of view, the opportunity to explore the prospective client's exact needs during the RFI stage is critical to success in the RFP stage. However, the University insisted on following the stringent rules of the RFI-RFP process and stonewalled all attempts Vendor S made to learn more about the former's situation.

Vendor S explained the way the market worked: "You issue an RFI and get a short-list. The short list might be ten or so vendors. You go to the RFP and end up short-listed again. But a lot of organisations now are not allowing you to do much discovery and to have many meetings during the RFI and RFP process. So most of the time, an RFI and to a lesser degree an RFP is a qualification process for us as much as the person looking for the software. . ."

Yet the information given at the RFI stage is "nothing exciting" from the vendor's point of view: "It's just, oh, we need a set of financial software, and these are the requirements, and so on. . ." Clients putting out a Request for Information don't necessarily know what a vendor will need to hear to design the best software package for their needs.

A Critique Against the RFI-RFP Process

Clearly, a significant weakness of the RFI-RFP process is its failure to provide prospective vendors with the opportunity to undertake a thorough discovery process.

Vendor S explained: "We do that just about in every case, not just [for the University of Australasia]. We will make contact with a prospect and ask for a meeting to discuss. .. what they really want to achieve and what the real problems are with their current solutions. Some will have the meeting. But these days, most will not. .. because. .. if they do it with us, they will have to do it with the twenty other vendors. And. .. they don't want to spend a lot of time with the vendors."

He also cited the perception that accommodating prospective vendors would "somehow... unduly influence the process." Attempting to articulate the client's point of view, he added, "This is very much a process to get a decision and things like meetings and informal discussions are seen as subverting the process."

There was clearly a conflict of mindsets at work when Vendor S declined to participate in the RFP. While the executive sponsor believed that the formal procurement process assured the University would end with the best enterprise system for its needs and that bending the rules for any vendor would unfairly sway the results, Vendor S perceived the same structured process as too rigid for vendors to compete with a good chance of winning. There was also a conflict in the way Kevin was perceived: to Craig, his experience was invaluable to the University; to Vendor S, his employment at Providence Consulting meant a greater likelihood that its competitor would land the account. These strikingly different takes on the same situation should be taken into account by anyone studying.

Yet the University had no scruples about changing one of the RFI-RFP rules at its own discretion. As Vendor S bluntly recalls, "They changed the process. The original process was they would review the RFP and then they would select two to do the demonstration. Once [Vendor N] pulled out, they decided to do a presentation with three vendors. That is a bad thing for us because we are now investing more money with a thirty percent chance."

This decreased chance of winning a bid was a huge consideration for Vendor S, which was already pessimistic about its chances, for other aforementioned reasons. In the end, it is the *relationship* that counts.

Craig was so preoccupied with a possible inability to backfill his staff that he left too much of the RFI-RFP process in Kevin's hands. By not cultivating a better relationship with Vendor S, on the grounds of avoiding "undue influence," he might have ended up costing the University a software package better suited to its unique needs than that of Vendor F.

Vendor S gave other critical insights which any company considering a structured RFI-RFP-BCD process should note: Firstly, he pointed out that customer-organisations usually ask vendors "to spend a big chunk of money so that they can get hold of a lot of information, which, quite frankly, are mostly irrelevant." It would be more beneficial, he seemed to be implying, for a customer to trust a prospective vendor's expertise in determining its system requirements rather than compelling a vendor to jump through unnecessary hoops before sharing essential information.

Asked to analyse the reason for this trend, he said, "People have gone this way because they are scared. In the past there had been too much buying of software based on relationships." The paradox is that while cultivating a good relationship with the right vendor should be a primary objective, already having a good relationship with a vendor does not mean he is the right vendor. Even Vendor S admitted that he was a loyal customer of a friend with whom he goes drinking and plays rugby, but maintained that structured procurement is just "another extreme" which puts too much emphasis on "cold hard facts" and too little emphasis on a relationship. "But the reality is, there needs to be somewhere in between, because you still need to look at people in the eye. You have to be talking and getting that nice feel that you can do business with this person."

Not surprisingly, Craig disagreed with the above analysis, saying that executive sponsors of large organisations were "not born yesterday," and would know better about relying too much on "cold hard facts" to the detriment of professional relationships. He claimed to be aware that the responses to the RFI and RFP might not be perfectly accurate, but stood by the process anyway. This is another example of a conflict of mindsets between a customer and a vendor during the crucial matchmaking stage.

The question remains: given the imperfections of the RFI-RFP process, did the matchmaking achieve its objective to pinpoint the right vendor for the University's needs?

EVIDENCE, REASONS, AND CONCLUSION

Deploying new enterprise systems or replacing old enterprise systems is a problem-solving intervention (Thull, 2005), under conditions of incomplete information (Taleb, 2007). A small trigger may result in large unexpected consequences (Ormerod, 2005; Rogres, 2003).

The possible presence of defensive reasoning and behaviour might mean that rational behaviour is actually irrational behaviour in disguise (Argyris, 2004; Luhmann, 2006). There is the possibility that the University of Australasia's risk-averse mindset, which influenced its very rational decisions about the procurement process, might ultimately have worked against it. The critiques of Vendor S point out the disadvantages of relying on such a highly structured procurement process and giving too much control to a third party agent.

All innovations depend on reasoned action for success. Yet "reasoning" is often based on a subjective set of beliefs and motives held by the executives sponsor, the opinion leaders, and other supporters of the innovation. These beliefs, motives and other assumptions are called a "mindset," and the study shows that they play a bigger role in procurement than the rigid structure of the process would suggest. A pro-innovation bias is always implicit when an organisation starts the procurement process.

While interacting with the opinion leader, change agent, prospective vendors, and other people, an executive sponsor further develops his awareness of the innovation and its attributes, and whether his organisation has a true need to deploy a radically new system. He can also better anticipate both desirable and undesirable consequences and more convincingly "sell" the innovation to other users in his organisation, before any big decisions are made. That Craig went to these lengths supports his reasoned approach to innovation and his pro-active mindset.

The structured approach in the matchmaking stage definitely highlights the reasoned actions on the part of the customer-organisation (Argyris, 2004; Luhmann, 2006), but it represents only one mindset. It does not take the mindsets of the prospective vendors into consideration. This is a mistake, in the light of matchmaking's objective to help an organisation cultivate a good relationship with the right vendor. An alignment of both mindsets is necessary before the Matchmaking stage can give way to the Business Case Development stage (FIGURE 1).

To conclude, a pro-active mindset among people involved in the innovation process is a positive influence for fostering relationships with opinion leaders, change agents, project team members, and other stakeholders. The same mindset influences some vendors to develop relationships with prospective customers and to understand more thoroughly their needs. Successful matchmaking depends on an alignment of these mindsets.

REFERENCES

Agarwal, R., & Prasad, J. (1998). A conceptual and operational definition of personal innovativeness in the domain of information technology. *Information Systems Research, 9*(2), 204–215. doi:10.1287/isre.9.2.204

Ajzen, I. (1985). From intentions to action: A theory of planned behavior. In Kuhl, J., & Bechmann, J. (Eds.), *Action Control: From Cognition to Behavior* (pp. 11–39). New York: Springer Verlag.

Ajzen, I. (1991). The theory of planned behavior. *Organizational Behavior and Human Decision Processes, 50*, 179–211. doi:10.1016/0749-5978(91)90020-T

Apte, U. S., & Karmarkar, U. M. (2007). Current research on managing in the information economy: Introductory note. In U. S. Apte & U. M. Karmarkar (Eds.), Managing in the Information Economy: Current Research Issues. New York: Springer Science+Business Media, LLC.

Argyris, C. (2004). *Reasons and Rationalizations: The Limits to Organizational Knowledge*. Oxford: Oxford University Press.

Auh, S., & Menguc, B. (2005). Top management team diversity and innovativeness: The moderating role of interfunctional coordination. *Industrial Marketing Management, 34*(3), 249–261. doi:10.1016/j.indmarman.2004.09.005

Avery, J. (1997). *Progress, Poverty and Population: Re-reading Condorcet, Godwin and Malthus*. Portland, Oregon: Frank Cass Publishers.

Bell, W. (1996). An overview of future studies. In R. A. Slaughter (Ed.), The Knowledge Base of Future Studies (Vol. 1, Foundations, pp. 26-56). Hawthorne, Victoria: DDM Media Group.

Bottomore, T., & Nisbet, R. (Eds.). (1978). *A History of Sociological Analysis*. New York: Basic Books.

Caldeira, M. M., & Ward, J. M. (2002). Understanding the successful adoption and use of IS/IT in SMEs: an explanation from Portuguese manufacturing industries. *Information Systems Journal*, *12*(2), 121–152. doi:10.1046/j.1365-2575.2002.00119.x

Chase, S. E. (2005). Narrative inquiry. In Denzin, N. K., & Lincoln, Y. S. (Eds.), *Handbook of Qualitative Research* (3rd ed., pp. 651–679). Thousand Oaks, CA: Sage Publications.

Christensen, C. M., & Shih, W. C. (11 Dec 2007). Successful Innovation: The Intersection of Theory and Practice Retrieved 30 Jan 2009, from http://www.hbs.edu/centennial/conversation/successful_innovation/

Clegg, S. R., Kornberger, M., & Rhodes, C. (2004). Noise, parasites and translation: Theory and practice in management consulting. *Management Learning*, *35*(1), 31–44. doi:10.1177/1350507604041163

Clough, P. T. (2000). Comments on setting criteria for experimental writing. *Qualitative Inquiry*, *6*, 278–291. doi:10.1177/107780040000600211

Corbett, J. (2005). Toresten Hägerstrand: Time Geography Retrieved 14 September 2005, from http://www.csiss.org/classics/content/29

Davis, F. D. (1986). *A technology acceptance model for empirically testing new end-user information systems: Theory and results.* Unpublished Doctoral dissertation, Massachusetts Institute of Technology.

Davis, F. D. (1989). Perceived usefulness, perceived ease of use, and user acceptance of information technology. *Management Information Systems Quarterly*, *13*(3), 319–340. doi:10.2307/249008

Davis, F. D., Bagozzi, R. P., & Warshaw, P. R. (1989). User acceptance of computer technology: A comparison of two theoretical models. *Management Science*, *35*(8), 982–1003. doi:10.1287/mnsc.35.8.982

Ellet, W. (2007). *The Case Study Handbook*. Boston, Massachusetts: Harvard Business School Press.

Ellul, J. (1964). *The Technological Society*. New York: Alfred A Knopf, Inc.

Gatian, A. W., Brown, R. M., & Hicks, J. O. (1995). Organizational innovativeness, competitive strategy and investment success. *The Journal of Strategic Information Systems*, *4*(1), 43. doi:10.1016/0963-8687(95)80014-H

Greenhalgh, T., Robert, G., MacFarlane, F., Bate, P., & Kyriakidou, O. (2004). Diffusion of innovations in service organizations: Systematic review and recommendations. *The Milbank Quarterly*, *82*(4), 581–629. doi:10.1111/j.0887-378X.2004.00325.x

Habermas, J. (1970). Knowledge and interest. In Emmet, D., & MacIntyre, A. (Eds.), *Sociological Theory and Philosophical Analysis* (pp. 36–54). London: Macmillan.

Habermas, J. (1974). *Theory and Practice*. London: Heinemann.

Hirschman, E. C. (1981). Innovativeness, novelty seeking, and consumer creativity. *The Journal of Consumer Research*, *7*(4), 63–71.

Holly, K. (4 Feb 2008). The politics of change. *BusinessWeek*. Retrieved from http://www.businessweek.com/innovate/content/feb2008/id2008024_250194.htm

Houghton, L., & Kerr, D. V. (2004). *Understanding Feral Systems in Organisations: A case study of a SAP implementation that led to the creation of ad-hoc and unplanned systems in a large corporation.* Paper presented at the 9th Asia-Pacific Decision Sciences Institute Conference, Seoul, Korea.

Houghton, L., & Kerr, D. V. (2006). A study into the creation of feral information systems as a response to an ERP implementation within the supply chain of a large government-owned corporation. *International Journal of Internet and Enterprise Management*, *4*(2), 135–147.

Hult, G. T. M., & Hurley, R. F. (1998). Innovation, market orientation, and organisational learning: An integration and empirical examination. *Journal of Marketing*, *62*, 42–54. doi:10.2307/1251742

Hult, G. T. M., Hurley, R. F., & Knight, G. A. (2004). Innovativeness: Its antecedents and impact on business performance. *Industrial Marketing Management*, *33*(5), 429–438. doi:10.1016/j.indmarman.2003.08.015

Lefebvre, E., & Lefebvre, L. A. (1992). Firm innovativeness and CEO characteristics in small manufacturing firms. *Journal of Engineering and Technology Management*, *9*(3-4), 243–277. doi:10.1016/0923-4748(92)90018-Z

Lincoln, Y. S. (1997). Self, subject, audience, text: Living at the edge, writing in the margins. In Tierney, W. G., & Lincoln, Y. S. (Eds.), *Representation and the Text: Re-framing the Narrative Voice* (pp. 37–55). Albany: State University of New York Press.

Lincoln, Y. S. (2000). Narrative authority vs perjured testimony: Courage, vulnerability, and truth. *Qualitative Studies in Education*, *13*, 131–138. doi:10.1080/095183900235654

Marx, K., & Engels, F. (1973). *Karl Marx: On Society and Social Change*. Chicago: University of Chicago Press.

Mazis, M., Ahtola, O., & Kippel, R. (1975). A comparison of four multi attribute models in the prediction of consumer attitudes. *The Journal of Consumer Research*, *2*, 38–53. doi:10.1086/208614

McLuhan, M. (1964). *The medium is the message Understanding Media* (pp. 7–23). London: Routledge and Kegan Paul.

Midgley, D. F., & Dowling, G. R. (1978, March). Innovativeness: the concept and its measurement. *The Journal of Consumer Research*, 4.

Mills, C. W. (2000). *The Sociological Imagination*. New York: Oxford University Press.

Mintzberg, H., Raisinghani, D., & Théorêt, A. (1976). The structure of "unstructured" decision processes. *Administrative Science Quarterly*, *21*, 246–275. doi:10.2307/2392045

Öberg, S. (2005, June). Hägerstrand and the remaking of Sweden. *Progress in Human Geography*, *29*(3), 340–349.

Ormerod, P. (2005). *Why Most Things Fail: Evolution, Extinction and Economics*. London: Faber and Faber Limited.

Orwell, G. (1949). *Nineteen eighty-four*. London: Secker.

Pennings, J., & Buitendam, A. (Eds.). (1987). *New Technology as Organizational Innovation*. Cambridge, Massachusetts: Ballinger.

Porat, M. U., & Rubin, M. R. (Eds.). (1977). *The Information Economy*. Washington, DC: US Department of Commerce/Office of Telecommunications Special Publication.

Rogers, E. M. (1962). *Diffusion of Innovations*. New York: The Free Press of Glencoe.

Rogers, E. M. (2003). *Diffusion of Innovations* (5th ed.). New York: Free Press/Simon & Schuster, Inc.

Sohal, A. S., & Fitzpatrick, P. (2002). IT governance and management in large Australian organisations. *International Journal of Production Economics*, *75*(1-2), 97–112. doi:10.1016/S0925-5273(01)00184-0

Subramanian, A. (1996). Innovativeness: Redefining the concept. *Journal of Engineering and Technology Management*, *13*(3-4), 223–243. doi:10.1016/S0923-4748(96)01007-7

Taleb, N. N. (2007). *The Black Swan: The Impact of the Highly Improbable*. London: Penguine Books Ltd.

Thull, J. (2005). *The Prime Solution*. Dearborn Trade Publishing.

Toffler, A. (1970). *Future Shock*. New York: Bantam Books.

Van de Ven, A. H., Polley, D. E., Garud, R., & Venkataraman, S. (1999). *The Innovation Journey*. New York: Oxford University Press.

Venkatesh, V., & Davis, F. D. (2000). A theoretical extension of the Technology Acceptance Model: Four longitudinal field studies. *Management Science*, *46*(2), 186–204. doi:10.1287/mnsc.46.2.186.11926

Weill, P., & Ross, J. W. (2004). *IT Governance: How Top Performers Manage IT Decision Rights for Superior Results*. Boston, Massachusetts: Harvard Business School Press.

Willcocks, L., Feeny, D., & Olson, N. (2006). Implementing Core IS Capabilities: Feeny-Willcocks IT Governance and Management Framework Revisited. *European Management Journal*, *24*(1), 28–37. doi:10.1016/j.emj.2005.12.005

Woodside, A. G. (2005). Firm orientations, innovativeness, and business performance: Advancing a system dynamics view following a comment on Hult, Hurley, and Knight's 2004 study. *Industrial Marketing Management*, *34*(3), 275–279. doi:10.1016/j.indmarman.2004.10.001

Chapter 6
The Development of Emerging Medical Devices:
The Lead-User Method in Practice

Brian O'Flaherty
University College Cork, Ireland

John O'Donoghue
University College Cork, Ireland

ABSTRACT

This case study explores the application of the Lead-user method in the development of medical applications based on Wireless Sensor Network (WSN) technology by three independent research teams. This exercise produced surprising results, with the emergence of diverse WSN technology product concepts applied to Geriatric Falls Detection & Analysis, Sport Cardiac Screening and Critical Care Vital signs within accident and emergency environments. This case highlights the segmented nature of medical areas and the difficulty in applying a generic WSN technology to meet the functional requirements of the broader individual medical domains. It questions the appropriateness of applying 'total' highly functional technologies broadly across highly specialised niche medical areas.

BACKGROUND

This case outlines the experiences of three Postgraduate Innovation teams, students on a one-year taught masters programme that are required to 'build products and services that don't exist yet.' The Masters programme, which targets technology graduates, includes a significant innovation component requiring the teams to validate market existence and develop a prototype and business plan with the assistance of an industry mentor. The students respond very well to the Lead-user Method, the 3M case study and the accompanying videos on Eric Von Hippel's website. These research innovation teams were created to explore the potential role of wireless sensor network (WSN) technology in the medical area. The teams independently focussed on three distinct areas, namely: 1) Geriatric Falls Detection & Analysis; 2) Sport Cardiac Screening; and 3) Critical Care Vital signs within accident and emergency environments. Each of the teams operated independently of each other as to not taint or indirectly alter one another's perceptions of their individual application areas. Each team consisted

DOI: 10.4018/978-1-61520-609-4.ch006

Figure 1. The process of the lead-user method

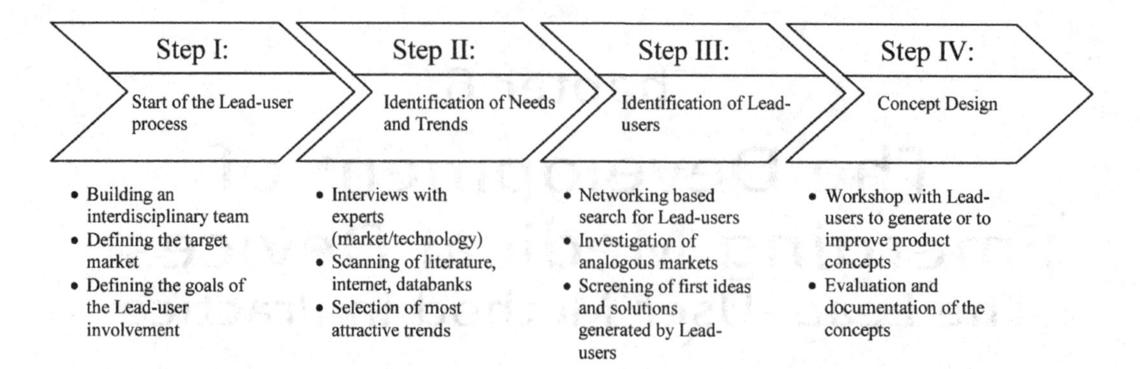

of five members with diverse backgrounds from commerce, electric and electronic engineering, and computer science. Subsequently each team was assigned an industry mentor to help guide them on a viable commercial path. Finally all three teams were lead by two project managers Dr. Dan Nielsen (technical lead) and Dr. Miyagi (product innovation).

The Lead-user process has been successfully adopted within a diverse range of application domains i.e. development of medical equipment technology (Lettl, et. al., 2006), medical infection control devices in 3M (Von Hippel, 1999), weblog technology (Kaiser, et. al., 2008) and extreme sports communities (Schreier, et. al., 2007). It was selected as the process to help guide each of the three teams in developing potentially successful commercial products/services (Von Hippel, 1998), (Franke, et. al., 2006). The 'functional' source of innovation provides a good starting point for innovation teams to explore the relationship between innovator and innovation. (Von Hippel, 1998) defines innovation as anything new that is actually used ("enters the marketplace"), whether major or minor. A distinction is made between a 'user' and 'manufacturer' innovation. With 'user' innovation the developer expects to benefit by using it and in the case of a 'manufacturer' innovation the developer expects to benefit by selling it. The Lead-user method has developed into a four stage approach (cf. Figure 1), which includes I) Start of Lead-user process, II) Identification of Needs and Trends, III) Identification of Lead-users & IV) Concept Design using Lead-expert workshops (Lettl, et. al., 2006).

The high failure rate or lack of commercial success of a number of innovate products/services in the market place are a great cause of concern (Hassan, 2008). Non-Lead-users tend to be technology driven (to maximise their current skill or resource sets) this results in a lost opportunity as they struggle to grasp the true functional requirements of the Lead-user and build what they perceive is the correct product or service. This in turn results in a large number of unsuccessful product/services entering the market place.

The work of Eric Von Hippel (1986, 1988) makes a number of significant contributions to technology entrepreneurship and innovation. The first and less highlighted is the critical natures of correctly sourcing innovation depending on the domain area in question, which is an important first step in any opportunity recognition exercise. The next contribution is the Lead-user method, which built on innovation sourcing, identifies specific individuals in domains, such as, extreme sports, scientific instruments, etc., who have unmet needs and are in a position to address them.

Lead-users have two specific characteristics which can lead to a novel product or service. These are 1) they have general market place needs, which they have strongly identify well in advance and 2) they have the ability to develop solutions to meet their particular needs. Therefore Lead-users have three key characteristics, namely 'ahead of the market', 'level of expected benefit' and 'high level of innovation' (Morrison, 2004).

Vast arrays of medical devices have emerged over the years to help improve the overall delivery and quality of patient care. The driving force behind such devices stem from medical practitioners (expert users) who have experienced a number of early unmet needs during their day to day activities. The Lead-user method (Von Hippel, 1988) was deemed appropriate in assisting each of the three groups, as it provided them with the necessary structure to take expert user functional requirements and step by step identify and evaluate their commercial merit.

This case study is based on the findings of all three teams to determine lessons learned and potential pitfalls in the area of early development of WSN within a medical context. The initial presumption is, 'if three teams apply the same method to the same technology then they will end up with similar outcomes'? The results contradict this assumption and also give insights into the segmented nature of innovation within the medical profession. Lessons learned from this case highlight that innovation research teams have resourcing issues in implementing the complete method, as described in the 3M case study (von Hippel, 1999). But, even a limited application of the method provides a change in their attitude to opportunity identification and effective new product development.

RESEARCH METHOD

This case study explores the experiences of three project teams that applied the Lead-user method to the medical application of Wireless Sensor Networks (WSN), with each team depicted a single case study. All three groups studied in the same Masters course and applied the same Lead-user method of analysis. The case study method was applied (Yin, 1994), with multiple sources of data, including learning journals, business plans, feedback from mentors and presentations (Patton, 1990). The patterns were examined across the three cases (Miles, 1994), which is used in the analysis of the case outcomes.

Technology Overview

A WSN is made up of a collection or a network of miniaturised computing nodes (e.g. Tyndall 25mm (O'Flynn et al., 2006)) which are able to sense a variety of real-world characteristics such as light, sound, movement, and moisture. Apart from sensing, these nodes are able to autonomously evaluate their sensor readings. This helps to create intelligent environments which can react to a variety of sensor readings in a controlled manner. Once a node is within radio range it is capable of communicating to neighbouring nodes to help achieve an accurate picture of its entire environment (e.g. what is the overall room temperature)? All sensor readings residing within the WSN may also be polled periodically or if a particular event occurs (e.g. temperature level has exceeded a recommended level i.e. potential fire) the WSN nodes may be programmed to communicate with a central base station or server where further analysis on aggregated node datasets may be carried out (cf. Figure 1). All nodes within a WSN may be dynamically reprogrammed to: 1) Sense particular artefacts (i.e. assuming the sensor is attached to that node); 2) updated with new internal logic algorithms; or 3) reconfigured with new communication or power management protocols.

WSNs have seen a tremendous expansion over the last few years; this has resulted in low cost devices, high processing capabilities, advanced dynamic network protocols and MEMS

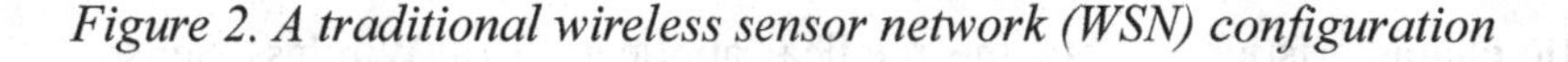

Figure 2. A traditional wireless sensor network (WSN) configuration

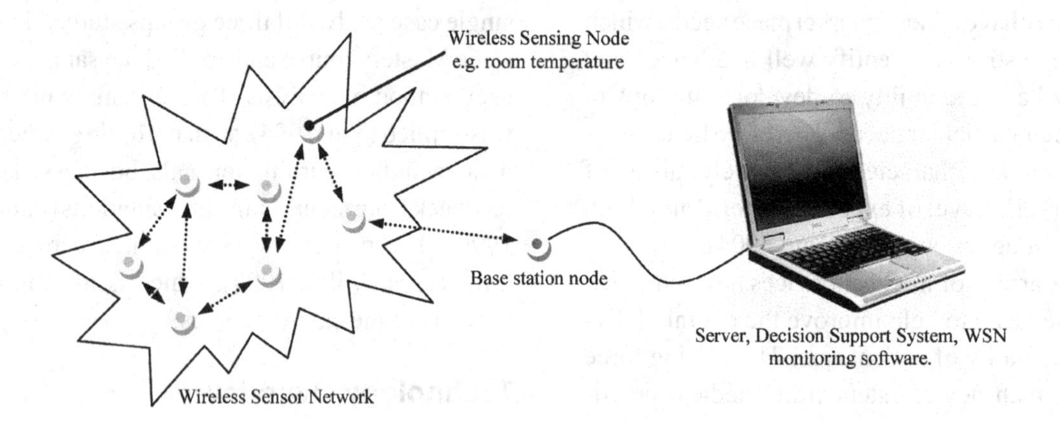

(Micro-Electro-Mechanical Systems). Power management solutions have begun to make WSN applications a viable and cost affective option for numerous application domains. One particular business domain has demonstrated significant potential through the usage of WSNs namely Body Area Networks (BANs) or Body Sensor Networks (BSN). A BAN (i.e. a wearable WSN) operates in a similar fashion to a WSN in that it can intelligently interact with its environment. BANs are considered a key element for patient-centric healthcare services for the future as it can provide a very effective way to collect, monitor and manage patient's physiological parameters such as glucose levels, blood pressure, heart rate, calorie burning, daily step numbers and foetal movement among others. A number of these devices have been developed over recent years to monitor specific patient vital signs e.g. (Lorincz, 2004; O'Flynn, 2006; Thiemjarus, 2005).

A major challenge within WSNs is the capturing and processing of relevant dataset as effectively as possible without eating into valuable resources (e.g. battery power). To make matters even more exigent, sensors within the WSN need to cope with high degrees of interference. If not managed correctly, poor sensor readings may lead to inaccurate reports. Levels of interference are a substantial issue within BANs as the motes are attached to the patient resulting in substantial levels of physical disturbance. This has major repercussions within cardiac and critical care vital-sign monitoring environments. When the erroneous datasets are identified and removed, access to accurate and relevant real-time patient information greatly enhances a medical practitioner's productivity level.

Currently, a large proportion of patient records and lab results are documented in written format, often kept in an unstructured manner. Cases frequently arise where patient medical records are unobtainable for many days and in some instances are lost. Moving from a complete paper based system to a complete electronic based environment does not guarantee a higher quality of service. A combination of the two may provide the optimal approach in handling patient datasets. The DMS in conjunction with the Tyndall-BAN (cf. Figure 3) is designed to manage context aware variables which may include patient and staff location, in union with real-time patient vital sign readings. The primary benefit of BANs within a medical setting is that it opens up staff resources to pay greater attention in delivering hands on patient care

Figure 3. A body area network layout to capture patient vital-signs and orientation

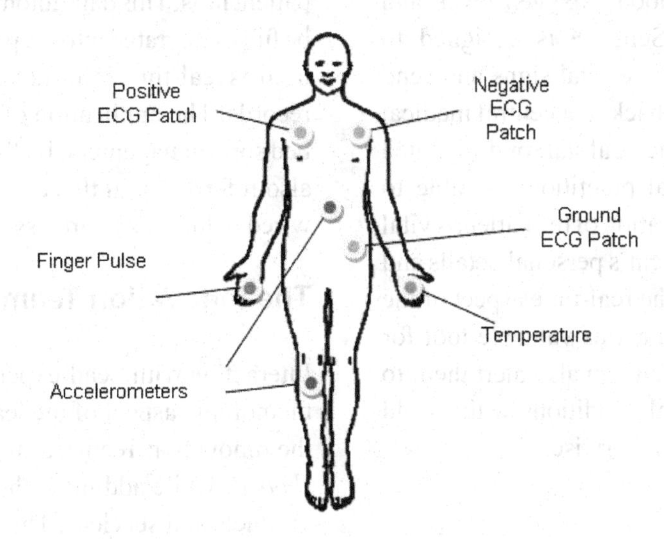

rather then consuming a great deal of their time monotonously sampling patient vital signs.

SETTING THE STAGE

Dr. Dan Nielsen recently completed his PhD in Body Area Networks (BAN), in the area of medical based wearable Wireless Sensor Networks (WSN), and started as a technical lead for a postgraduate innovation module. He wished to continue his work on the 'Total body area network (Total-BAN) concept' and was interested in working with the Innovation Module leader, Dr. Miyagi, who had a reputation for developing radical new product concepts. Dr. Nielsen, a pure technologist, highlighted some of the application assumptions that his previous research group held, which included the inappropriateness of WSN applied to Accident and Emergency or Critical-Care. Dr. Miyagi found this interesting and questioned the underlying motivations for these assumptions. They both decided that Dan's work and 'application myths' could be tested and developed by a number of the research teams in the Innovation Unit. This case study focuses on the experiences of these three teams, who collectively applied the Lead-user method to discovery of new applications of WSN in the medical domain.

CASE DESCRIPTION

Project Descriptions

Each of the three groups set about identifying opportunities for wireless sensor networks (WSN) using the Lead-user method, with Dr. Nielsen, providing technical guidance. All three groups developed a fictitious business and they were required to develop a business plan, in order to support their market research findings and also gauge commercial viability of their product. A description of each developed company's is described as:

Vital Solutions

Vital Technology Solutions developed a wireless sensor patient monitoring system designed to monitor patients' vital signs in critical medical environments such as the Emergency Department and the Intensive Care Unit. The V-Sense™ solution records and monitors a patient's heart rate, blood pressure, blood / oxygen level, and body temperature. V-Sense™ is designed to intelligently evaluate these vital signs and send key information points back to a central medical workstation to inform medical staff or designated caregivers. The medical practitioner is able to view "real-time" information of the patient's vital signs, as well as the patient's personal details and their medical history. The real-time aspect of the V-Sense™ solution can act as valuable tool for collection patient data but can also alert them to potentially fatal medical conditions with would not have been detected otherwise.

SensIT

Red Cross estimates that 13 people die every day in Ireland as a result of Sudden Cardiac Death (SCD), which is approximately 5,000 every year, 100 of which are individuals under the age of 35 (Dept of Health & Children Report, 2006). The only reliable means of preventing these deaths is through widespread heart screening. However, at present this is not possible due to insufficient personnel and technology. The SensIT Solutions' PreCG 'black boxes', the technology and the process of monitoring heart rhythms, allow for quick, painless, cost-effective and widespread remote screening of individuals so that existing conditions can be addressed before they deteriorate, ultimately saving lives.

Health Sense

Health Sense is a wireless sensor network system designed to assist independent living in residences of Nursing Homes, offering continual monitoring of residents and their behaviour while maintaining their privacy and dignity. Health Sense records and monitors patients' daily living activities in the nursing home and automatically summons help if it detects a possible problem. The product records and monitors changes in temperature and light and will also detect limb movement to sense if a patient falls. This data automatically collected can be fully integrated into the patient care record to be used as real-time information for patient hospital records. The application of WSN to the geriatric bed sore management, by the Com4care group, is also referred to in this case, but the team used a wired solution as wireless was not necessary.

The Innovation Team Experience

Interacting with Lead-experts proved to be a very memorable aspect of the learning experience for the Innovation Teams, as this gave the students a *'boost'*, while adding to the *'credibility'* of their products and services. The students found meetings with their Lead-users quite intense. Normally, large amounts of information were generated, which required significant analysis with the potential for a major impact on the outcome of their project. The Lead-user had an important role in defining the appropriateness of technology, but most importantly they determined when solutions *'would not work'* from both business and technical perspectives.

The postgraduate teams were required to keep a learning journal, which described their experiences in working with mentors, sourcing innovation, and their experiences interacting with various lead-experts on technology innovation projects. The learning journal documented the stages of the process of the Lead-user method. The documentation included the identification of consumer needs and industry trends; as well as, the screening of ideas as posted by the experts.

Needs Recognition

Lead-user involvement played an important role in identifying and validating the product needs. Several student quotes about the experience include:

'The whole concept behind the PreCG was an evolution from feedback we received from Lead-users,

who had recognised the need for such a device. 'We had a customer/Lead-user from initiation and this was very beneficial to gain an understanding of requirements and needs.' (A team member from SensIT Group)

'We had a customer/Lead-user from the beginning and this was very beneficial to gain an understanding of requirements and needs.' (A team member for Vital Solutions)

The learning journal entries from the students confirmed that the Lead-user interviews played an important role in validating the feasibility of the product concepts.

'This role involved interviewing identified Lead-users to discover the feasibility of our concept. This interview process meant meeting high ranking officials in various medical institutions, such as hospitals, nursing homes and other health care facilities.' (A team member from Com4care Group)

Sourcing Lead-Users

The teams found sourcing Lead-users or experts a challenging aspect of the process. They made numerous on site visits that accommodated the Lead-users' schedule. One student summarized the activities performed in working with a Lead-user.

'One of my main responsibilities was sourcing potential Lead-users to our project and conducting interviews with those various Lead-users. It also entailed making on-site visits to the Emergency Department & Intensive Care Units in order to gain hands-on insight into these environments. (A team member for Vital Solutions)

Some of the innovation team members found sourcing the Lead-user difficult, but fortunately they received significant help from their industry mentors. Several students summarized the importance of having a mentor in achieving their project objectives. These included:

'The team mentor played a pivotal role in securing meetings with many professionals and allowing us to use his networking contacts to get important information and obtain Lead-users and expert support for our project.' (A team member from SensIT Group)

'Our mentor also contacted an important Lead-user for us that we would have not been able to speak to without his assistance.' (A team member from Vital Solutions)

Lead-user groups typically involved experts that had a stake in technology innovation in terms of not only improving their work environment but also enhancing quality and longevity of life for their patients. The student team for the Vital signs project interviewed a range of experts in the Trauma (Accident and Emergency) area including a Ward Sister, Head of General Practice, a Lecturer in Emergency Services, Advanced Nursing Practitioner, Associate Professor in Orthopaedics & Rehabilitation and an Accident and Emergency Consultant. The breadth of participation, by experts in the field, provided extensive feedback on viable aspects of technology innovation.

Networking, Idea Validation, and Evolution

The Lead-users also assisted with referrals to research institutes or companies. One student summarized below the importance of Lead-users in providing access to additional resources.

'One of these Lead-users was a Medical academic and a physiology expert based in our university, who pointed us towards specific WSN technology, as a means of detecting heart arrhythmias. He arranged a meeting with AD Instruments, who provide software used by cardiologists when analysing results from ECG devices.' (A team member from SensIT Group)

Many of these resources would be difficult if not impossible for students without such Lead-user intervention. The example cited above allowed the students to meet with Lead-users in the field who provide real-world feedback on the viability of technology solutions.

The innovation teams also benefited from support by industry mentors, who are experienced technology entrepreneurs. One mentor emphasised the benefits of having a Lead-user forum.

'After a few meetings our mentor really wanted to have a Lead-user forum and he was adamant that this was the best thing we could do to make the project a success. He had used this method before for developing idea's and had proved to be most beneficial.' (A team member from Health Sense)

The lead forum concept was applied by the student teams in bringing together Lead-users with a range of expertise. The outcomes were similar to the mentor's comments in early stage evaluation of the viability of technology solutions.

Another mentor emphasised the importance of managing the expectations of the Lead-users.

'Our mentor's mantra was to "under promise and over deliver". Initially it was easy to be lured into promising our Lead-users many things to be developed into the project.' (A team member from Health Sense)

This was important information for the student teams, as many have little experience in working in real-world environments of building technology solutions. What may seem as a simple enhancement or modification to the project may turn out to be significant in the development of viable solutions.

One student expressed the team's frustration in realising that their original concept is invalid, which is similar to real-world experiences in technology innovation.

'It was also very tough to be positive after experiencing quite a few setbacks in the earlier stages, where numerous ideas for products failed, as our original idea was written off by Lead-users.' (A team member from Com4care Group)

The Lead-user involvement was also responsible for total changes in direction and even abandoning specific product ideas. The mentor provided valuable guidance on moving forward given these types of innovation scenarios.

Innovation Team Conclusions

The team experience was significant from a learning perspective. The journal entries reflected the frustration and excitement associated with technology innovation and building viable solutions. These types of experiences are difficult to simulate in a classroom setting.

The innovation team members found the Lead-user interview process initially intimidating.

'I had to organise and attend meetings with Lead-users in a number of hospitals and nursing homes throughout the city. It was quite daunting for me at the beginning as I never before had to meet and interview a total stranger previously; however, after the first few meetings the process became less intimidating and even an important skill I developed.' (A team member from Com-4care Group)

Two different team members emphasised both the broad applicability of the Lead-user method and also the systematic and methodical nature of the Lead-user method.

'I enjoyed learning the importance of researching a business idea. This included the Lead-user research method which I believe can be applied to most projects.' (A team member from SensIT Group)

'The experience of researching different aspects of innovation such as the Lead-user process, suggests that innovation can be created through a methodological process.' (A team member from Vital Solutions)

In addition to learning about developing solutions that are technology-driven, they gained skills and knowledge for example on: effective teamwork, interaction with a diverse Lead-user group, and dynamic problem-solving.

CURRENT CHALLENGES AND ANALYSES

Dr Nielsen and the Dr. Miyagi sat down to review their experiences of supervising three innovation teams. Nielsen's original assumptions that Wireless Sensor Networks (WSN) had a limited role in the Trauma (Accident and Emergency area) were explored. Nielsen found this difficult to clarify, but presumed that everyone in the '*WSN medical applications*' team assumed that the trauma area is well served by technology and that there was no need to propose new product offerings.

This was contradicted by the findings of the Vital team, *'When a patient comes into the Emergency Department, their vital signs are taken immediately and repeatedly (manually by a nurse / triage staff). A medical professional cannot tell how serious a patient's condition is until their vital signs are taken'* (Vital, Business Plan). Likewise a Professor of General Practice believed that, *'Intuitively, the ability for V-Sense™ to be wireless allows for a greater quality of care for the patient during their hospital stay.'* All indications are that a wireless vital signs device is ideal in the trauma emergency medicine area. Nielsen concluded that, *'You must be careful about assumptions, which must always be tested.'*

Myagi reminded Nielsen about his 'Total Body Area Network' concept and started scribbling on an index card. The three projects focused on applying one technology to a number of different medical areas, *'Why did this exercise generate different answers from the individual teams?'* Myagi asked Nielsen. *'Again this concept has an assumption that one adaptable technology Wireless Sensor Network technology will address multiple applications, but lets consider the project outcomes. Did the teams have any Lead-users in common?'* Hesitantly Nielsen replied, *'Well actually, no!'*

Myagi finished drawing and presented Nielsen with the index card (Figure 4) and the broad statement that, *'One technology, which is used in three respective medical areas, namely Cardiac Screening, Critical Care Vital Signs and Independent Living, with falls analysis, is used by three respective and independent groups of medical specialists.' 'Yes, that makes sense!'* responded Nielsen. *'What does this tell us about Medical device areas,'* asked Myagi in a probing manner. *'That the various medical areas are very segmented, like silos,'* replied Nielsen. *'Yes'*, responded Myagi, *'Unlike open source software development were you have large horizontal peer groups of Lead-users'* (Von Hippel, 2007).

Myagi hypothesised about the Total Body Area Concept and started scribbling on another index card. *'Technologists, by their nature, focus on what is possible, but find it hard to determine what is needed?'* (Stefik, 2004). Dr. Nielsen quietly took offence from this remark and declined to reply

Figure 4. WSN technology versus medical applications

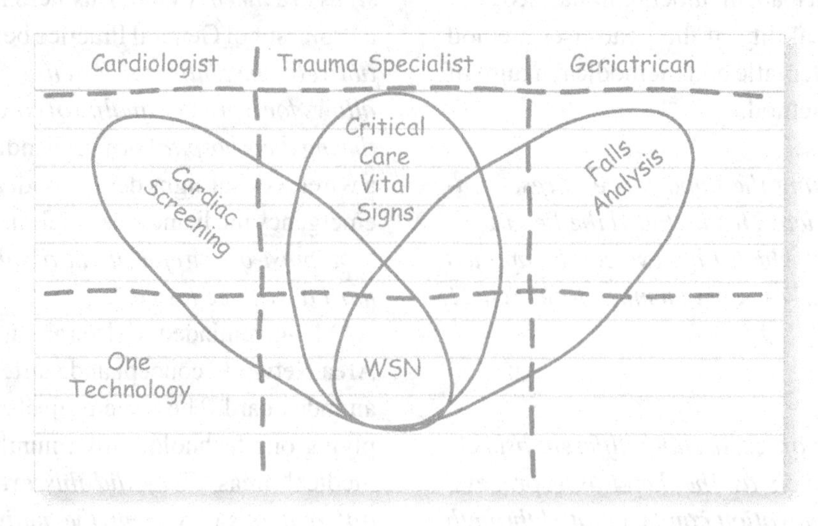

with the comment, *'Are you saying that we do not know what is needed?'*

Myagi muttered, *'Does one technical WSN solution fulfil all the broad non-invasive medical signs data for all the proposed medical applications?'* Continuing with the diagram, he asked a number of questions, *'What medical areas were addressed and their vital signs?'* Myagi finally asked Nielsen if he thought that the Innovation team exercise had covered all possible vital signs and medical applications. Nielsen replied, *'Definitely not!'*, and was presented with the second index card (Figure 5).

Nielsen studied the index card and was surprised at differences of each of the proposed systems and that no single wireless sensor network device would measure all the required vital signs across the specific applications areas. They joked about the ideal customer that would need the 'Total Body Area Network' system, *'a bed ridden geriatric athlete prone to bed-sores, who is in a coma.'*

In summarising, Myagi asked Nielsen, *'Can you see how some technologists can focus on a total technology solution, but taking needs analysis perspectives can produce different results?'*

LESSONS LEARNED

The critical learning outcome from the student's point of view is that it encouraged them to look for sources of innovation in their chosen domain area. In this case, it was medical applications of WSN. This involved sourcing and interviewing domain experts who can identify needs ahead of the general market. The cross case analysis of the application areas indicated that there is significant divergence of product ideas, when applications are sought in niche medical areas. This divergence is shown in Figure 3, which highlights the technology, the medical application area, and the appropriate Lead-experts that were approached as part of the studies.

In terms of the technology, not all groups found the WSN technology commercially viable. Com4care, the group that studied bed-sore management, selected a wired sensor solution for their prototype, as mobility was not a critical issue in a cost-sensitive market, such as, nursing homes, who found it hard to justify the additional expense. The other three areas, namely falls assessment, cardiac screening and critical care vital signs, all would benefit from a wireless solution

Figure 5. Cross case grid index card

Vital Signs Vs Medical Area	ECG	Blood Pressure	Other Medical Signs	Physical Orientation	Spectral Analysis	Pressure Sensing	Beat 2 Beat Blood Pressure
Cardio Screening	✓		....?....		✓		
Critical Care	✓	✓	....?....		✓		
Falls Analysis	✓		....?....	✓	✓		✓
Other Areas	....?....	....?....	....?....	....?....	....?....	....?....	
Bedsore Prevention			....			✓	

as patient mobility added flexibility to the care of patients.

An assumption underlying this study was that one-wireless sensor device could have multiple applications across a range of medical areas. This selection of medical areas clearly highlighted the insular nature of the Lead-experts within each area, the lack of communication across these areas, and their differing medical sensory requirements. Falls assessment is highly specialised requiring cardiac readings, blood pressure and spatial orientation data to support the assessment of the patient. Critical care and vital signs involved monitoring heart-rate, blood pressure, blood/ oxygen levels, body temperature and ECG. Sports cardiac screening involves heart rate and ECG monitoring, with customised Spectral Analysis software. This cross-case analysis therefore clearly highlights the niche and specialised nature of non-invasive sensory medicine and that no single product device would support generic medical needs. This exercise also highlights some interesting challenges facing the transition from wireless sensor network applications to actual medical devices.

Trialling the Lead-user method in an academic environment allows inexpensive experimentation, which would be difficult to justify in a business setting. This approach allows multiple teams to go through multiple iteration of the Lead-user process in a relatively short period of time. The counter argument to this point is that student teams find in difficult to get access to top expert Lead-users and do not have the resources to facilitate international meetings with world experts, but generally international experts are general with advice via email. The method is useful in teaching technology entrepreneurship, as it sensitises the students to alternative sources of innovation and encourages them to interact with domain experts in niche areas. The Lead Use method is not applied correctly as many of the experts interviewed are not in fact Lead-users, but are better described as Lead-experts. The process described by Von Hippel (1999), can be expensive to implement and the logistics of getting a group of experts together for a think tank forum is beyond the resources of most student groups.

REFERENCES

Department of Health & Children. (2006, March). *Reducing the risk: A strategic approach - the report of the Task Force on Sudden Cardiac Death*. Public Report, Department of Health & Children, Ireland.

Franke, N., von Hippel, E., & Schreier, M. (2006). Finding commercially attractive user innovations: A test of Lead-user theory. *Journal of Product Innovation Management*, 301–315. doi:10.1111/j.1540-5885.2006.00203.x

Hassan, S. T. G. (2008). Bringing Lead-user innovations to the market: Research and management implications. *SAM Advanced Management Journal*, 51-58.

Herstatt, C., & Von Hippel, E. (1992). From experience: Developing new product concepts via the Lead-user method: a case study in a 'low tech' field. *Journal of Product Innovation Management*, *9*(3), 213–221. doi:10.1016/0737-6782(92)90031-7

Kaiser, S., & Müller-Seitz, G. (2008). Leveraging Lead-user knowledge in software development—the case of Weblog technology. *Industry and Innovation*, 199–221. doi:10.1080/13662710801954542

Lettl, C. H. (2006). Users' contributions to radical innovation: evidence from four cases in the field of medical equipment technology. *R & D Management*, 251–272. doi:10.1111/j.1467-9310.2006.00431.x

Lorincz, K., Malan, D., Fulford-Jones, T., Nawoj, A., Clavel, A., & Shnayder, V. (2004). *Sensor networks for emergency response: Challenges and opportunities. IEEE Pervasive Computing*. Oct/Dec.

Lüthje, C., & Herstatt, C. (2004). The Lead-user method: Theoretical-empirical foundation and practical implementation. *R & D Management*, *34*(5), 549–564.

Morrison, P. D., Roberts, J. H., & Midgley, D. F. (2004). The nature of Lead-users and measurement of leading edge status. *Research Policy*, *33*(2), 351–362. doi:10.1016/j.respol.2003.09.007

O'Flynn, B., Angove, P., Barton, J., Gonzalez, A., O'Donoghue, J., & Herbert, J. (2006). Wireless bio-monitor for ambient assisted living. In *Proceedings International Conference on Signals and Electronic Systems*, (ICSES'06), Lodz, Poland (pp. 257-260).

Schreier, M., Oberhauser, S., & Prügl, R. (2007). Lead-users and the adoption and diffusion of new products: Insights from two extreme sports communities. *Marketing Letters*, 15–30. doi:10.1007/s11002-006-9009-3

Stefik, M., & Stefik, B. (2004). *Breakthrough: Stories and Strategies of Radical Innovation*. Cambridge, MA: MIT Press.

Thiemjarus, S., Lo, B. P. L., & Yang, G.-Z. (2005). Body sensor network – a wireless sensor platform for pervasive healthcare monitoring, *Adjunct Proceedings of the 3rd International Conference on Pervasive Computing*, (PERVASIVE 2005) (pp. 77-80).

Thomke, S., & von Hippel, E. (2002). Customers as innovators: A new way to create value. *Harvard Business Review*, *80*(4), 74–81.

Von Hippel, E. (1988). *The Sources of Innovation*. Oxford: Oxford University Press.

Von Hippel, E. (2005). *Democratizing Innovation*. Cambridge, MA: MIT Press.

Von Hippel, E. (2007). Horizontal innovation networks - by and for users. *Industrial and Corporate Change*, 293–315. doi:10.1093/icc/dtm005

Von Hippel, E. (2008). Users as sources of invention. In Hall, B. H., & Rosenberg, N. (Eds.), *Handbook of Economics of Technological Change*. New York: Elsevier B.V. Press.

von Hippel, E., Thomke, S., & Sonnack, M. (1999). Creating breakthroughs at 3M. *Harvard Business Review*, *77*(5), 47–57.

Yin, R. K. (1994). *Case Study Research: Design and Methods* (2nd ed.). Thousands Oaks, CA: Sage Publishing.

Chapter 7
Paradigms, Science, and Technology:
The Case of E-Customs

Roman Boutellier
ETH Zurich, Switzerland

Mareike Heinzen
ETH Zurich, Switzerland

Marta Raus
ETH Zurich, Switzerland

ABSTRACT

This chapter explores the concept of paradigms, science, and technology in the context of information technology (IT). Therefore, the linear model of Francis Bacon and Thomas Kuhn's notion of scientific paradigms are reviewed. This review reveals that the linear model has to be advanced, and supports the adoption of Kuhnian ideas from science to technology. As IT paradigms transform business processes, a five-level concept is introduced for deriving managerial implications and guidelines. Within the case of e-customs, a European-funded project tries to ease border security and control by adopting a common standardized e-customs solution across the public sector in Europe. The rise of the IT paradigm within customs and its effect on business operations will be explained. This chapter contributes to the research in diffusion and adoption of innovation using science progress and the interplay of science and technology as dominant concepts.

"Give a little boy a hammer and everything looks like a nail." Abraham Harold Maslow (1908-1970), American psychologist

DOI: 10.4018/978-1-61520-609-4.ch007

INTRODUCTION

"Give a little boy a hammer and everything looks like a nail." The quote of Abraham Harold Maslow (1908-1970), an American psychologist, guides the

reader through each section. In analogy to the case of electronic-customs (e-customs), which will be presented later, the quote could be interpreted as giving Internet technology to a customs office, while that office's export processes remain paper-based.

The boy instinctively relates what he already knows about the hammer to new problems and challenges. Only if major problems cannot be solved by this method and more problems accumulate, we begin to ask ourselves if there are other ways of looking at the situation. In Thomas Kuhn's words, this is the heart of the scientific progress.

Thomas Kuhn's notion of scientific paradigms (Kuhn, 1962) that have been early adopted as technological paradigms (Dosi, 1982), can be described as the punctuated nature of technological change combined with path dependency of innovation processes in times of incremental progress (Peine, 2008). When analyzing the quote, the path dependency of the innovation process shows that the boy will use the hammer very efficiently for nails, whereas in times of radical technological change, he will scrutinize the one-sided use of the hammer. A paradigm sets out an array of expected solutions to accepted problems (Lakatos & Musgrave, 1970).

The description of an advanced model of Francis Bacon (Bacon, 1605) supports the transferability from science to technological paradigms, which are exemplified with IT paradigms. Examples are software and programming developments, as well as the technical progress from mainframe to personal computers, are currently finalized in a technological paradigm of virtual resources accessible through the Internet.

Internet technology is the underlying technology of e-customs. Theories are applied to an e-customs project called Information Technology for Adoption and Intelligent Design for e-Government (ITAIDE), which has been launched for the adoption of standardized e-customs systems on an international level. A value assessment and a five level business transformation plan accompany the description of technical progress in IT.

SCIENCE AND TECHNOLOGY

While science is considered a knowledge base, technology is defined as the physical manifestation of that knowledge (Khalil, 1999). Some definitions of science and technology are:

- Science analyzes and looks for explanations in general terms. The benchmark is still the natural law and the periodic system for classification. Explanation is a natural law applied to specific boundary conditions. Technology is very much problem solving oriented, where problems come from practical considerations.
- Science lives on open questions – surprises. Every surprise is a potential candidate for a new research field and new results. Within a technology, side effects are objectionable; everything has to be controlled to avoid accidents and non-performance.
- Technology must have a goal to fix or build something, a reliable and sturdy bridge for example. The goal in science usually comes from within science. The goals of technology are defined by the needs of customers– short term solutions are better than solutions for the long term.

Indeed, science and technology are much closer today than in earlier times. A good example is computer technology. It developed in a logical path from Leibniz to Microsoft, yet only in hindsight. Another example is the X-ray technology invented by Konrad Röntgen in 1895.

Within one year after the detection of X-rays, more than 49 scientific books were on the market, including more than 1,000 scientific papers. X-ray technology spread with a speed that is amazing even for 21st century standards. Just two months

Figure 1. Linear model of Francis Bacon (1561 – 1626): 17th century and the scientists' optimism (Bacon, 1605)

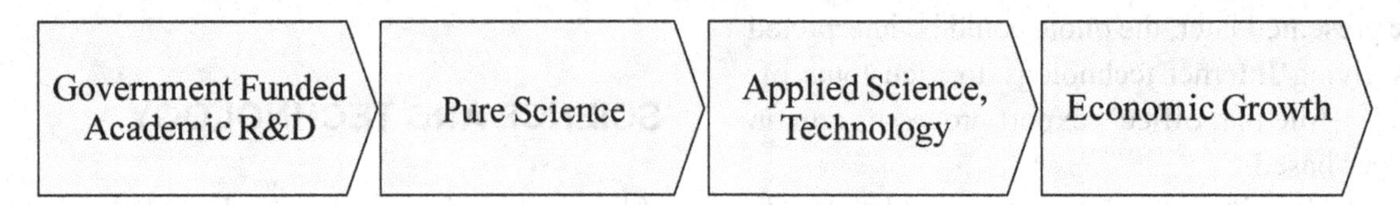

after invention, it was used for forensic evidence in an American court. Thus, fact that both science and technology developed parallel to each other, is evident within a historical perspective.

Linear Model

In ancient times, technology was simply a tool to survive. In Galileo's times, around 1650, this changed. Telescope and microscope helped to improve our scientific understanding of the world. Technology became more and more a tool of improving conditions of life.

In order to explain the interaction between science and technology, a simple model was created in the 17th century: The linear model of Francis Bacon (Figure 1). Basic science leads to applied science which leads to technology, and with that to growth and welfare (Bacon, 1605).

Until the late 19th century, technology and science developed in parallel, but well separated from each other. There is some evidence that technology fertilized science more than the other way around. Galileo did not understand the optics of his telescope but used it to prove his theories. Stevenson, the inventor of the locomotive, was illiterate and his first machines contradicted accepted thermodynamic theories of his time. The middle of the 19th century brought the turning of the tide: Justus Liebig applied science to improve fertilizers and storage of meat. Scientific technology began its victorious campaign.

Today, most people are strong believers of Francis Bacon's linear model: Because in current times, we try to explain everything with as much science as possible. For example, even in management mathematical models predominate with all the risks involved: Black-Sholes equations are the fundament of all structured financial products.

Advanced Linear Model

While the linear model has useful explanatory value, its output, discovery, and innovation are much less deterministic. As history shows, most innovations derived from small improvements in technology and were surprisingly random products. Sometimes innovation developed through the application of unspecified research results, like the ABS, airbag, calculator, and telephone. As (Bygrave, 1989) stated, "We should avoid reductionism in entrepreneurship research. Instead, we should look at the whole."(p.20).

An alternative model to consider is the interdependence of science and technology which includes technology, basic and applied research, innovation, and wealth (Figure 2). Not only basic research, but applied research as well can lead to technology and vice versa, and thus, to economic wealth. Innovation plays a big role too, as it is the outcome of science and technology and the major driver for economic growth and important for the future wealth of nations (Easterly, 2001).

Technology is certainly one of the most important drivers in today's economy. Scientific results push frontiers forward, engineers improve existing technologies and introduce new ones, and society is as receptive for new technologies as never before. In the middle of the 19th century, technology became more and more driven by science. Hence, today's society is very much driven by technology - progress has become technical progress. Neill

Figure 2. The multi-dimensional model of scientific progress: We have to look at the whole (Bacon, 1605 modified)

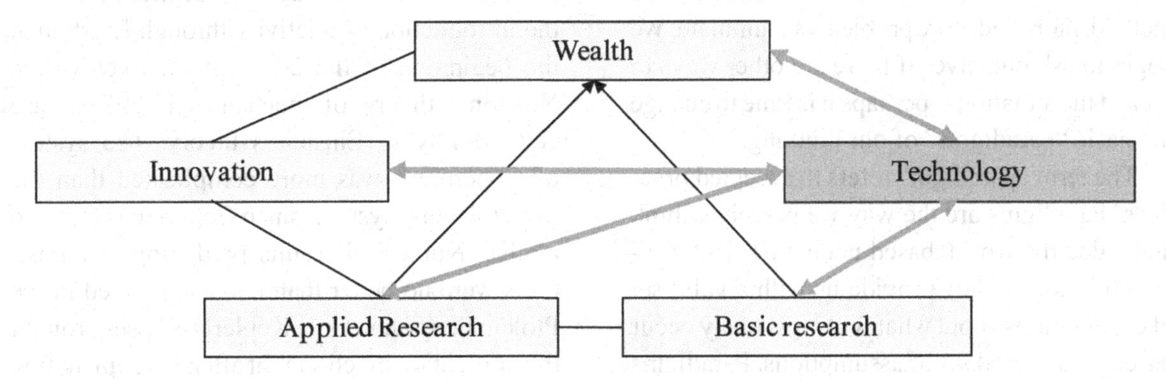

Figure 3. Mario Botta: The chapel of St. Mary of the Angel, Monte Tamaro

Postman, an American scholar of communication calls it "Technopoly", the submission of all forms of cultural life to the sovereignty of technique and technology (Postman, 1992).

That science can lead to technology is clear, but the reverse is as true as well: Without technology, no modern science could exist. Just think about CERN, the European organization for Nuclear Research, in Geneva or medical imaging technology like Computed Tomography Angiography (CTA) and Magnetic Resonance Imaging (MRI). Science tries to explain existing objects, technology and engineering want to create new objects, artifacts, and objects created by human beings.

The Sunniberg Bridge near Klosters, Switzerland, could be built only by an experienced engineer inspired by strong aesthetical thinking. The engineer behind it, Christian Menn from ETH Zurich, has succeeded in harmonizing technology, material functionality, environment, and usage. The bridge is a masterpiece of simplicity and functionality, just like Mario Botta's church on Monte Tamaro/Ticino (Figure 3). Mario Botta used only two local materials. An integral innovation; however social choices are driven by human values, aesthetics, which influence science and technology.

The fact that science, technology, and values are highly interwoven is the basis for the following section about scientific paradigms, with its underlying thesis that scientific problems influence technological problems and vice versa.

Such multi-dimensional models have greater explanatory value in the development and application of information technologies, which is illustrated in the case study.

THEORY OF PARADIGMS

One of the ways we solve problems is by instinctively applying what we already know to new problems and challenges. However, if over time major problems in a field cannot be solved by this method, more and more problems accumulate. We begin to ask ourselves if there are other ways to look at the situation—perhaps it is time to change the basic "paradigms" of our thinking.

The term "paradigm" refers to accepted practices. Paradigms are the way we perceive, think and value the world, based upon a particular vision of reality. They provide us with a valid set of expectations about what will most likely occur based on a shared set of assumptions. Paradigms establish boundaries and define how to succeed within the boundaries. When we are in the middle of a paradigm, it is difficult to imagine any other paradigm. Conditions are ripe for a "paradigm shift" when a sufficient number of people agree that the old ways no longer solve important problems and that new ways are needed.

Of course, not everyone in a field does change because a new paradigm comes on the scene. Some people are so committed to particular paradigms that they will never change. It is not always easy to know when to change and when to step around those who do not. According to Max Planck (1949), "A new scientific truth does not triumph by convincing its opponents and making them see the light, but rather because its opponents eventually die, and a new generation grows up that is familiar with it" (p.596).

Thomas Samuel Kuhn, a historian of science, gave the word "paradigm" its contemporary meaning when he adopted it to refer to the set of practices that define a scientific discipline during a particular period of time. According to Kuhn (1962), a scientific paradigm is as fundamental belief which is signified by a consensus on problems and methods within a field of research. From time to time, it is followed by systemic innovations, what Kuhn termed "paradigm shifts". Then, the period of the old paradigm is over and a new one begins.

Kuhn's model is one of the most often cited philosophical results of the 20th century. Historic examples are the struggle between the Ptolemaic systems and Copernicus' heliocentric system or the introduction of relativity through Einstein at the beginning of the 20th century. Even today, Newton's theory of mechanics is still in use concurrently to Einstein's theory. The system of Copernicus was more complicated than the old Ptolemaic system, since Copernicus stuck to circles. Numerical results predicting star positions were no better than results obtained in the Ptolemaic system. Only Kepler`s ellipses brought the unification much sought after. An explanation in the sense of a law was given later by Newton. However, Newton had to introduce the notion of gravitational force working at infinite speed without any intermediaries. A notion perceived as an unacceptable setback by many contemporary scientists (Koestler, 1989).

Kuhn confirmed the notion that science develops in a non-linear pattern and not within a linear model. Each generation solves its own problems - not because the problems have changed, but because the knowledge and mindsets of each generation evolve after each scientific revolution (Lakatos & Musgrave, 1970).

Therefore, science and technology are highly interacting, and thus, leading to innovation. Several authors adopted Kuhn's notion of scientific paradigms as technological paradigms (Dosi, 1982; Granberg & Stankiewicz, 1981; Johnston, 1972), using Kuhnian ideas describing technological progress. But also profound progress in medicine, transportation, the physical sciences, labor laws, social attitudes, and the way we work are taking place within a continuous strain, broken up by disruptive innovations from time to time (Christensen, 1997).

An example of management trend triggered by earlier paradigm shifts includes the Total Quality movement that began in Japan in the late 1950s, spreading to manufacturing all over the world. A technological paradigm shift was Sony's invention of the Walkman in the field of personal entertainment. The Walkman was the direct predecessor of the concept of mobile communications embodied by today's cellular telephones.

As Kuhn pointed out, changes in paradigms have a predictable pattern. Understanding this pattern improves our ability to use it for problem solving. Those who recognize that a paradigm shift is taking place will benefit from consequences of the shift. Many become leaders in their industry or field. For example, Albert Einstein was a youthful beginner in the field of science when he first defined his theory of relativity. The founders of FedEx and of Apple Computer were also new to their fields when they conceived new technical and business paradigms.

Applying Abraham Maslow's quote again, the boy who will not use only the hammer when he sees nails will create a new paradigm.

Thomas Kuhn's Scientific Paradigms

Diving deeper into Thomas Samuel Kuhn's work, he differentiates between 'normal science' and 'scientific revolutions' (Kuhn, 1962). Normal science is a process of puzzle-solving, where scientists work with proven methods to achieve expected solutions on accepted problems. The puzzles within these problems are perceived as scientifically interesting and attract an enduring group of adherents from competing modes of scientific activity. They also have to be open-ended leaving problems for their group of practitioners (and their students) to resolve. For example, in the Ptolemaic astronomical model, the earth was at the center of the universe. This scientific model gave unsatisfactory explanations to the retrograde motion of the outer planets until the Middle Ages, when Copernicus presented a heliocentric framework to explain planetary movement.

Students study these paradigms in order to become members of the specific scientific community in which they will later practice. Because students largely learn from and are mentored by researchers who learned the bases of their fields from the same models, there is seldom disagreement over fundamentals. Scientists, whose research is based on shared paradigms, are committed to the same basic rules and standards for their scientific practice. This way, the paradigm implicitly drives how scientists solve problems. Scientists automatically prefer problems that can be solved with these rules leading to a repetitive application of methods, tools, and routines through which scientists become efficient. The more positive results we get, the higher the acceptance of this theory – a self-reinforcing circle. This is Kuhn's second definition of a paradigm (Kuhn & Neurath, 1970), where paradigms embody tacit norms and rules. It even refers to the priority of paradigms: "Paradigms may be prior to, more binding and more complete than any set of rules for research that could be unequivocally abstracted from them" (p. 24).

According to Kuhn, scientists will apply the current paradigms' methods until anomalies or surprising discoveries emerge – the beginning of 'scientific revolutions' (Kuhn, 1962). The emergence of a new theory is generated by the persistent failure to solve puzzles within normal science. Failure of existing approaches is the prelude to a search for new ones. In early stages of a new paradigm, many alternatives are explored. Once a paradigm is entrenched (and the tools of the paradigm prove useful to solve the problems the paradigm defines), alternative theories are consolidated. Crises provide the opportunity to retool. Like one of the most intriguing aspects in Einstein's theory the Clock Paradox, this started a big discussion among scientists.

Technology Adoption seen as Paradigm Change

Kuhn's scientific paradigm has been applied to technological progress (Dosi, 1982; Granberg & Stankiewicz, 1981; Johnston, 1972). Johnston ascribed the punctuated nature of technological change to the internal structure of technology, meaning a "set guiding principles accepted by

Figure 4. Kuhn's scientific theory adapted to technological paradigms (based on Kuhn 1970)

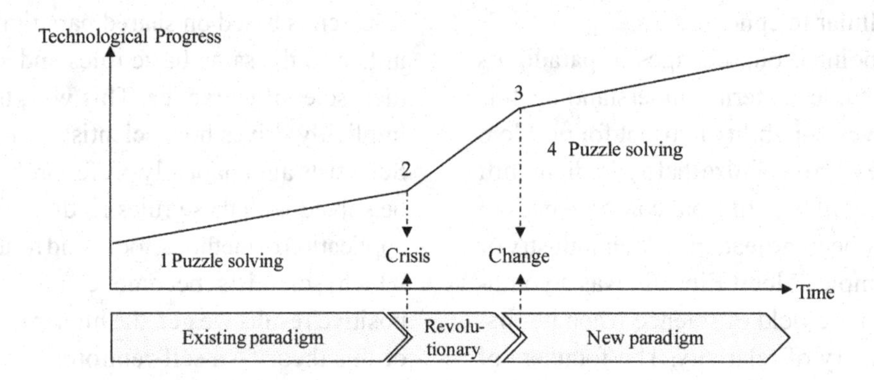

practitioners in a particular field of technology" (p. 122). With his definition, he focused especially on the paradigm shift. In contrast to Johnston, Granberg and Stankiewicz elaborated the activities of the community of technologists, such as technological research and functional analysis, in more detail (p. 216). Also, Dosi described that phases of incremental technological change are alternating with punctuated changes of radical shifts.

"In broad analogy with the Kuhnian definition of a 'scientific paradigm', we shall define a 'technological paradigm' as 'model' and a 'pattern' of solutions of selected technological problems, based on selected principles derived from natural sciences and on selected material technologies" (Dosi, 1982, p. 152). Further, Dosi defined two drivers of technological progress: 1) the material part of a technology, which is the exemplary artifact representing the function of a paradigm; and 2) the disembodied part of a technology containing expertise, experiences and practical knowledge.

Peine (2008) described the adoption of the Kuhnian framework of scientific paradigms to technological change as still underexposed, and Vincenti (1995) highlighted the instructive impact of Kuhn's work in understanding the "technical shape of technology."

Thus, this research applies Kuhn's idea of scientific paradigms to technological progress that moves also in revolutionary steps and long eras of small improvements. Both scientific and technological progress needs four steps in three phases (Figure 4) for the birth of a new paradigm: (1) puzzle solving within an existing paradigm, (2) anomalies, leading to disputes among scientists, (3) change of paradigm and change of culture, and (4) puzzle solving within a new accepted theory.

However, it is important that material and the disembodied part of a technology are considered separately (Dosi, 1982). Staying with the example of the Sony Walkman, the material technology of personalized entertainment had a different pace of progress than the expertise, experience and practical knowledge developing that technology.

Technological and Computing Paradigms

A big technological revolution was the development from analog to digital systems. At some point, analog signals were too limited in bandwidth, and hence, a better system had to be developed. The big advantages of a digital system are that operations can be conducted more frequently and are less error-prone, and therefore, copies can be made indefinitely. Without digital systems, computing paradigms would have developed differently.

A computing paradigm emerged due to modularization in software development. Until

1980, procedural programming was in place. After that time, three industry layers evolved: 1) a generic layer that is responsible for generic software components, 2) an application layer that is responsible for special industry applications, and 3) a customer layer that is responsible for customer requests. Such layer modules are applied by companies like SAP, Oracle, Sony and Ericsson. Further, procedural programming was redeemed by object-oriented programming, using "multiple inheritances" within the concept of classes and instances – a new programming paradigm. Object-oriented programming developed as the dominant programming methodology during the mid-1990s, largely due to the influence of C++.

Another big information technological progress has been reached through the development of personal computers. Before the introduction of the microprocessor in the early 1970s, computers were generally large, costly systems owned by large corporations, universities, government agencies, and similar-sized institutions. Until that time, none in the information technological community thought computer systems could be of interest for individuals. In 1977, Apple Computers introduced the Apple II as the world's first PC. From that time on, computers were developed for household use. A new paradigm had evolved, convincing the community of computer scientists that computers would be demanded by individuals. Since that time, computer scientists have kept working on processor performance, speed and miniaturization, resulting in ever smaller, highly productive PCs including laptops, net books, table PCs and Pocket PCs.

Another paradigm in IT-Management goes a little bit further: Cloud computing offers the customer virtualized resources provided as a service over the Internet. Cloud computing customers do not generally own the physical infrastructure. Instead, they avoid capital expenditure by renting usage from a third-party provider. They consume resources as a service and pay only for resources that they use. Some of the vendors providing cloud computing services are Google, Amazon, Microsoft and Yahoo. However, opinions of specialists about cloud computing differ. Some call it the biggest revolution since the Internet, others criticize cloud computing as being a regression in limiting both freedom and creativity of the user. Thus, cloud computing cannot be called a paradigm yet – the revolution is still up in the air.

The Paradigm Effect on Business Transformation: Managerial Guidelines

Because most business processes today are IT related, computing paradigms have had and will continue to have an impact on business transformation. IT has become a fundamental enabler in creating and maintaining flexible business networks. Thus, understanding these paradigms, together with the knowledge of Kuhnian non-linear concepts, provides guidance to managers.

Venkatraman (1994) breaks the IT-enabled business transformation into five levels with two dimensions: the range of IT's potential benefits and the degree of organizational transformation (Figure 5). The central underlying hypothesis is that benefits of IT progress are lower if organizational strategies, structure processes, and culture are not adapted to IT development.

The first level is called "Localized Exploitation". In this stage, managers often respond to an operational problem with a localized isolated system; for example, a customer order-entry system or a toll-free customer service system. Learning among managers is minimal and thus limits potential benefits. Competitors can easily imitate these standard technical applications.

The second level, "Integral Integration", is characterized by two types of integration: 1) technical interconnectivity through a common IT platform; and 2) business process interdependence across different functions within the organization. Venkatraman (1994) observed that managers allocate their interest more to techni-

Figure 5. Five Levels of IT-enabled Business Transformation (Venkatraman, 1994, p. 74)

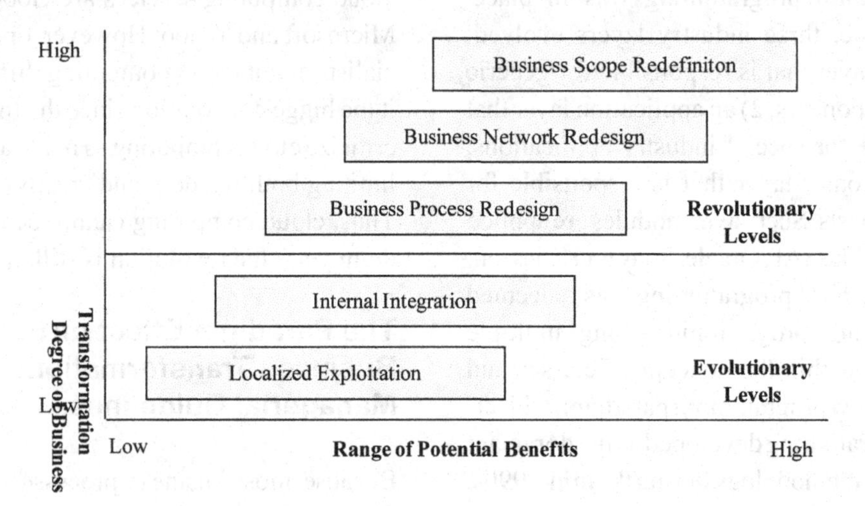

cal interconnectivity than to business process interdependence.

The first two levels are called "evolutionary" because only small changes to the business processes are required; whereas, the subsequent three higher levels are "revolutionary", since they require fundamental changes in organizational routines. The higher the level of transformation, the greater the benefits of organizational routine changes.

With "Business Process Redesign", IT is fully integrated in the organization. However, benefits of IT functionality cannot be fully realized because general concepts of business processes are still valid, such as centralization vs. decentralization, functional specialization and administrative mechanisms.

The three lower levels focus only on a single organization, whereas the next higher level, "Business Network Redesign", expands interconnections to include external businesses, such as suppliers, buyers and other intermediaries with regard to IT deployment. Potential benefits, besides administrative and operational efficiency, are leveraging competencies due to IT-enabled strategic alliances and enhanced learning.

Within the top level "Business Scope Redefinition", IT deployment enables or facilitates the repositioning of a firms' business scope. Strategy concepts that led to increased emphasis on vertical integration would be replaced by newer concepts, such as joint ventures or virtual business networks.

Thus, an organization should first identify the transformational level that is appropriate for its situation. This evaluation will depend on managers' perception as to whether IT capabilities are seen as an opportunity or a threat to the status quo. It can be helpful to to assess where other leading companies are positioned in order to create organizational awareness of limitations and to get commitment.

In summary, we have shown that the linear model of Francis Bacon has to be enlarged; and also that Kuhn's notion of scientific paradigms, with its emphasis on progress in steps, can be applied not only on science, but also to technology and management. After presenting scientific and technological examples for theory clarification, we focused on computing paradigms. As IT paradigms enable business transformations, we introduced a five level framework for deriving managerial guidelines and implications. This theoretical background will be used to get a better understanding of the Case of e-customs.

THE CASE OF E-CUSTOMS

One of the interesting effects of information technology is that it changes organizational paradigms. As problems with an existing paradigm surface, technological solutions may be applied, resulting in a paradigm shift. This can be illustrated by a case concerning border control, where the old paradigm of government customs officers processing paperwork, as a way of ensuring border control and security, has been shown to be inadequate. With the application of information technologies under the auspices of e-customs, a new paradigm in border security has emerged. The technology for e-customs exists, using the Internet as a service-tool. However, business processes have not been aligned yet to the new technology available in Europe and worldwide.

Therefore, the EU currently funds the project called the Information Technology for Adoption and Intelligent Design (ITAIDE, IST-02789) involving stakeholders from academy, industry, and governmental institutions to support European member states with e-Custom solutions. Within that project, one initiative of the European Commission to create a simple and paperless environment for customs and trade is the Multi-Annual Strategic Plan (MASP) setting the framework for the following case. The main goal of MASP is to share vision, objectives, strategic framework, and milestones for an e-customs implementation.

The Advanced Linear Model Within E-Customs

The boundaries of science and technology are hard to establish, as many computer scientists find their research in algorithms or programs immediately applicable as a technology. This is the case in e-customs, where progresses in scientific and technology interact with one another. Scientific progress started with the development and diffusion of information and communication technology (ICT). The invention of the Internet revolutionized ICT and enabled electronic business concepts, where individuals or groups do not deal face to face, but conduct business remotely, regardless of location and time.

In addition to traditional business-to-business (B2B) and business-to-consumer (B2C) technologies, business-to-government (B2G) technologies recently have incorporated transactions from public sectors. According to Wassenaar (2000), e-Government describes an information exchange which can be a transaction or a contract between companies, governments, customers, suppliers or other partners. E-customs is an innovation within e-Government and focuses mainly on border control and security. Its underlying technology is ICT using electronic data networks, such as provided by the Internet. With the diffusion and adoption of a common standardized e-customs solution across Europe, EU hopes to mainly reduce administrative burden and strengthen the European economy. As the application of ICT in customs transforms business processes, a new field of research about organizational change has evolved. This is demonstrated by the vast research literature put forth by authors, such as, Baida, Liu, & Tan (2007), Baida, Rukanova, Wigand, & Tan (2007); Bjørn-Andersen, Razmerita, & Henriksen (2007), Boyd, Hobbs, & Kerr (2003), Henriksen & Rukanova (2008), Henriksen, Rukanova, & Tan (2008), Kuiper (2007), and Raus, Kipp, & Boutellier (2008).

As scientific results push frontiers forward, engineers continue to improve existing technologies and introduce new ones. Society is unprecedently receptive to the introduction of new technologies. As such, the project of ITAIDE is a perfect example of Francis Bacon's advanced model (Figure 6), as explained in the following paragraph.

In modern research, engineers and industry are both involved in research projects from the beginning. This is the case with the ITAIDE project. These types of research projects, funded by the EU, may be considered as partnerships between research universities and industries. Pure Science,

Figure 6. The e-customs EU-funded project ITAIDE is a good example of an advanced model of Francis Bacon

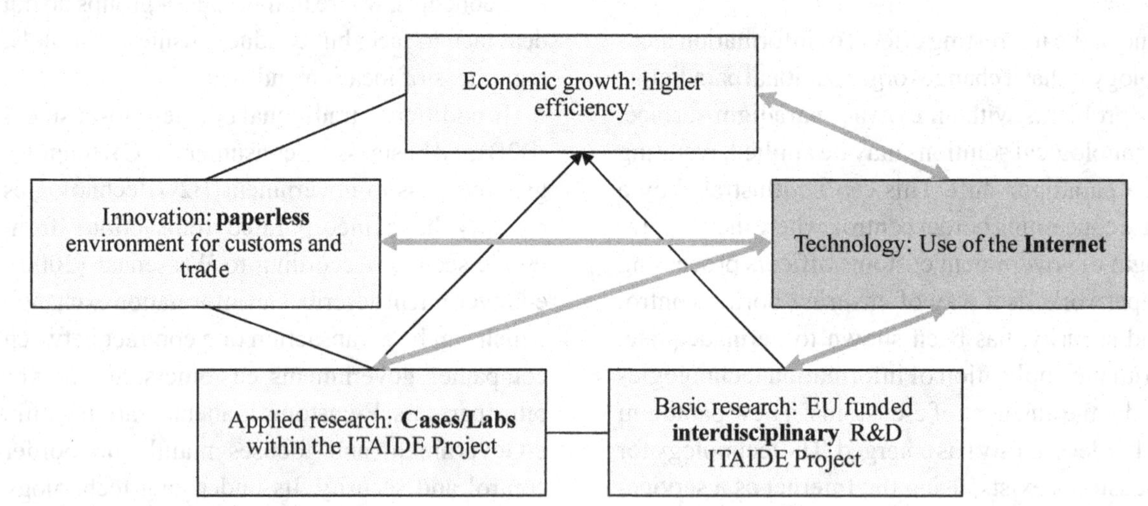

applied science and technology go hand in hand, working in workshops, projects and at universities together. This is unlike Bacon's linear model where one typically follows the other. Even though science and technology have different objectives and methods, they can enrich each other, which leads to innovation and thus, to economic growth.

An intensive integration of distinct research disciplines: social sciences, information technology, and political science lead to balanced solutions, higher efficiency and hopefully wealth in the future.

The Technological Paradigm of E-Customs

As basis for studying the application of the theory of technological paradigms to the case of e-customs, this section will focus on the export of goods from EU countries to non-EU countries. This is a pilot project within ITAIDE concentrating on a food industry company, exporting from Denmark to Russia (ITAIDE 2009). Important in this regard is that Denmark already has a national e-customs system in use, which in comparison is different from Germany or Italy whereby ERP-systems such as Atlas or Aida are used.

The following sections will describe the adoption of e-customs as a technological paradigm by means of Kuhn's four steps: 1) the old paradigm of customs, 2) issues, questions and problems with the old paradigm, 3) technological opportunities, and 4) the new paradigm.

The Old Paradigm of Customs

E-customs focuses on international trade, where "International trade is characterized not only by the physical movement of goods across national boundaries but by voluminous paperwork that captures information pertinent to identification, delivery, and government control of transported goods"(Teo, Tan, & Wei, 1997). Although today's information technology offers modern e-Solutions, e-customs management is still associated with administrative processes, filling documents, multiple data entry on paper, and electronically processed forms. It is not seen as innovative as product design or manufacturing automation (Raus, Flügge, & Boutellier, 2008). The customs authorities still have a bureaucratic, administrative, and old-fashioned image – a paradigm that sets the rules of the game: At present, EU customs offices communicate by telephone, emails, and fax, or by periodic meetings. Data, despite being virtually identical, are processed via separate IT systems. Personnel still feel they are more efficient with their old tools in comparison to applying and learning new ones. Metaphorically speaking, *they still use the hammer for fixing the nail.*

The case of the exporting food company in Denmark confirms the old paradigm described above. The export procedure from Denmark to Russia is currently paper based. This has been examined by an as-is analysis with the help of interviews and a value modeling tool, where all communication, transportation and transaction paths are documented. Throughout the export process, several value exchanges have to be arranged by the Danish food company, such as getting money from the importing Russian company, paying the logistics or shipping costs, and obtaining various certificates for export to Russia. For these value exchanges, various documents are required. An example is the GOST certificate, which is an acronym for state standards of Russia and is obligatory for export to Russia and has to be submitted in original form along with 20 to 30 copies. It takes at least two months to get a GOST certificate and costs about 2000 Euro for the exporter. Another example is the packaging list, which is a Microsoft Excel spreadsheet created manually. More than 12 other documents are required, most of them paper-based, with mandatory stamps and signatures.

Obviously, the technological paradigm of a common Internet-based customs solution is not sufficiently interesting to attract market entry by the customs community in the competing paper-based customs process. Therefore, EU has set up an EU-Directive that makes the implementation of a standardized common e-customs system mandatory by 2013. In spite of this directive, the old paradigm is still ruling. Kuhn (1970) said that paradigms can be more binding than any rules.

Although information and Internet technologies are developing rapidly, basic e-customs solutions are currently available, and Internet usage is growing worldwide, there is an international resistance to the adoption of e-customs. Kuhn's explanation to such resistance is the loss of efficiency by implementing new technologies.

Issues, Questions and Problems With the Old Paradigm

The first big change in thinking about the traditional paper-based border control and security came with the September 11th terrorist attacks in the United States as well as subsequent attacks in other places. These events led to higher legal standards and safety regulations in trade and customs. With this crisis, the necessity of new global standards and transparency evolved. As Internet technologies in general reduce corruption through transparency and harmonization of rules, e-customs received a new wave of attention: The perception became that the globalized world trade is to be treated with corresponding techniques – IT solutions. Metaphorically speaking, *nails changed and they cannot be treated with a hammer anymore. The old paradigm runs into difficulties.*

That political developments can have a big influence on scientific or technological developments has been stressed by Kuhn (1970) in the section entitled, "The nature and necessity of scientific revolutions" where he states, "This genetic aspect of the parallel between political and scientific developments should no longer be open to doubt." (p. 93) and also, "One aspect of parallelism must already be apparent … that existing institutions have ceased adequately to meet the problems posed by an environment they have in part created." (p. 92).

After the September 11th terrorist attacks, the US customs wanted to increase safety regulations by requiring one common and standardized European e-customs system. Then, the US would have a universal European customs partner instead of several national trading partners with different systems.

A big issue within the Danish Russian export example is the co-existence of trade infrastructure, which makes the different actors dependent on one another. The reason is that data sets are not standardized, multiple authorities have to be approached for one and the same commercial transaction, and economic zones have different certification programs.

In the European Union, national e-customs declaration systems are already used in many countries (e.g., Atlas in Germany, Aida in Italy or Sagitta Entry in the Netherlands). However, the European member states are self-reliant on how to adopt an individual e-customs system, hampering trade and transactions on the European market when compared to American, Asian or Pacific businesses.

Technological Opportunities

The technological opportunities for an e-customs system have been in existence since the Internet started a vigorous campaign that lead to e-Business solutions. This is the first dimension in our transformation model, shown in Figure 5. As previously mentioned, some EU members are taking advantage of these opportunities. Although countries, such as Denmark, already use e-customs, a common standardized electronic customs system is still missing. For this reason, the EU is funding the project ITAIDE for trade facilitation as well as securing import and export by the use of an e-customs system to be implemented no later than 2013.

Two e-customs concepts of the ITAIDE project will simplify the adoption of a new standardized e-customs system in Europe: 1) the Single Window Access (SWA) and 2) the Authorized Economic Operator (AEO). Both topics are addressed by the EU initiative to reduce the administrative burden of trade transactions and increase security and control mechanisms. A single window is "a facility that allows parties involved in trade and transport to lodge standardized information and documents with a single entry point to fulfill all import, export, and transit-related regulatory requirements. If information is electronic, then individual data elements should only be submitted once" (Dedrick & West, 2003; United Nations Economic Commission for Europe, 2005). This common solution would harmonize differences in both systems and regulations. In addition, the automation would help to accelerate the export execution process and decrease entry errors supporting both government and business companies in a faster, more efficient collaboration. With the AEO status, reliable operators can be certified, including those that are also compliant in respect of security and safety standards, and thus, can be considered secure traders.

A redesign of the Danish-Russian export example would incorporate the idea of a common European customs coordinator, which is implemented as a web service. The scenario builds on an elimination of papers in the customs process and harmonization of European customs processes through the introduction of a common data model (ITAIDE, 2009).

Several authors have summarized facilitators for the diffusion of e-customs solutions (Henriksen & Rukanova, 2008; Raus, Flügge, & Boutellier, 2009), which will support the paradigm shift:

1. E-customs leads to time and financial savings, as well as higher accuracy in data processing. Time savings have been realized due to executing procedures faster and diminishing multiple manual data entries. Financial savings were the result of streamlining operations and the computerization of repetitive tasks. New puzzles can be solved.
2. The implementation of precertification of organizations and accessing data prior to the arrival of the shipment helped to eliminate irrelevant process steps and the redesign of processes, so that tasks can be carried out in parallel.

3. The formulation of new laws or changing current ones are a key requirement to stimulate environment for the development of e-customs in the countries.
4. E-customs will define common process patterns. Considering the variety of stakeholders involved in the export process, the standardization of the process steps, such as ordering, export declaration, and delivery, would ease the supply chain flow.

In Kuhn's words, these enablers represent a typical paradigm shift: If the technology of e-customs is established, processes get more and more standardized, efficient and routine. The access of registration-related information worldwide will broaden the scientific community and will support the paradigm shift, as rules are visible for everybody. This is especially the case if IT standards are capable of spreading on a global basis. In order to reach a large community, the determination, definition, and specification of standard characteristics are key factors.

The New Paradigm

As globalization, statutory provisions, IT modernization of public authorities and growth of world trade continue to increase, IT standards in the field of customs management will expand worldwide. The vision of e-customs is a paperless environment for customs and trade. Internet technology in customs has changed practices and regulations and leads to new ways of working between government and businesses.

On the European path to this new paradigm, EU has introduced a strategic action plan to connect IT-systems and procedures of business units with e-customs systems of all European member states. Due to business networks or statistical specifications, non-EU members are affected by this plan as well. At an international level, the Kyoto ICT Guidelines[1] demand to apply common standards for imports, exports and documentation worldwide.

According to Kuhn, these treaties are the first step to a paradigm shift, as they set new rules for the environment. (2004) confirm: "regulation is the intentional activity of attempting to control, order or influence the behavior of others" (p.332). The definition outlines a common understanding of regulation as a steering mechanism in society.

As international e-customs solutions are not in use yet, the paradigm has not shifted completely. However, some approaches have been made in the last years to stimulate common e-customs system worldwide. For instance, the case project ITAIDE in which e-custom solutions are being tested includes actual business scenarios. The development phase is conducted in so-called living laboratories that are characterized by pre-defined industry specific or cross-industrial trade scenarios with one exporting country and one importing country. Resolutions take place in specific places and move on much later, like the pilot project of using a national e-customs system in Denmark and the exemplary Danish Russian export case.

Denmark, Sweden and Switzerland are examples of successful implementation of parts of e-customs solutions fostering the paradigm's diffusion.

Denmark reported in 2009 almost all export declarations electronically. As there is no standardization across Europe, Denmark had to directly computerize the previous manual and paper-based system. Thus, it is still possible to have paper equivalents for electronic messages in situations where companies do not want to submit export data electronically (such as in the case of an export to Russia).

In the Danish-Russian export case, an information bridge between Danish and Russian customs will be linked by extra information, the "e-Import" information, which is directly derived from the "e-Export" information achieved from the Danish customs. With this linking between the two customs offices, Russian customs will know about

Figure 7. E-customs is currently still on the edge between invention and adoption due to barriers like unclear regulations and missing cultural change

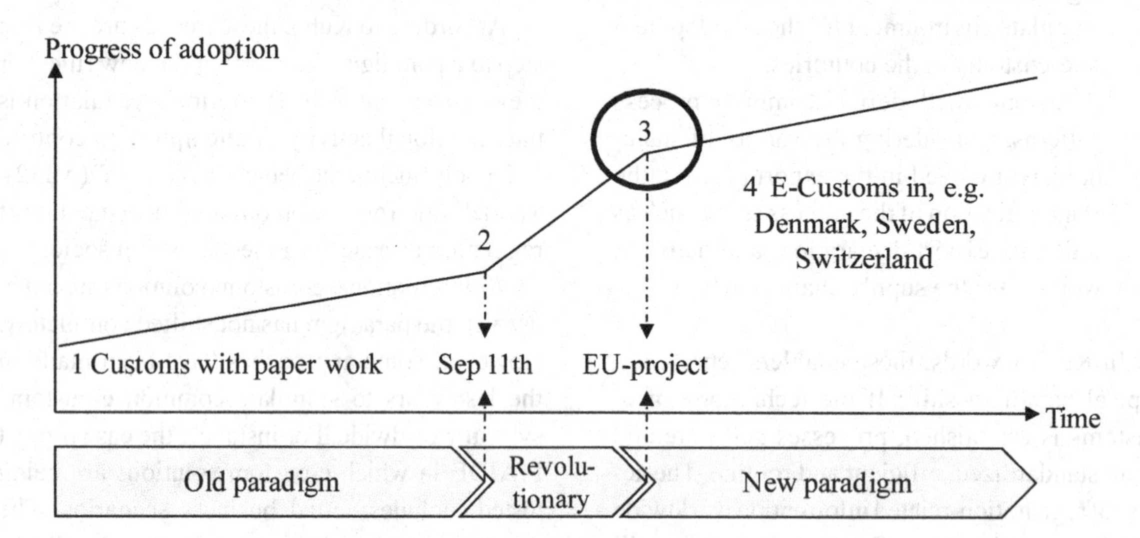

the status of the goods before the goods actually arrive. Import tax fraud with double invoicing can be minimized. The same is true regarding certificate handling. An "e-Certificate" will replace the current paper and stamp based certificate. No posting and human handling will be necessary, and all the certificates will be verified electronically by certification authorities and sent directly to the Russian customs. Risk of certification fraud is thus minimized (ITAIDE 2009). This is a typical example of loss of efficiency during a Kuhnian revolution.

On a conceptual level, export processes from Sweden and Denmark are very similar and follow the same steps. For example, Sweden introduced digital signatures in their customs procedures and replaced paper-based and stamped document checks. Consequently, they can comply with the requirement of the four eye-principle[2] by requesting visible control checkpoints across organizations (Tullverket, 2006).

Switzerland is a special example in demonstrating the importance of facilitated legislations and e-customs systems as a small non-EU country inside the EU. It needs specific customs regulations for every single trade within Europe. For being compliant with EU regulations and especially with the e-customs initiative, the Swiss Federal Customs Administration launched in June 2007 a project called IDEE (**Id**eale **e**lektronische **E**xporteurlösung, ideal solution for exporters). The goal of this project is the replacement of its old, paper-based VAR system (VAR = Vereinfachte Ausfuhrregelung, a simplified set of export rules), originated from the 1970's, with a new e-customs system called e-dec[3] (electronic declaration), which will be binding from July 1, 2009. After that date, Swiss companies can declare their exports electronically. Consequently, approximately 900 Swiss companies using VAR today have to adapt their export procedures to electronic ones.

That e-customs indeed drives value in the public sector shows value assessment in a Danish food company, belonging to the ITAIDE project (Raus, 2009). Four major goals with an e-customs adoption have been identified: 1) security, 2) reduction of administrative burden, 3) compliance and 4) communication.

In the Danish food company case, the common e-customs system reduced monthly safety irregularities by 50-75%. This accelerated the order execution and reduced personnel cost by

25-50%. Due to the reduction of administrative burden, export process cycles increased about 25-50%. Even though the company is already compliant with current regulations, e-customs directives, like Single Window, make it easier to be compliant to all regulations being processed and tested electronically. It is now easier to move to higher levels of business transformations: New puzzles may be solved like business network redesign.

Figure 7 summarizes this study. The figure concentrates on the diffusion of e-customs in general using Kuhn's notion of scientific progress with the mentioned four steps of: (1) puzzle solving, (2) crisis, (3) change and (4) new puzzle solving. The paradigm shift from a paper-based customs administration to electronic customs can be confirmed to a certain degree given e-customs is viewed as mandatory in international trade. EU has a committed time schedule for 2013 e-customs implementation.. Thus, the paradigm shift to a common e-customs solution may be considered predetermined.

Globalization and the complexity of trade and political events such as terror attacks, led to stronger regulations that supported the necessity of a transparent and safe international trading system. With the technological opportunity of the Internet and IT modernization this could be realized by IT standards. Consequently, most of the rules of the paradigm are set through treaties, guidelines, and technical standards. In spite of these rules, there is still resistance within the customs community.

Problems, Issues and Questions Within the New Paradigm of Border Control

The diffusion of a standardized common e-customs system across European countries turns out to be difficult, as the procurement of information and technology and the implementation of e-customs solutions is not specified across countries (Raus, Flügge, & Boutellier, 2009). Rules and guidelines mandated by governments and EU are still unclear and unspecified for some members of the customs community, such as companies and Custom authorities. This uncertainty leads to a resistance within the "old scientific community" to switch to the new paradigm.

The three major barriers of the adoption of e-customs are (Henriksen & Rukanova, 2008; Raus, Flügge, & Boutellier, 2009):

1. Regulations can be a powerful instrument on a national level for e-customs implementation, the power of regulation fades to recommendations and soft laws on an international level. If customs aim to fulfill the principle Pacta Sunt Servanda (agreements must be kept), one of the oldest principles in international law (Henriksen, Rukanova, & Tan, 2008), they have to stick to their old instruments, as the new one is missing procedural details and direction.
2. Speaking in Kuhn's words, regulations of an international e-customs system do not involve the whole customs community, as not all standards are applicable to all industries. SMEs have difficulties in implementing and maintaining different standards or systems. Even though the EU aims for establishing one common e-customs system for all, countries still have to adopt the systems individually. Thus, some countries might not deploy the outlined roadmap to its full extent (Raus, Flügge, & Boutellier, 2009).
3. While e-customs support citizens and businesses through more convenient and accessible services (Burn & Robins, 2003), governmental activities might change or become obsolete. This leads to an intermediate loss of efficiency. The fear of losing jobs or not having the right skill-set to cope with the new technical environment constitutes a crucial issue. The risk of resistance within organizations is still big, due to cultural diversity across the European member states. The

theory of e-customs is still not valid enough to even take away their fears of adopting a new theory.

Jeyaraj, Rottman, and Lacity (Jeyaraj, Rottman, & Lacity, 2004) support these findings affirming that the relative advantage will be perceived if the innovation can be adapted to the individual situation of the adopters. In addition, the authors explain that the behavior of individual adopters to accept a new solution is very much linked to the social network the adopter is embedded in. This leads back to Kuhn's concept of the scientific community. The community will only accept a new paradigm if its members are strong believers and involved in the process. (Henriksen, Rukanova, & Tan, 2008) state that "an interesting trend is the tendency to emphasize that regulation is created by technocrats but it is the practitioners which have to deal with regulation. Thus, the need to manage change is one of the major outcomes of the analysis in the case of diffusion and adoption of e-customs system innovations."(p. 17).

A change in e-Government initiatives was developed by Guha, Grover, Kettinger, and Teng (1997), and later modified by Burn and Robins (2003) based on the statement of Kalakota, Oliva, and Donath (1999), "top management support is essential" in taking an "active role organizationally to shape their firms' policies and standards.".

Although the paradigm has changed, the culture has not. As the barriers showed, some involved parties still fear new technology and change. Nevertheless, an information technological paradigm shift from customs to e-customs is irrevocable.

Implications for Business Processes and Customs Officials

Shifting from paper-based to paperless processes does not necessarily involve significant changes in the processes. However, in order to make full use of the optimizations incurred by the reduction of paper, a process redesign can yield further significant optimizations in time, complexity and money. New puzzles may be solved in a different way.

According to Venkatraman's (1994) five-level model (shown in Figure 8), the EU has already defined the transformational level that would be appropriate in 2013: Business Network Redesign where interconnections are expanded to external businesses, like suppliers, buyers or other intermediaries with regard to IT deployment. This is the goal the EU is trying to achieve with ITAIDE, not for competitive reasons like a company, but for facilitating trade and transactions worldwide and enhancing operational and financial efficiency in the public sector.

Currently, the adoption of e-customs appears to be located in levels two or three. In some countries, such as, Denmark, Sweden and Switzerland, IT is fully integrated in the customs area. This places these countries in level three "Business Process Redesign". General business principles in these countries avoid optimized IT integration in their e-customs systems. Countries, such as Russia, relying on paper-based importing and exporting, may be located in level two "Internal integration", where technical interconnectivity surely takes place due to IT modernization. However, the export processes have not been adapted to the new technology.

Thus, one managerial implication for customs officers is to continuously develop their country's e-customs systems. As such, they further push for a common standardized e-customs system. The formal paradigm shift will occur with the EU-Directive saying that the 27 member states have to adopt the system by 2013. As such, the customs community, including all exporting companies, customs officials and customs workers, has to be convinced entirely to adopt the e-customs system. The revolution will be over in 2013. Thereafter, companies may differentiate themselves with business process redesign, network redesign and business scope redefinitions.

Figure 8. Different countries are on different levels and seek level four (according to Venkatraman, 1994, p. 74)

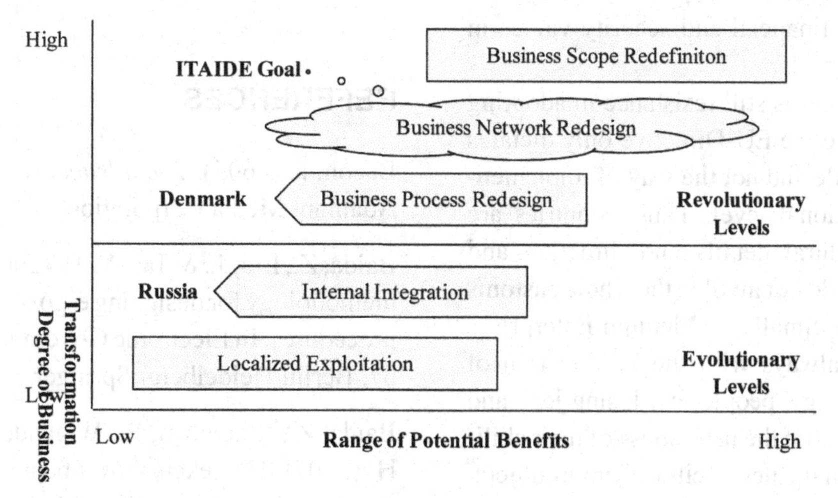

CONCLUSION

History has shown that the linear model of Francis Bacon had to be advanced with feedback mechanisms that include organizational issues. This led to the notion of technological paradigms associated with Thomas Kuhn's concept of scientific paradigms and proved that Kuhnian ideas could also describe technological progress. Here, Kuhn's idea of a community forming a paradigm is still valid: If a majority of engineers working on one technology are convinced of a need for change, paradigms shifts will occur.

Historical examples of information technological (IT) progress are the modularization of software development, the move from mainframe to personal computers, and currently finalized in a technological paradigm of virtual resources accessible through the Internet.

The underlying technology in the Case of e-customs is the Internet, starting with e-Business (B2B and B2C) and then also reaching the public sector with e-Government solutions.

The technological progress of the adoption of a common standardized e-customs system in Europe has been explained with Kuhn's four steps of a paradigm: (1) puzzle solving, (2) crisis, (3) change and (4) new puzzle solving.

About 30 years ago, an export process within customs was fully paper-based and required approximately 200 documents. Even though the Internet evolved, the adoption of e-customs system did not take place. With globalization, increasing trading complexity and political events, such as the terrorist attacks that occurred in the US on September 11th, new legal standards and safety regulations in trade and customs are required. As the Internet in general reduces corruption through transparency and harmonization of rules, e-customs got a new wave of attention:

The US required one common and standardized European e-customs system with the intention of having a universal European customs partner instead of several national trading partners with different systems. As a result, the EU currently funds the project ITAIDE for achieving trade facilitation as well as to secure import and export with a common European e-customs system to be implemented by 2013. With this EU-Directive and an implemented strategic action plan, the technological paradigm shift to a common use of e-customs system is predetermined. Some pilot projects have already taken place in Denmark,

Sweden and Switzerland. The Danish example shows that an e-customs adoption would significantly increase financial and security values in the public sector.

However, there is still resistance in adopting e-customs, since the EU-Directive only dictates the time schedule and not the way of implementation on a national level. Thus, countries are missing procedural details and direction and some standards do not involve the whole customs community, e.g. Small and Medium Enterprises (SMEs). Like always with the introduction of Internet technology, people fear losing jobs and are concerned about the usefulness of their skill-sets to cope with the new technical environment. Although the paradigm has changed, the culture has not. Nevertheless, an information technological paradigm shift from customs to e-customs is irrevocable.

The EU has set its managerial goals with its EU-Directive, aiming at a "Business Network Redesign" where interconnections are expanded to external businesses, like suppliers, buyers or other intermediaries with regard to IT deployment. However, with countries being still at the level of "Internal Integration" regarding their process transformation, like Russia, the EU has to find the commitment of the whole customs community in giving them clear and standardized directions for e-customs implementation. Propositions are to introduce one common European customs coordinator and to strengthen the value added due to e-customs adoption. For customs officers it means to further develop the e-customs system to convince the whole customs community.

ACKNOWLEDGMENT

The presented case study is based on the integrated project ITAIDE (Nr.027829), funded by the 6th Framework IST Programme of the European Commission (see www.itaide.org). The ideas and opinions expressed by the authors do not necessarily reflect the views/insights/interests of all ITAIDE partners.

REFERENCES

Bacon, F. (1605). *The advancement of learning*. Adamant Media Corporation.

Baida, Z., Liu, J., & Tan, Y.-H. (2007). Towards a methodology for designing e-government control procedures. In Electronic Government, 4646, 56-67. Berlin/Heidelberg: Springer.

Baida, Z., Rukanova, B., Wigand, R., & Tan, Y. H. (2007). Heineken shows benefits of customs collaboration. *Supply Chain Management Review*, *11*(7), 11–12.

Bjørn-Andersen, N., Razmerita, L. V., & Henriksen, H. Z. (2007). The streamlining of cross-border taxation using IT: The Danish eExport solution. In Makolm, J., & Orthofer, G. (Eds.), *E-Taxation: State & Perspectives: E-Government in the Field of Taxation: Scientific Basis, Implementation Strategies, Good Practice Examples* (pp. 195–206). Linz, Austria: Trauner Verlag.

Boyd, S. L., Hobbs, J. E., & Kerr, W. A. (2003). The Impact of Customs Procedures on Business to Consumer E-commerce in Food Products. *Supply Chain Management: An International Journal*, *8*(3), 195–200. doi:10.1108/13598540310484591

Burn, J., & Robins, G. (2003). Moving towards e-government: A case study of organisational change process. *Logistics Information Management*, *16*(1), 25–35. doi:10.1108/09576050310453714

Bygrave, W. D. (1989). The entrepreneurship paradigm (I): A philosophical look at its research methodologies. *Entrepreneurship Theory and Practice*, *14*(1), 7–26.

Christensen, C. (1997). *The innovator's dilemma: when new technologies cause great firms to fail*. Harvard Business School Press.

Dedrick, J., & West, J. (2003). Why firms adopt open source platforms: A grounded theory of innovation and standards adoption. In J. L. King & K. Lyytinen (Eds.), *Proceedings of the Workshop on Standard Making: A Critical Research Frontier for Information Systems,* (pp. 236-257). Seattle, WA, USA.

Dosi, G. (1982). Technological paradigms and technological trajectories. *Research Policy, 11*(3), 147–162. doi:10.1016/0048-7333(82)90016-6

Easterly, W. (2001). The elusive quest for economic growth: Economists' adventures and misadventures in the tropics. Cambridge: MIT Press. *The Economist (2000), "Growth is good", The Economist, May 27*, 82.

Granberg, A., & Stankiewicz, R. (1981). The development of generic technologies - the cognitive aspects. In Grandstrand, O., & Sigurdson, J. (Eds.), *Technological and Industrial Policy in China and Europe* (pp. 196–224). Lund: Research Policy Institute.

Guha, S., Grover, V., Kettinger, W. J., & Teng, J. T. C. (1997). Business process change and organizational performance: exploring an antecendent model. *Journal of Management Information Systems, 14*(1), 119–154.

Henriksen, H. Z., & Rukanova, B. (2008, April 23-25). *Barriers and Drivers of eCustoms Implementation: Never Mind IT.* Paper presented at the 6th Eastern European eGovernment Days, Prague, Czech Republic.

Henriksen, H. Z., Rukanova, B., & Tan, Y.-H. (2008). Pacta Sunt Servanda but Where Is the Agreement? The Complicated Case of eCustoms. In Wimmer, M. A., Scholl, H. J., & Ferro, E. (Eds.), *EGOV 2008* (pp. 13–24). Berlin, Heidelberg: Springer-Verlag.

ITAIDE. (2009). *Report on redesign of administrative processes, interoperability and standardization.* Retrieved from http://www.itaide.org

Jeyaraj, A., Rottman, J. W., & Lacity, M. C. (2004, December). *Understanding the Relationship between Organizational and Individual Adoption of IT Innovations: Literature Review and Analysis.* Paper presented at the Diffusion Interest Group in Information Technology, Washington D.C., USA.

Johnston, R. (1972). The Internal Structure of Technology. In Halmos, P., & Albrow, M. (Eds.), *The Sociological Review Monograph 18 - The Sociaology of Sciences* (pp. 117–130). J.H. Brookes Printers Limited, Keele.

Kalakota, R., Oliva, R. A., & Donath, B. (1999). Move Over, E-Commerce. *Marketing Management, 8*(3), 22–32.

Khalil, T. M. (1999). *Management of technology*. McGraw-Hill Science/Engineering/Math.

Koestler, A. (1989). *The Sleepwalkers*. Arkana/ Penguin.

Kuhn, T. S. (1962). The Structure of Scientific Revolutions Chicago. University of Chicago Press. the problem of induction 1(1), 5.

Kuhn, T. S., & Neurath, O. (1970). *The Structure of Scientific Revolutions: International Encyclopedia of Unified Science*. University of Chicago Press.

Kuiper, E. J. (2007). *Convergence by Cooperation in IT – The EU's Customs and Fiscalis Programmes*. Delft, The Netherlands: Delft University of Technology.

Lakatos, I., & Musgrave, A. (1970). *Criticism and the Growth of Knowledge*. Cambridge Univ Press.

Parker, C., Scott, C., Lacey, N., & Braithwaite, J. (2004). *Regulating law*. Oxford University Press.

Peine, A. (2008). Technological paradigms and complex technical systems—The case of Smart Homes. *Research Policy, 37*(3), 508–529. doi:10.1016/j.respol.2007.11.009

Planck, M. (1949). A scientific biography. In O. U. Press (Ed.), The Oxford dictionary (2004 ed., p. 596).

Postman, N. (1992). *Technopoly: The Surrender of Culture to Technology*. New York: Knopf.

Raus, M. (2009). *Value Assessment of Business-to-Government IT Innovations: a Case Study. 22nd Bled eConference eEnablement: Facilitating an Open, Effective and Representative eSociety*. Slovenia: Bled.

Raus, M., Flügge, B., & Boutellier, R. (2008). Innovation Steps in the Diffusion of e-Customs Solutions. In S. A. Chun, M. Janssen & J. R. Gil-Garcia (Eds.), *ACM International Conference Proceeding Series* (Vol. 289, pp. 315-324). Montréal, Canada: Digital Government Society of North America.

Raus, M., Flügge, B., & Boutellier, R. (2009). Electronic Customs Innovation: an Improvement of Governmental Infrastructure. *Government Information Quarterly*, *26*(2). doi:10.1016/j.giq.2008.11.008

Raus, M., Kipp, A., & Boutellier, R. (2008). Diffusion of e-Government IT Innovation: a Case of Failure? In Cunningham, P., & Cunningham, M. (Eds.), *Collaboration and the Knowledge Economy: Issues, Applications, Case Studies*. Amsterdam: IOS Press.

Teo, H. H., Tan, B. C. Y., & Wei, K. K. (1997). Organizational transformation using electronic data interchange: The case of TradeNet in Singapore. *Journal of Management Information Systems*, *13*(4), 139–165.

Tullverket. (2006). *Säkerhetsfragor I Tullverkets EDI-System* (1.0 ed., pp. 1-9).

United Nations Economic Commission for Europe. (2005). *Recommendation and Guidelines on establishing a Single Window - Recommendation No. 33*. Geneva: UN/CEFACT.

Venkatraman, N. (1994). IT enabled business transformation. *Sloan Management Review*, *35*(2), 73–78.

Vincenti, W. (1995). The Technical Shaping of Technology: Real-World Constraints and Technical Logic in Edison's Electrical Lighting System. *Social Studies of Science*, 553–574. doi:10.1177/030631295025003006

Wassenaar, A. (2000). *E-governmental value chain models-E-government from a business (modelling) perspective*.

ADDITIONAL READING

Drews, J. (1998). Die verspielte Zukunft - Wohin geht die Arzneimittelforschung? Basel, Boston, Berlin: Birkhäuser. A. Lightman The best American science writing, Harper 2005, p.76.

Fukuyama, F. & C, S. (2003). *Our posthuman future*. New York: Picador New York. Kealey, T. (1997). *The economic laws of scientific research*. Palgrave Macmillan. Kuhn, T. (1996, July 13). The nature of science. *The Economist*.

Snow, C. P. (1964). *The two cultures*. Cambridge University Press.

Wilson, E. O. (1998). *Consilience: The unity of knowledge*. New York: Knopf.

Yoshikawa, H. (1995). President Tokyo University. In Technology's New Horizon (p. 123). Oxford.

ENDNOTES

1 The Kyoto ICT Guidelines suggest that Customs should review their current procedures and processes prior to adopting any IC technology tools (www.wcoomd.org – World Customs Organization).

[2] For assuring integrity of information and communication, the four-eye principle guarantees that all communications should generally be checked and counter-signed by a person who was not involved in drafting it.

[3] E-dec designates a cargo processing IT product, developed by the Swiss federal customs administration, which is to standardize existing IT-supported (cargo processing) procedures (www.ezv.admin.ch).

Chapter 8
Growth Trajectories of SMEs and the Sensemaking of IT Risks:
A Comparative Case Study

Arvind Karunakaran
The Pennsylvania State University, USA

Jingwen He
The Pennsylvania State University, USA

Sandeep Purao
The Pennsylvania State University, USA

Brian Cameron
The Pennsylvania State University, USA

ABSTRACT

Our case will describe two small to medium enterprises which are located within the same region and sharing the broad industry sector but at a different 'growth stage' perceive the role of Information Systems differently. We describe how these two firms, at different growth stages and at different levels of maturity with respect to their information systems, perceive the usefulness of information systems differently. We extend the interpretations to discuss sub-sections within SMEs, which are at different stages of growth, and how the nature of information systems' risks is likely to differ depending on these growth stages. We emphasize the importance of owner/manager's "sensemaking of risks" as a key variable that influences the demarcation between entrepreneurs and small business owners, beyond the oft-discussed variables such as "achievement motivation," "risk-taking propensity," and "preference for innovation." We conclude with the proposition that SMEs should not be considered as unitary entities; and suggest that there are likely to be different varieties of risks that SMEs face, and suggest the growth stage and organizational filters as key determinants of the owner/managers' understanding of these risks.

DOI: 10.4018/978-1-61520-609-4.ch008

EXECUTIVE SUMMARY

Small and Medium Enterprises (SMEs) are considered to be the engines of growth and innovation of the modern economy (SBA, 1998, 2004c). According to Office of Advocacy estimates, small firms with fewer than 500 employees represent 99.9% of the 27.2 million businesses (SBA, 2004a, 2004b). However, two-thirds of them survive two years, 44 percent survive for four years, and 31 percent survive for seven years (SBA, 2004a). The reason for their failure is attributed to various factors like the institutional environment, financial crisis, owner/manager motivation, risk-taking propensity, competition, market conditions etc (Miner & Raju, 2004).

Our proposed case will describe two small to medium enterprises – Bedrock Manufacturing and VPro, Inc,- which are located within the same region and sharing the broad industry sector but at a different 'growth stage' (Churchill & Lewis, 1983) perceive the role of Information Systems differently. Bedrock Manufacturing designs and manufactures construction equipments for dozers, loaders, excavators, and motor graders. It has about 190 employees and annual revenue of 20 million dollars. VPro, Inc. manufactures industrial process furnace and provides custom vacuum and thermal process equipments for various industries. It has 8 employees with annual revenue of around 1 million dollars. Both are located in the I-99 Innovation Corridor of Central Pennsylvania and fit the U.S. Small Business Administration's definition of a "Small Business". We describe how these two firms, at different growth stages and at different levels of maturity with respect to their information systems, perceive the usefulness of information systems differently. We extend the interpretations to discuss sub-sections within SMEs, which are at different stages of growth, and how the nature of information systems' risks is likely to differ depending on these growth stages. We emphasize the importance of owner/manager's 'sensemaking of risks' as a key variable that influences the demarcation between entrepreneurs and small business owners, beyond the oft-discussed variables such as 'achievement motivation', risk-taking propensity', and 'preference for innovation'. We conclude with the proposition that SMEs should not be considered as unitary entities; and suggest that there are likely to be different varieties of risks that SMEs face, and suggest the growth stage as a key determinant of the owner/managers' understanding of these risks.

THE REGION

I-99 Corridor As A Legacy Industrial Era Region

The I-99 Corridor (Interstate-99) is a partially-completed intrastate interstate highway (i.e. interstate highway located within a single state) in Central Pennsylvania, linking other cross-state corridors like I-80 and I-76 (Dytche & Warren, 2007a). The current southern terminus is at the north of Bedford, while the northern terminus is near Bellefonte (Interstate 80). It also passes through Altoona and State College (home of The Pennsylvania State University). The full route of the corridor is part of "Corridor O" under the Appalachian Development Highway System, which runs from Interstate 68 near Cumberland in Maryland onto Bedford in Pennsylvania.

The Corridor is home to large number of towns whose economies are fairly fragile, which could be classified as "legacy industrial era regions". Legacy Industrial era regions are those regions which are still undergoing a transition in adjusting to the post-industrial society (Bell, 1973; Masuda, 1980). Economic activity in these regions forms a significant portion of the heavy industry and manufacturing sectors of the U.S. economy (Faberman, 2002). Sometimes referred to as "Rust Belt", these regions signify the steep decline of the manufacturing industry throughout the late 1960s and 1970s, which lead to signifi-

Figure 1. Legacy Industrial Era Regions – Comparison Statistics - Education

Education	I-99 Corridor	National	NewYork	Illinois	Ohio	Michigan	Wisconsin	Pennsylvania	Indiana	West Virginia
Bachelor's Degree or higher (2000)	3%	27.20%	27.37%	26.06%	21.09%	21.76%	22.42%	22.35%	19.40%	14.84%
High School Diploma (2000)	85%	84.60%	79.06%	81.43%	82.97%	83.41%	85.09%	81.90%	82.13%	75.21%

Figure 2. Legacy Industrial Era Regions – Comparison Statistics – Economic Diversity

Economic Diversity	I-99 Corridor	National	NewYork	Illinois	Ohio	Michigan	Wisconsin	Pennsylvania	Indiana	West Virginia
Manufacturing	18%	5%	8%	13%	17%	17%	20%	13%	21%	12%
Services	23%	41%	32%	31%	28%	31%	23%	29%	26%	24%
Trade, Transportation & Utilities	28%	22%	20%	22%	21%	20%	22%	22%	22%	23%
Construction	4%	12%	4%	5%	5%	4%	5%	5%	5%	5%
Government & Other	27%	20%	35%	22%	28%	28%	30%	31%	26%	36%

cant loss of jobs in those regions (Kahn, 1999). Between 1969 and 1996, when the manufacturing employment of the country grew by 1.4%, manufacturing employment in the Rust Belt fell by 32.9% (Kahn, 1999). Also, when the nation as a whole has shifted toward a service economy, these regions failed to catch up with the changes that were happening around them and are still in the process of undergoing the transition to post-industrial society (Bell, 1973).

I-99 Corridor share some distinct characteristics with other similar 'legacy industrial era regions'. For example, the percentage of population who has a Bachelor's degree in these regions is tends to be lower than the national average (See Figure 1), and is sharply lower for the specific case of the I-99 corridor (3% compared to the national average of 27%).

Also, the region has not yet made a stride into the Service economy, and continues to hold on to the Manufacturing and related sectors (contributing only 23% of economic activity, as compared to the national average of 41%), a trend similar to but worse than the other regions shown (see Figure 2).

Clearly, the I-99 corridor presents a difficult and intriguing challenge even within the general category of legacy industrial era regions.Recent research suggests that due to the geographical remoteness of the region and due to the lack of tangible as well as intangible infrastructures, the I-99 region is at an inherent disadvantage with respect to technology transfer and new business development (Dytche & Warren, 2007b; Warren, Hanke, & Trotzer, 2008). The tangible and intangible infrastructure available to this region is relatively low, compared to other highly developed regions. One of the metrics which a recent report used for measuring the strength of such infrastructures was the value of Venture Capital (VC) investments within a fifty mile radius. The density of VC investments tended to be high within 50 mile radii of the large research universities. For example, regions around California and along the East Coast had higher densities due to concentration of a number of large research universities (Dytche & Warren, 2007a, 2007b). However, the I-99 Corridor, in spite of the presence of The Pennsylvania State University, continues to show a much lower density of VC investments (Dytche & Warren, 2007b) suggesting the presence of causes other than availability of high calibre research outcomes, researchers and student pool. The lack of these potential enablers or catalysts may help explain the slow transition from the manufacturing base to a service economy. The continued lag in the key indicators (e.g. Education levels) shown in Figures 1 & 2 above suggest what some of these enablers may be. It is against the above backdrop that we describe the two cases in this chapter.

Table 1. Key indicators for the two organizations

Name	Industry	Annual Revenue	Number of employees	Age (in years)	Location
Bedrock Manufacturing	Custom Construction Equipment & Plate Work Manufacturing	$20 million	190	60+	Bedford, PA
VPro, Inc	Industrial Process Furnace and Oven Manufacturing	$1 million+	8	50+	Everett, PA

THE ORGANIZATIONS

Bedrock Manufacturing Company

Bedrock Manufacturing is located in Bedford, Pennsylvania. The I-99 corridor passes to the west of Bedford, providing interstate access to I-80. Bedrock Manufacturing design and manufacture construction equipments for dozers, loaders, excavators, and motor graders and specialize in the production of custom construction equipment. Their products are referred to as 'attachments for heavy equipment'. Their external image is that of the makers of "allied equipments for construction machinery'. However, they consider themselves as manufacturers of"heavy equipment attachments" which are designed for "specific applications" that would enable better machine productivity from dozers, loaders, excavators and motor graders. Their equipments are available for almost every major brand in the construction machinery industry.

Bedrock Manufacturing has been in operation for more than sixty years and has expertise in designing and manufacturing custom construction equipments. Over the years, they have made several unique contributions to the construction equipment industry. In the late 50's, they designed and patented the first ever land-clearing rake with reversible and adjustable teeth – considered to be a major advancement at the time. The construction equipment industry was "startled" when Bedrock Manufacturing offered a lifetime guarantee to their land-clearing rake against any failure from defective design, materials, or workmanship. During the early 60's, in a joint venture between two other companies, they designed and built the world's largest dozer blade. They also designed and manufactured the first-ever spade nose rock bucket to be put in a quarry, which helped them in proving their assertion that wheel loaders could handle shot rocks effectively.

Currently, they have around 190 employees, with annual revenue of 20 million dollars. Their customer base covers much of the continental US, with a few international customers.

VPRO, INC

VPro, Inc. is located in Everett, which is a borough in Bedford County, Pennsylvania. VPro, Inc. is a design, manufacturing, service, engineering, research, and installation entity, which provides custom vacuum and thermal process equipments for various industries. They are in operations for over 50 years. Predominantly, they see themselves as "custom vacuum and thermal process equipment providers", with a focus on providing complete turnkey solutions leading to "process and equipment improvement". Their products and services include Turn Key Thermal & Vacuum System Manufacturing, Hot Zone Repair, Parts, Vacuum Pumping Systems, Pumps, Repair, Service, Maintenance and Service Supplies. Their customer base is centered on Continental U.S, with few international customers. They would fit into the "Industrial Process Furnace and Oven Manufacturing" segment.

As compared to Bedrock, they are relatively small. Currently, they have 8 employees and they are expecting to expand upto 20 employees by next year. Their current annual revenue is 1 million dollars. Both these organizations would fit into the U.S Small Business Administration (SBA) definition of a 'Small Business' (SBA, 2004a).

RELATED PRIOR WORK

This section reviews related research that can help understand and better describe the cases that follow. It is not meant to be a complete literature review; instead, it provides terminology and is meant to sensitize the readers to constructs explored by prior research.

Small and Medium Enterprises and Information Technology

Small and Medium Enterprises (SMEs) are considered the engines of growth and innovation of the modern economy (SBA, 1998, 2004c). According to U.S Small Business Administration, SMBs represent 99.7 percent of all employer firms and pay nearly 45 percent of total U.S. private payroll (SBA, 1998, 2004b, 2004c). They hire as many as 40 percent of high tech workers such as scientists, engineers, and computer workers (SBA, 1998, 2004a, 2004b, 2004c). According to Office of Advocacy estimates, there were 27.2 million businesses in the United States; of these, 99.9% are small firms with fewer than 500 employees (SBA, 2004b). However, two-thirds of them survive two years, 44 percent survive for four years, and 31 percent survive for seven years (SBA, 2004a). The reasons for failure can include institutional environment, financial crisis, owner/manager motivation, risk-taking propensity, competition, market conditions etc (Miner & Raju, 2004).

The volume of IT spending from these Small to Medium Businesses tends to be comparable to that of Large Enterprises. The SMEs form the so-called 'Long Tail' (Anderson & Andersson, 2006) of IT spending. Together, they represent 49% of the overall IT spending, making it a lucrative market for the IT Service providers (Speyer, Pohlmann, & Brown, 2006). However, these SMEs seldom have access to quality IT service providers who can help them with their day-to-day IT operations. Anecdotal evidence suggests that the IT providers are often too mired in technology details and solutions (e.g. 'my router is not working'), and rarely, if ever, consider aiding the business owners to find 'Business-IT synergies' or in using IT towards facilitating business growth (e.g. 'moving to strategic uses of IT with new software functionalities'). These SMEs do not have the internal resources to forecast their future Information Systems needs and also, have very little access to quality IT Service providers who could assess their IT needs and could forecast, plan and mitigate risks related to the evolution of their information systems platforms (Ballantine & Levy, 1998; Levy, Powell, & Yetton, 1998).

Also, researchers tend to overlook the phenomenon of 'Managing IT Risks within SMEs'. Past studies on 'Risk Management' under the "Information Systems" (IS) and "Information Technology" (IT) stream, focus on large enterprises (see, for example, (Alter & Sherer, 2004a; Stoneburner, Feringa, & Goguen, 2002)). Not many cases have been recorded which describe the risks faced by SMEs when they introduce innovative information systems into the organization.

SME Growth Theories

Theories on SME growth provide us with a framework that lets us distinguish between the various attributes and dimensions of those two companies (Bedrock Manufacturing and VPro, Inc.) (O'Farrell & Hitchens, 1988) categorize theories on SME growth, proposed within the disciplines of Management and Economics, into four major groups:

1. Static equilibrium theory
2. Stochastic models of firm growth
3. Strategic management perspectives on growth
4. Stages-of-development theory

Static equilibrium theory is derived from the field of industrial economics and is preoccupied with attainment of economies of scale and minimization of long-run unit costs (Nelson & Winter, 1982). It operates under the overarching assumption that large firm is the ultimate stable outcome of growth. According to the stochastic models of firm growth, there are many factors that affect growth and, therefore, there is no dominant variable which determines the rate of growth. However, it does operate under the assumption that business growth rates are independent of enterprise size (Nelson & Winter, 1982; O'Farrell & Hitchens, 1988).

Strategic management perspective considers the owner/managers on growth and fulfillment to be the dominant factor which influences the growth of a firm. It operates under the assumption that not all SME owner-managers have the desire to grow their business due to various personal reasons like inclination to a particular life-style, disinclination to surrender control etc. (Henderson & Venkatraman, 1999; O'Farrell & Hitchens, 1988; Wernerfelt, 1984)

Stages of development theory offer several Enterprise life-cycle models, which explain the growth of a firm. Hanks, Watson, Jansen, & Chandler (1993) reviewed and synthesized 10 such enterprise life-cycle models and they define a life-cycle development stage as "a unique configuration of variables related to organization context or structure". The enterprise life-cycle model has two dimensions:

- *Contextual Dimension:* enterprise size and age, growth rate, and focal tasks or challenges faced.
- *Structural dimension*: formalization, centralization, vertical differentiation, and number of organizational levels.

Churchill & Lewis (1983) came up with a Five Stage model of small business growth – Existence, Survival, Success-Disengagement and Success-growth, Take-Off and Resource Maturity, which is one of the widely used models to classify SMEs based on their growth stage. These models of growth provide us with the necessary dimensions, such as, size, age, growth rate, formalization, centralization, organizational levels etc., that would further enable us to exemplify the differences between the two organizations studied and would let us examine the different varieties of IT risks that SMEs are likely to face at different growth stages.

Risk Management

"Risk Management" literature within the IS/IT stream conceptualizes risk in a variety of ways. The simple and elegant definition of risk is the "probability of negative outcomes (sometimes weighted by loss)" (Alter & Sherer, 2004b). Risk management literature within the IS/IT stream could be classified under two major categories based on their research objectives. The first category is mainly concerned with the risks that arise during systems development process and about ways for controlling the process. These risks are called 'System Development Risks' and works in this area focuses only on risks arising in the system analysis, planning and design stages. The other category of the literature focuses more on IT integration risks. The focus of this stream of research is on the larger organization and its associated work processes and practices. These risks are otherwise referred to as Enterprise Integration Risks. Enterprise systems integration is a phrase that is often used to describe the efforts undertaken by an organization to achieve cross-functional integration of its information, processing and work practices. Risk in enterprise integration projects are emergent because problematic situations do not occur suddenly but rather, generate momentum and become a significant occurrence only over time.

However, most of the studies under the "System Development Risks' literature as well as the 'Enterprise Integration Risks' literature were conducted within the context of 'Large Enterprises'. As briefly discussed in previous sections, past studies on 'Risk Management' within the IS/IT stream tend to overlook the contextual and structural dimensions like size and age, growth rate, formalization, centralization, organizational levels, thereby, diluting the differences between a large enterprise and a SME (see, for example, Alter & Sherer (2004a) and Stoneburner, et al. (2002)).

Sensemaking

The term "sensemaking" has been widely used in various fields, including organizational science, communications, education, computer-supported cooperative work and information systems. The process of sensemaking involves finding structure in a seemingly unstructured situation (Corley, 2002; Dervin, 1992; K. E. Weick, 1993). According to Weick, sensemaking consists of seven major aspects - grounded in identity construction, focused on extracted cues, retrospective, enactive, social, ongoing and driven by plausibility than accuracy (K. Weick, 1995). Sensemaking, at its core, involves three tasks (Corley, 2002; K. Weick, 1995).

1. Scanning the environment to collect information (about the actual or potential changes in the environment)
2. Interpreting the information collected, and
3. Taking action based on those interpretations.

Sensemaking process is not considered to be sequential and unidirectional but as interactive, recursive, and sometimes, even discontinuous, as participants make sense of the dynamic environments around them. On the other hand, the term *sensemaking* is not only about *scanning the environment, interpreting the information & taking actions based on the interpretation*, (i.e. a person retrospectively making sense of what has already happened), but also about being *enactive of sensible environments* (i.e. participating in the evolution of the environment, and then making sense of the environmental events that resulted from the participation).

However, the process of sensemaking does not happen in a vacuum and the organizational context has a significant impact on the sensemaking process (Corley, 2002; J Dutton & Duncan, 1983; Gioia & Thomas, 1996). Context serves as a filter and provides a heuristic frame towards directing attention towards some issues (Corley, 2002; J Dutton & Duncan, 1983; Gioia & Chittipeddi, 1991). 'Context', in this case, should not be considered as being 'external' to the firm. Past research has shown that internal context plays a critical role in the sensemaking process (Gioia & Chittipeddi, 1991; Thomas, Clark, & Gioia, 1993). Due to the capacity limits of individuals and organizations towards processing vast amount of information, the organizational context provides a good frame towards selectively attending to certain critical problems and actively constructing a reality towards which they would like to streamline their efforts. Organizational filters are also considered to be a function of the an organization's strategy(JE Dutton & Duncan, 1987; Gioia & Chittipeddi, 1991; Thomas, et al., 1993), organizational growth stage, organizational memory, in terms of the past crisis/response it encountered through out its growth stages and the rules, routines and best practices it acquired over a period of time(Cohen & Levinthal, 1990; Corley, 2002; Dervin, 1992; Dutton & Duncan, 1987; Gioia & Chittipeddi, 1991).

Thus, the growth stage an SME is at influences its organizational filter, which in turn affects its sensemaking process. Hence, it is important to consider the growth trajectory of SMEs as a critical variable towards influencing the process of "Sensemaking IT Risks".

CASE DESCRIPTIONS

Both Bedrock Manufacturing and VPro, Inc. are looking forward to growth in the next few years. The owner/managers of both the companies feel that their current IT infrastructure might not be sufficient to cater to their future needs. Both are in the process of evaluating their existing information systems and are looking for resources that would help them in the process. However, due to the inherent disadvantages of location, they do not have easy and direct access to quality IT service providers – e.g. someone who could who could understand the 'big picture' and offer a holistic solution instead of ad-hoc, 'bandaged' solutions. This has resulted in possible scalability and integration concerns. Although they continue to work with the current IT systems and platforms, they are unable to evaluate the possible emergent risks that might occur if they encounter failure related to upgrading existing information systems, including integrating with organizational processes and work practices. They are not aware of the formal mechanisms which are available that would let them evaluate their existing information systems, identity the possible risks, assess the severity of those emergent risks and come up with a possible strategy to address those risks. Neither do they have access to quality IT service providers who could do this for them.

Even the existing "Risk Management Frameworks" which were put forth by leading IT Consulting firms focus on large enterprises, and they largely ignore the unique issues faced by SMEs like Bedrock Manufacturing and VPro, Inc. At this unique juncture, a student consulting team from a nearby land-grant, space-grant public research university was brought in to address the problem which confronted both these organizations. The students are trained as a part of the larger NSF-funded project called AESOP (Augmenting Education of System of Systems Professionals), which prepares them to work in multi-disciplinary teams to understand organizational work practices and processes better. The intent of engaging the student teams with those SMEs is twofold:

- To nurture future computing professionals with the needed skill sets that would enable them to better understand the organizational context in order to bridge the gap between IT solutions and work practices. Through the process of engaging the student teams in real-time projects, experiential learning is achieved.
- SMEs could work with student consulting teams to evaluate their existing information systems, identity possible risks which are unique to their organization, assess the severity of those emergent risks, and come up with a possible mitigation strategy. By this way, the SMEs could leverage research-intensive university's knowledge in addressing the problems (in this case, their IT problems) for which they are confronted.

The student consulting team, aided by a senior faculty member and a graduate research assistant, did a thorough assessment of their existing information systems. They did this through a mixture of face-to-face meetings and multiple qualitative, semi-structured, telephonic interviews. The interviews were designed to understand the current and future business and IT needs of those companies. The existing IT infrastructure of those companies was assessed, in order to understand the gaps between the current and future needs. Based on the preliminary analysis of the initial interview, further interview questions were developed to understand how the perception of the usefulness of information systems differed among SMEs at different stages of growth; and how this in turn impacts the owner/manager's 'sensemaking of emergent risks'. A risk-item, specific to the organizational IT, was developed and the severities of those risks were assessed. Based on the final analysis, recommendations' were made to these SMEs.

BEDROCK MANUFACTURING COMPANY – AN ASSESSMENT

Bedrock Manufacturing Company had a rich history of coming up with innovative work practices and processes within their industry segment. They were the first allied equipment manufacturer to provide a toll free 800 number for their customers. They wanted to develop their IT infrastructure in such a way that would let them handle their business needs effectively. However, they are currently concerned about keeping up with growth and expansion which they are expecting within the next few years.

IT Infrastructure Support

Bedrock Manufacturing employs a mixture of "in-house" and "third party" support in addressing their everyday IT needs. Currently, their in-house IT department consists of only one employee. The role of the IT staff is to provide support for all on-site tasks that must be routinely performed and to trouble-shoot and handle other on-the-job issues which require immediate attention. They depend upon third party IT service providers for addressing other emergent needs – like replacing or repairing their existing workstation. Most of the software applications they use are not developed in-house, with the sole exception of an application that generates the 'bill of materials' for the engineering department. Since it is very basic software, they have had little trouble in maintaining it. For troubleshooting all the other off-the-shelf products, they again need to depend on the vendors of those products or their third party IT service providers.

Workstations and Servers

Bedrock has roughly 70 to 75 workstations stations. All of them are connected to a server and have access to internet via wired connections. They do have few laptops with wireless capabilities. The servers that Bedrock employs are currently being phased out in favor of a more efficient and faster one. However, even with the older servers, they had little productivity issues. The servers that are being used currently run on Windows 2003 Server (32 bit).

Backup and Intra-Company Communication System

Bedrock manufacturing is aware of the risks related to 'system failure' and the consequential impacts of 'data loss'. They have implemented a backup plan to protect their data. Every night, there is an onsite backup of all important data onto a 4 TB drive. In addition to that, they also do offsite backups every week to ensure that critical information is not lost. Since they feel that communications plays a big role within their company, they have implemented a company cell phone plan through Verizon. This lets their employees be in touch with each other at a much discounted price. They use Exchange Server 2003 Pro for their email needs and Message Pal for instant messaging. However, they are unhappy with current their email and communication software applications. They tend to crash often, exhibiting unpredictable behavior that leads to discontinuous communication and loss of productivity.

Table 2. Comparison of IT infrastructure at the two organizations

Name	3rd party providers	Use custom developed software	Number of workstations	Server	ERP System	On-site backup
Bedrock Manufacturing	PC Works Plus	Some	75	Windows 2003	Vantage	*Onsite*: Each night; Offsite: once a week
VPro, Inc.	N/A	N/A	6	N/A	N/A	*Onsite*: Accounting data daily, everything else, weekly *Offsite*: No

ERP System Integration

Bedrock has purchased an ERP System called 'Vantage'. Before they made this purchase decision, they speculated whether to custom develop an ERP system instead of getting an off-the-shelf product. However, in order to make an informed decision on this, they felt that they needed to determine what would better fit their needs. When they began this process of determining the correct 'fit', they were confused in assessing and elaborating their actual needs. Thus, they decided to purchase a bundled package consisting of the 'off-the-shelf product' and the 'customization services' that came along with the product. But, they were unable to make sure whether the application could be sufficiently customizable to fit into their business requirements. Since they were unsure of this, they kept postponing their decision. Ultimately they decided to choose 'Vantage'. They later found out that the application could not be customized to an extent that would let them fulfill their business needs. An early beta version was rolled out with incomplete features. Initially, most employees were unwilling to use the new system. However, over a period of time, employees started using the system more frequently. Currently, the ERP System is used in all departments at Bedrock except in the Human Resources Department. Human Resources uses Microsoft Excel to meet their spreadsheet and data management needs, instead of the ERP System and they were unwilling to adopt the new system.

The senior management of Bedrock wanted to integrate Human Resources Department into the ERP System, but was unable to do so because of internal resistance within the HR department. In order to appease the employees within the HR department, they thought of customizing the application further to make sure that the application was compatible with the HR personnel's whims and fancies, but were afraid to do so since they felt these changes might cause dissatisfaction and reduced use with employees from other departments.

VPRO, INC – AN ASSESSMENT

VPro, Inc addresses all its IT needs in-house and does not use any third party IT services. They based their decision on rationale that since there aren't many IT functions needed for their business, no third party services are necessary.

IT Support and Development

Currently, their in-house IT department comprises of only one employee, who addresses the needs of eight other employees. Since they do not employ any other third party services, the IT personnel is expected to cater to each and every need of the other eight employees within the company. There

Table 3. Comparison of communication system used at the two organizations

Name	Mobile	Email	Instant Messaging
Bedrock Manufacturing	Verizon	Exchange Server 2003 Pro	Message Pal
VPro, Inc.	Yes	Outlook	Yes

are no specific responsibilities assigned to the IT department, apart from responding to issues raised by employees and resolving them. They feel that it is manageable so far, since their current employee head count is just eight. However, they do have plans for expansion and will likely be around 20 members strong by the end of next year. Thus, they are not sure whether the current system could cater to the future needs and they did not think much about exploring the possibilities of using the services of a third party. Also, VPro, Inc. currently does not use any custom developed software, since they find it difficult to integrate it with other software systems they are using currently. They are considering the possibilities of purchasing an off-the-shelf ERP system, but they feel that none of the existing systems could offer what they really need.

Workstations and Servers

VPro, Inc. currently has six workstations, but they do not have a server. Five of their workstations are connected to the Internet. Due to this lack of central storage area, employees are facing two major problems – 1) they are unable share information effectively. 2) They are unable to access information remotely. Due to this lack of central storage area, employees follow a round-about process of transferring the order forms to make needed modifications and get approvals from key stakeholders. Currently, employees either e-mail the order forms to each other, or they pass around the form in a thumb drive, so that three to four people can edit and sign off on the order form. Thus, there are multiple instances and versions of the same order form and this leads to sub-optimal co-operation and poor customer service. A central storage location, which could in the form of a file server, could solve the problem and stream line this process.

Most of the employees who travel to client locations want remote access to VPro, Inc's network. These employees often forget to take the necessary documents with them and so, they must either call or e-mail their colleagues at office to request for those needed documents. Thus, remote access to a central file repository would solve this problem and offer the much needed flexibility employees. All of their workstations run on Windows XP Professional. There are some performance shortfalls in their workstations, since they were not upgraded for long. They are reluctant to explore new software applications, as they are afraid that it would be incompatible with their workstations. VPro Inc's office workstations are connected using a wired connection, but their shop floors are connected using wireless connections. They are using a DSL connection for Internet and they feel that it is adequate for now. They are not sure if this would hold true in the near future.

Backup and Intra-Company Communication System

VPro Inc's accounting data is backed up daily while all other data is backed up once a week. They use telephones, voice mail, and e-mail for their telecommunications but they do not use instant messaging. Their e-mails are backed up under the same system that they previously mentioned weekly. They believe that their communications are functioning smoothly due to the use of mobile, e-mail, and normal phone services. They use MS

Office, Outlook, Adobe, and DWG for drawings, DWG readers, and QuickBooks Pro 2006. VPI does not use any ERP systems, as they feel that none of the existing systems could offer what they really need. They also use QuickBooks Pro 2006 for accounting and financial purposes. VPI feels that all of their software works well independently as monoliths, but not as a whole. They feel that their systems could be better integrated, but they are not sure how to go about the process of integration.

THE SENSEMAKING OF IT RISKS

"Sensemaking of IT risks" is not only about making sense of the IT risks that emerges within the current environment, but also about enacting a future environment, envisioning the risks that might emerge in such an environment, and planning accordingly to avoid those risks. To put this in a slightly different way, sensemaking of IT risks is not only about the risks which emerges due to the *presence* of various artifacts, tools and technologies, and their interactions, but also about the risks that might emerge due to the *absence* of such artifacts, tools and technologies. It is about the process of constructing a hypothetical mental model of a future situation and how it could possibly evolve over time (Dervin, 1992; Weick, 1993).

Traditionally, researchers have focused on the former, i.e. IT risks that emerge within complex environments, with a multitude of interacting systems. However, not many studies have been conducted based on the latter, i.e. IT risks that emerges due to the *absence* of tools & technologies, due to a failure to construe a mental model of a future situation (business growth, personnel growth, and the corresponding IT needs). Thereby, Bedrock Manufacturing and VPro, Inc. provides an apt setting to understand this understudied phenomenon.

After the initial assessment conducted by the student team, a detailed analysis was carried out to envision the nature of IT risks which confronts these two organizations. Further interview questions were developed to understand how the perception about the role of information systems differed among SMEs which are at different stages of growth and at different levels of maturity with respect to their information systems. To accomplish this, the student teams conducted semi-structured, qualitative interviews with the owner/manager's of these two companies.

Mr. Daniel Lewis is the Director of Product Marketing for the Bedrock Manufacturing in Bedford, PA. Mr. Lewis also takes care of the IT department within Bedrock and he is the key decision maker when it comes to planning and managing Bedrock's IT operations. Mr. Lewis feels that Information Systems' are critical to their day-to-day operations, and thus, he continually emphasized the importance of streamlining their IT function they to a level that would meet their needs and the expectations of their client. He also feels that his organization has reached a certain level of maturity with respect to aligning their IT organization with the Business organization. He values the learning and the resultant knowledge which the company had acquired over a period of time, which enabled them to streamline their IT operations in an effective and efficient manner. However, Mr. Lewis feels that there is still room for growth and expansion. Though Mr. Lewis feels that third party services sufficiently meets the requirements and handles the problems quickly and efficiently, he is worried whether this would remain efficient if the company continues to grow. He is actively brainstorming with the senior management of Bedrock to think of alternative action plans to address their future IT needs. He is cognizant of the consequences of the lack of action with respect to envisioning and planning Bedrock's future IT needs. When asked how he went about envisioning the future IT needs of Bedrock, Mr. Lewis replied that he thought about the failures that could possibly occur, if the organization continued to grow and if its information systems remains the same. This in turn let him foresee the possible risks that might arise as a consequence of "inaction" in the present. Through this process of constructing a hypothetical mental model of the current situation and how it could possibly evolve over time, Mr. Lewis could make sense of an ambiguous situation that might occur in the near future. This let him plan accordingly to meet the future demands.

Mr. Jack Norton, the president of VPro, Inc. was also interviewed as a part of the student consulting team. Their IT functions are all supported in-house and they do not use any third party services. Neither do they intend making use of third party services in the near future. When asked for the reason, Mr. Norton replied, *"We decided against it since there aren't many IT functions in our business"*

Currently, their IT department has a single member. Sometimes, ad-hoc requests and upgrades and installations are handled by Mr. Norton himself. Though Mr. Norton sees that this could pose a problem when the company expands, he had not thought much about this. Also, VPro, Inc. currently does not use any custom developed function-specific software, though they would like to. The reason for this is attributed to their inability to integrate custom developed software with their existing Information Systems. Also, they often considered moving to a mid-market, off-the-shelf ERP system, but they feel that none of the vendors showed what they really need. Mr. Norton says:

"They have not shown us what we want! The existing ERP packages are either too costly or they do not have the functionalities which we would like to have".

They are reluctant to try out new software applications, since they are afraid that it might not be compatible with their existing workstations. They currently have six workstations with no server. In addition to that, they are not sure how to go about handling the 'central storage repository' and 'remote access' issue. Though Mr. Norton feels the criticality of sorting out these issues quickly, he did not think about it much. Also, Mr. Norton feels that all of their existing software works properly but could be better integrated. When asked whether they tried to integrate those applications, Mr. Norton replied that they have not tried to integrate their discrete applications, since they do not know how to do it without interrupting the day-to-day operations. In addition to that, he feels that the main reason why VPro, Inc. is reluctant to invest more in their Information Systems and IT infrastructure is that they feel that their business would not benefit enough from those investments. Though he feels that it does hinder their business growth, it is not a major area of concern for them currently. When asked how he went about envisioning the future IT needs of VPro, Mr. Norton said that he did not think about it much either since he feels that there are other "larger" operational and customer-related issues which he needs to handle first, before he could think about his IT needs.

THE INFLUENCE OF GROWTH TRAJECTORIES ON SENSEMAKING OF IT RISKS

From the conducted exploratory study, we observed that SMEs are not unitary entities and even with them, there are sub-sections, each at a different growth stage and at a different level of maturity with respect to their information systems. We observed linkages between the growth stage a SME is at, the perceived role of information systems and the sensemaking of IT risks. While Bedrock Manufacturing emphasized the critical role of information systems to their business, VPro, Inc. did not hold the same view. In addition to that, the nature of information systems' risks differs depending on what stage of growth the SME is at. While Bedrock was confronted with issues related to 'integrating discrete applications', 'employee resistance to adopt new technology' etc., VPro, Inc. faced rudimentary problems related to the 'remote access' and 'centralized repository'.

Also, the differences and overlaps existing among enterprise, entrepreneurship and small business is often much discussed (Bridge, O'Neill, & Cromie, 2003). Various approaches (Economic, Sociological, and Cognitive) and theories (Personality theories, Behavioral theories) try to distinguish between entrepreneurial firms and small businesses and they try to come up with a demarcation between the two (Carland, Hoy, Boulton, & Carland, 1984). Often, 'risk-taking propensity', 'achievement motivation' and 'preference for innovation' are considered to be some of the key variables which influence this demarcation. Entrepreneurs are said to exhibit a higher risk-taking propensity than small business owners (Stewart, Watson, Carland, & Carland, 1999).

From the conducted pilot study, we observed *'sensemaking of risks'* to be one of the key variables that influence the demarcation between entrepreneurs and small business owners. Owner/manager's of Bedrock Manufacturing were cognizant of the consequences of the lack of action with respect to envisioning and planning their future IT needs. They achieved this through a process of constructing a hypothetical mental model of a future situation and how it could possibly evolve over time (Dervin, 1992; Weick, 1993). Through this sensemaking process, they could evaluate and assess the risks that might occur in the near future and plan accordingly to mitigate/avoid them. On the other hand, owner/manager's of VPro, Inc were unable to construct this hypothetical mental model of a future situation. Thus, they were unable to foresee, evaluate and assess the risks related to their information systems.

Our proposition is that IT risks faced by a 15 member firm during their process of switching from Microsoft Excel to a simple Accounting package is different from a 60+ member firm who is undergoing a transition from a simple accounting package to a mid-tier ERP system. This in turn is different from the risks faced by the company, which uses a full blown ERP system, has a process consultant in its payroll and wants to move to the next level – standardizing their business processes using business process modeling tools, and the nature and criticality of these risks is proportionate to the stage of growth of an SME. In addition, the process of "Sensemaking IT Risks" is also influenced by the growth trajectory of SMEs, since the growth trajectory encompasses elements like past critical decisions made and strategies taken, rules, routines and best practices acquired, etc.

CONCLUSION

A number of tentative conclusions may be reached from this exploratory study. Risk Management studies within the IS/IT stream should not ignore the "organizational context" and should move beyond coming up with ideal frameworks for abstract organizations (Agell, 2004; Alter & Sherer, 2004a; Ballantine & Levy, 1998; Cragg & King, 1993). They should focus on contextual and structural dimension like the enterprise size and age, growth rate, formalization, centralization, number of organizational levels etc. Contextual and Structural differences between a 'large enterprise' and a 'SME' should be explicitly recognized. Also, SMEs should not be abstracted out as unitary entities. From our research study, we observe that there are different varieties of risks within SMEs and the nature of those risks is proportionate to the stage of growth an SME is at.

In addition to 'risk-taking propensity' and other variables like 'achievement motivation' and 'preference for innovation' (Carland, et al., 1984), we consider active 'sensemaking of risks' to be one of the key variables that influence the demarcation between entrepreneurs and small business owners. This process of sensemaking is in turn influenced by its 'organizational filters', which in turn is influenced by the organizational growth trajectory - in terms of the past crisis/response it encountered through out its growth stages and the rules, routines and best practices it acquired over a period of time. Hence, it is important to consider the growth trajectory of SMEs as a critical variable towards influencing the process of "Sensemaking IT Risks". As the firm traverses through different growth stages, their 'risk sensemaking' capability tends to vary.

Opportunities to extend the study would include the following. First, key variables could be identified to demarcate SMEs according to their growth stage. Based on the identified variables, theoretical sampling could be done, as per the SME growth stage and other key business growth indicators. Then, risks which are specific to a particular growth stage could be identified and validated through a mixture of qualitative and quantitative studies. Finally, based on the findings of the above studies, we could come up with constructs to measure the linkage between 'SME Growth Stage', 'Perceived Usefulness of Information Systems' and 'Sensemaking of IT Risks'.

ACKNOWLEDGMENT

The work reported has been funded by the National Science Foundation's PFI grant under award number 0650124. Any opinions, findings and conclusions or recommendations expressed in this material are those of the author(s) and do not necessarily reflect the views of the National Science Foundation (NSF).

REFERENCES

Agell, J. (2004). Why are small firms different? Managers' views. *The Scandinavian Journal of Economics, 106*(3), 437–452. doi:10.1111/j.0347-0520.2004.00371.x

Alter, S., & Sherer, S. A. (2004). A General, but readily adaptable model Of information system risk. *Communications of the Association for Information Systems*, (14): 1–28.

Anderson, C., & Andersson, M. (2006). The long tail. *Wired Magazine, 12*(10).

Ballantine, J., & Levy, M. (1998). Evaluating information systems in small and medium-sized enterprises: issues and evidence. *European Journal of Information Systems*, 7(4), 241–251. doi:10.1057/palgrave.ejis.3000307

Bell, D. (1973). The coming of post-industrial society. *Business & Society Review/Innovation* (5).

Bridge, S., O'Neill, K., & Cromie, S. (2003). *Understanding enterprise, entrepreneurship and small business*. Basingstoke, UK: Palgrave Macmillan.

Carland, J., Hoy, F., Boulton, W., & Carland, J. (1984). Differentiating entrepreneurs from small business owners: A conceptualization. *Academy of Management Review, 9*(2), 354–359. doi:10.2307/258448

Churchill, N. C., & Lewis, V. L. (1983). The five stages of small business growth. *Harvard Business Review, 61*(3), 30–49.

Cohen, W., & Levinthal, D. (1990). Absorptive capacity: a new perspective on learning and innovation. *Administrative Science Quarterly, 35*, 128–152. doi:10.2307/2393553

Corley, K. (2002). Breaking away: an empirical examination of how organizational identity changes during a spin-off. Unpublished doctoral dissertation, The Pennsylvania State University, Pennsylvania.

Cragg, P. B., & King, M. (1993). Small-firm computing: Motivators and inhibitors. *Management Information Systems Quarterly*, *17*(1), 47–60. doi:10.2307/249509

Dervin, B. (1992). From the mind's eye of the user: the sense-making qualitative-quantitative methodology. In Glazier, J. D., & Powell, R. R. (Eds.), *Qualitative research in information management* (pp. 61–84). Englewood, CO: Libraries Unlimited.

Dutton, J., & Duncan, R. (1983). *The creation of momentum for change through the process of organizational sensemaking*. Unpublished manuscript, New York University, New York.

Dutton, J., & Duncan, R. (1987). The influence of the strategic planning process on strategic change. *Strategic Management Journal*, *8*(2), 103–116. doi:10.1002/smj.4250080202

Dytche, J., & Warren, A. (2007a). *I99 Corridor Innovation Portal - summary: Farrell Center for Corporate Innovation and Entrepreneurship*. Pennsylvania: The Pennsylvania State University.

Dytche, J., & Warren, A. (2007b). *Penn State KIZ Innovation Grant Team I-99 Innovation Network Portal: Farrell Center for Corporate Innovation and Entrepreneurship*. Pennsylvania: The Pennsylvania State University.

Faberman, R. (2002). Job flows and labor dynamics in the US Rust Belt. *Monthly Labor Review*, *125*(9), 3–10.

Gioia, D., & Chittipeddi, K. (1991). Sensemaking and sensegiving in strategic change initiation. *Strategic Management Journal*, *12*(6), 433–448. doi:10.1002/smj.4250120604

Gioia, D., & Thomas, J. (1996). Identity, image, and issue interpretation: Sensemaking during strategic change in academia. *Administrative Science Quarterly*, *41*, 370–403. doi:10.2307/2393936

Hanks, S. H., Watson, C. J., Jansen, E., & Chandler, G. N. (1993). Tightening the Life-Cycle Construct: A Taxonomic Study of Growth Stage Configurations in High-Technology Organizations. *Entrepreneurship: Theory and Practice*, *18*(2), 5–30.

Henderson, J. C., & Venkatraman, N. (1999). Strategic alignment: Leveraging information technology for transforming organizations. *IBM Systems Journal*, *38*(2/3), 472–484.

Kahn, M. (1999). The silver lining of Rust Belt manufacturing decline. *Journal of Urban Economics*, *46*(3), 360–376. doi:10.1006/juec.1998.2127

Levy, M., Powell, P., & Yetton, P. (1998). *SMEs and the gains from IS: from cost reduction to value added*. Paper presented at Joint Working Conference on Information Systems, Helsinki, Norway.

Masuda, Y. (1980). *The information society as post-industrial society*. Tokyo, Japan: Institue for the Information Society.

Miner, J. B., & Raju, N. S. (2004). Risk propensity differences between managers and entrepreneurs and between low- and high-growth entrepreneurs: A reply in a more conservative vein. *The Journal of Applied Psychology*, *89*(1), 3–13. doi:10.1037/0021-9010.89.1.3

Nelson, R., & Winter, S. (1982). *An evolutionary theory of economic change*. Harvard University Press.

O'Farrell, P., & Hitchens, D. (1988). Alternative theories of small-firm growth: a critical review. *Environment and Planning, 20*(3), 1365–1383. doi:10.1068/a201365

SBA. (1998). *The new American evolution: The role and impact of small firms*. Small Business Research Report. Retrieved September 12, 2008, from http://www.sba.gov/advo/

SBA. (2004a). *Small business resources for faculty, student, and researchers*. Retrieved September 12, 2008, from http://www.sba.gov/advo/.

SBA. (2004b). *Small Firms and Technology: Acquisitions, Inventor Movement, and Technology Transfer, Small Business Research Summary No. 233*. Retrieved September 12, 2008, from http://www.sba.gov/advo/

SBA. (2004c). Top ten reasons to love small business. *Small Business Administration News Release, SBA 04-06 ADVO*. Retrieved September 12, 2008, from http://www.sba.gov/advo/.

Speyer, M., Pohlmann, T., & Brown, K. (2006). *IT spending in the SMB sector*. Cambridge, MA: Forrester Research.

Stewart, W., Watson, W., Carland, J., & Carland, J. (1999). A proclivity for entrepreneurship A comparison of entrepreneurs, small business owners, and corporate managers. *Journal of Business Venturing, 14*(2), 189–214. doi:10.1016/S0883-9026(97)00070-0

Stoneburner, G., Feringa, A., & Goguen, A. (2002). *Risk management guide for information technology systems* (Tech. Rep. No. SP800-30). National Institute for Science and Technology.

Thomas, J., Clark, S., & Gioia, D. (1993). Strategic sensemaking and organizational performance: Linkages among scanning, interpretation, action, and outcomes. *Academy of Management Journal, 36*(2), 239–270. doi:10.2307/256522

Warren, A., Hanke, R., & Trotzer, D. (2008). Models for university technology transfer: resolving conflicts between mission and methods and the dependency on geographic location. *Cambridge Journal of Regions. Economy and Society, 1*(2), 219–232.

Weick, K. (1995). *Sensemaking in organizations*. Sage Publications, Inc.

Weick, K. E. (1993). The collapse of sensemaking in organizations: The Mann Gulch Disaster. *Administrative Science Quarterly, 38*(4), 628–652. doi:10.2307/2393339

Weick, K. E., & Sutcliffe, K. M. (2001). *Managing the unexpected: Assuring high performance in an age of complexity*. San Francisco, CA: Jossey-Bass.

Wernerfelt, B. (1984). A Resource-based view of the firm. *Strategic Management Journal, 5*(2), 171–180. doi:10.1002/smj.4250050207

Chapter 9
Use of the Concern-Task-Interaction-Outcome (CTIO) Cycle for Virtual Teamwork

Suryadeo Vinay Kissoon
RMIT University, Australia

ABSTRACT

This chapter introduces the CTIO (Concern-Task-Interaction-Outcome) Cycle as a means of studying team member interaction using face-to-face and virtual interaction media in retail banking. The type of interaction is discussed in terms of different conceptual cycles having a linkage in the framing of the CTIO Cycle. In the past, routine teamwork using face-to-face communication was important. Today, with emerging technologies for retail banking organizations, teamwork through virtual communication has been gaining importance for increased productivity. This chapter addresses different problem-solving cycles, each of which relates to the mode of interaction medium (whether face-to-face or virtual) used by team members, facilitators, or managers to resolve problems in the workplace. The chapter focuses on understanding the relationship between face-to-face and virtual interaction variables. This is important to researchers in identifying retail banking trends using hybrid teams and virtual group networks with routine teamwork. Using virtual over face-to-face interactions in the different data life cycles linkages are gaining importance from the perspectives of data and information quality. This can be attributed to the increased use of technologies and virtual network features. Current trends are leading to the triangulation of continuous improvement, routine teamwork, and virtual teamwork in support of retail banking organizations achieving productive performance.

DOI: 10.4018/978-1-61520-609-4.ch009

Figure 1. The CIT model realized through the CTIO cycle (Kissoon, 2007)

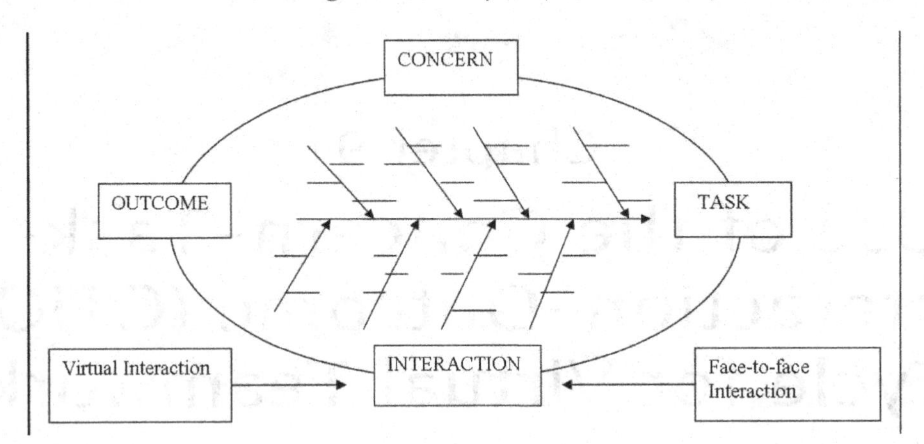

BACKGROUND

This chapter provides a background on an evolving approach to teamwork in the retail banking sector. The case refers to a new teamwork approach where routine teamwork is integrated with virtual teamwork using a continuous improvement initiative. The Concern-Task-Interaction-Outcome (CTIO) Cycle (Kissoon, 2007) is the continuous improvement initiative mapped from other conceptual problem-solving data life cycles.

The CTIO Cycle refers to the new evolving consultative, participative, virtual and interactive virtuous teamworking approach. It reflects the effect of employee interaction using both face-to-face and virtual interaction media in achieving productive performance for the organization (Kissoon, 2008a).

Routine teamwork is used in reference to face-to-face interaction, while virtual teamwork is used in reference to virtual interaction media. The CTIO Cycle was researched in a major Australian financial organization, which employs about 30,000 employees and 1,700 retail branches in all states of Australia.

The technology utilization of the financial organization relates to conferencing, teleconferencing, videoconferencing, voice mail, internet/intranet and many other networks features, as shown in Appendix 1. The players involved are the branch managers, team leaders, financial planners, home loan managers, personal bankers, customer service officers and tellers in retail banking branches.

CONTINUOUS IMPROVEMENT TEAMWORK (CIT) MODEL

The Continuous Improvement Teamwork (CIT) Model, demonstrated by stages in the CTIO cycle (shown in Figure 1), is a virtual teamwork approach. The circle in Figure 1 represents the continuous working towards resolution of a concern through face-to-face interaction or virtual interaction by team members. This approach is aligned with common organizational objectives of effective communication using emerging technologies (Kissoon, 2008a). The CIT Model is illustrative of an evolving participative, virtual approach to teamwork currently used by a major Australian banking organization with about 30,000 employees including its international branches. The company's objective for using the CIT Model is to achieve quality performance for its products and services.

The CIT Model is comprised of the following phases:

- **Concern (Issue):** A team member, or team members, identifies an issue related to organizational performance.
- **Task (Action):** The issue identified needs to be addressed as soon as possible through tasks by team members, facilitators, or managers working for the respective organization or as an external consultant.
- **Interaction (Involvement and Connection):** The various tasks are done through face-to-face interactions and/or virtual interactions through communication media between team members.
- **Outcome (Result):** The CIT is a continuous process in achieving successfully and productive outcomes for the benefit of the firm, team and stakeholders.

The CIT Model is part of the CTIO Cycle. The CIT Model is a classificatory framework (Kissoon, 2008a), comprised of the mapping of three knowledge domains: continuous improvement, teamwork, and e-teamwork leading to the emergence of the Continuous Improvement Teamworking (CTIO) model. The CTIO Cycle is framed from other conceptual data life cycles (such as PDCA (Plan-Do-Check-Act) Cycle, DMAIC (Define, Measure, Analyze, Improve, Control) Cycle, Data Evolution Life Cycle, NEAT Methodology Data Life Cycle, and Information System Life Cycle). These process improvement approaches will be later explained.

This case illustrates the use of the CIT Model within the context of the CTIO Cycle. The CIT Model, evidenced through the CTIO Cycle (Kissoon, 2008a), was researched for about three years in a major leading Australian's service sector organization using both a deductive and inductive reasoning approach. The results from this participant observation study showed that synchronous conferencing, Internet online functional services, continuous improvement, and team meetings form the essential four core elements of the CIT Model/CTIO Cycle. The study showed that the adoption of the CIT Model assists in improving retail banking operational activities and in achieving better performance (Kissoon, 2008b).

The CIT Model/CTIO Cycle is presented in this case based on observations made in the Australian retail banking sector. The use of the CIT Model/CTIO Cycle is appropriate for the banking sector, as interactions among team members is critical to operational effectiveness related to banking services.

Kissoon (2007) identified competition as a primary reason for the emergence of the CTIO Cycle and the adoption of the CIT Model. Without deregulation, which started in 1980`s, banking organizations would not have been so competitive. International banking organizations with expertise in retail banking were compelled to enter the domestic, Australian market due to competitive advantage. With the introduction of the Financial Services Reforms Act (FSRA) of 1988 by the Australian government, following the stock market crash in 1987, competition was encouraged (Hutley & Russel, 2005).

Increased competition means that efficiencies in operations and customer service are key to profitability. The CIT Model promotes effective communication internally among employees and externally with customers. With the increasing use of Internet and Web technologies, the CIT concept is integral to promoting both face-to-face and electronic teamwork (e-teamwork) in the efficient resolution of issues as well as meeting the needs of customers. The CIT Model is a continuous loop focused on problem-resolution. As part of the CTIO Cycle, it offers an organization the opportunity for effective communication within a process improvement loop.

Researchers and practitioners have identified the concepts of teamwork, e-teamwork, virtual communication, and continuous improvement as crucial parameters in the service sector. Each of these plays an important role in addressing customer concerns effectively. They are also part of the Total Quality Management (TQM). TQM is comprised of management practices that are applicable throughout the organization. TQM promotes organizational consistency such that customer needs are met (or exceeded). TQM also promotes process metrics and control mechanisms in order for an organization to continuously improve.

TQM continues to grow in popularity in organizational areas of service quality, data quality, information quality, and performance management. Each of these areas of TQM is supported by effective communication within and external to the organization. E-teamwork and virtual communication, in particular, allow for customer interaction transcending time and location boundaries.

A participant study was conducted by Kissoon (2008a) over a nine month time period. The results of the study showed that the CIT approach addresses organizational issues associated with the smooth, operational activities of a bank through the use of face-to-face interaction; as well as, through virtual interaction media used to communicate with internal and external customers.

The CTIO Cycle is different than other performance cycles; such as, the PDCA Cycle, NEAT Methodology data life cycle, Benchmarking Cycle, Kolb's Cycle, and RADAR Life Cycle, among others, as it involves the triangulation of traditional teamworking, virtual teamwork and Continuous Improvement (CI). Continuous Improvement is part of the management of all systems and processes (Evans & Dean, 2003). Interaction mode of team members in a Continuous Improvement approach is important when service quality is initiated in an organization. Continuous Improvement, from a virtual teamwork perspective, means that team members in performance cycles uses both routine teamworking integrated with CI and virtual teamworking integrated with CI. The knowledge contribution is the integration of traditional teamworking and virtual teamworking with CI in the performance and problem-solving cycles to achieve productive performance.

The concept of virtual teams was added to the concepts of TQM and teamwork in developing the existing CIT Model. This integration of concepts is referred to as the Continuous Improvement Teamwork (CIT) approach whereby the concepts of teamwork, virtual teamwork, and continuous improvement are amalgamated to foster better productive performance and improve customer service (Kissoon, 2008a). The key measures, for both face-to-face interaction and virtual interaction, are presented in Appendix 1.

Many other concepts and performance cycles also have a linkage to the CIT Model/ CTIO Cycle, as briefly illustrated in this chapter. These performance cycles are mainly focused in the service sector. Deming Planning, Doing, Checking and Acting (PDCA) Cycle; Six-Sigma Defining, Measuring, Analyzing and Improving (DMAIC) Cycle; Root Cause Hypothesis Analysis Cycle; Data Evolution Life Cycle; NEAT Methodology Data Life Cycle; Information System Life Cycle; Resulting, Approaching, Deploying, Assessing and Reviewing (RADAR) Cycle; Acceleration of Innovation ideas to Market (AIM) innovation Life Cycle, Ethnographic Research Cycle, Action Research Cycle, among others, have a linkage with the CIT Model which is realised through the CTIO Cycle.

CTIO CYCLE LINKAGES TO OTHER CONCEPTUAL CYCLES

Deming PDCA or Deming PDSA Cycle

The variant PDSA Cycle of the traditional Deming-Shewhart PDCA Cycle is a simple methodology for continuous improvement. The PDCA cycle helps team members solve problems. It provides the structure for work improvements so that a team can: (1) use process tools logically, (2) identify and analyse problems, (3) develop workable solutions, and (4) solve problems and ensure that they will not happen again (STA, 1996). This learning

Figure 2. The PDCA cycle problem solving approach (Quality System, 1996)

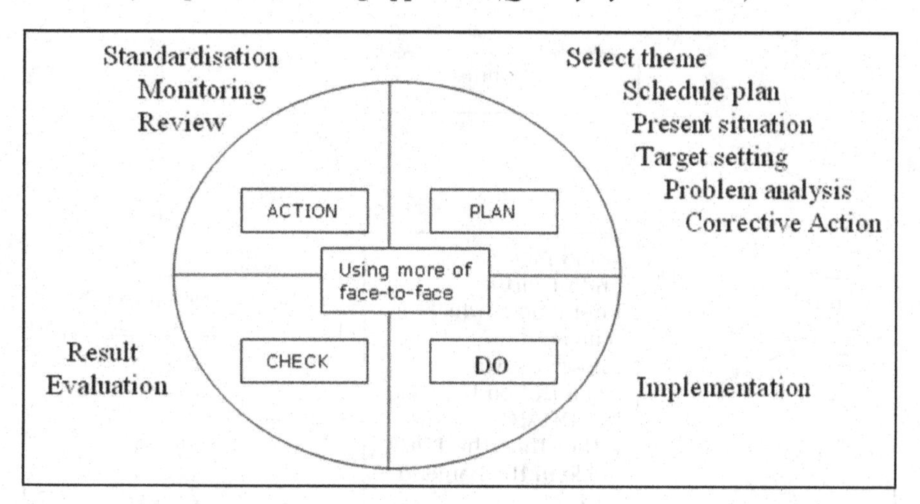

loop of planning, doing, checking, and acting is shown in Figure 2.

The CIT approach adopts the philosophy of the PDCA Cycle with its systematic problem-solving method. The CIT approach integrates teamwork into the process improvement cycle inclusive of traditional (routine) teamwork and virtual teamwork.

With globalisation, technological advances, virtual organizations, deregulation, and organizational competitiveness, the amalgamation of teamwork with virtual teamwork, as a hybrid team, is being envisaged. The use of hybrid teams has only recently emerged given technological and societal advances. However it is important to note that the original quality circles, which in the 1980's used the PDCA problem-solving approach, placed increasing emphasis on teamwork in organizational importance.

The Six-Sigma Methodology (DMAIC Cycle)

DMAIC Cycle, shown in Figure 3, is a problem-solving cycle related to improvement and achieving better customer service in the service sector. It is used by organizations to continuously monitor customer requirements and assess process performance. Six Sigma is a derivative of TQM (Total Quality Management) emerging in the late 1980`s as a way for an organization to solve quality problems and maintain improvement. These process improvement techniques were first adopted by Motorola.

The successful completion of a Six Sigma team project is achieved through the effective use of teams. Team members, both virtuously and virtually, work together as a team to achieve organizational outcomes (e.g., process efficiencies, quality customer service). The benefit of combining the CIT Model with Six Sigma formal processes includes team efficiencies in job performance, effective use of virtual interaction media, convergence of team members into high performance, business-to-enterprise (B2E) teams, information quality, virtual working arrangements (e.g., telecommuting), e-training (electronic training), e-learning (electronic learning), and effective use of collaborative tools and other computer mediated communication. This breadth of benefits is typically not achievable by face-to-face interaction of team members working on process improvement associated with a project. The CIT approach illustrates the interaction of team members face-to-face and virtually with managers and

Figure 3. The Kaizen DMAIC six-sigma cycle

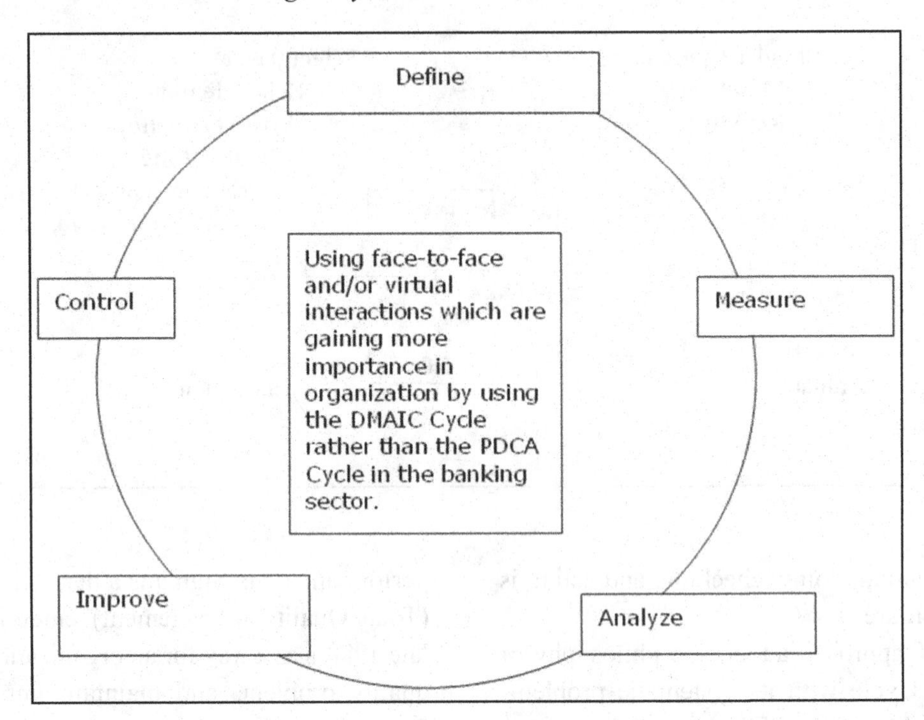

other team members to work better as a dynamic team on any available project.

DMAIC Cycle Application in the Financial Services

The results of research with senior representatives from eleven financial services organizations showed that these financial institutions are leading exponents of lean, Six-Sigma and business Process Management Methods within the financial industry (Hayler & Nichols, 2007). These financial institutions are American Express, Bank of America, Credit Suisse, Dresdner Kleinwort Wasserstein, First Data Resources, JP Morgan Chase, Lloyds TSB, MBNA Consumer Finance, Merrill Lynch, Overseas Chinese Banking Corporation and UBS. Pande, Newman and Cavanagh (2000) identified Six Sigma as the organizational quality management system helping leading international organizations in saving millions of dollars by producing more satisfied customers. All four major Australian banking organizations are using Continuous Improvement and Six-Sigma problem-solving methodology (Figure 3) for achieving high performance of their organizations.

The Root Cause Hypothesis Analysis Cycle

The Root Cause Hypothesis Analysis Cycle (Hayler & Nichols, 2007) is illustrated in Figure 4 where the "Analyze" stage from any problem-solving cycle can be applied in process improvement as a cycle using both face-to-face interaction and virtual interaction of team members working on the project. The cycle is driven by generating and evaluating "hypotheses" (or "educated guesses") as to the cause of the problem. The causes are evaluated by using available data obtained from processes. Each of the hypotheses are refined or rejected according to the severity of the causes. The vital causes leading to the root causes are

Figure 4. The root cause hypothesis analysis cycle (Hayler & Nichols, 2007)

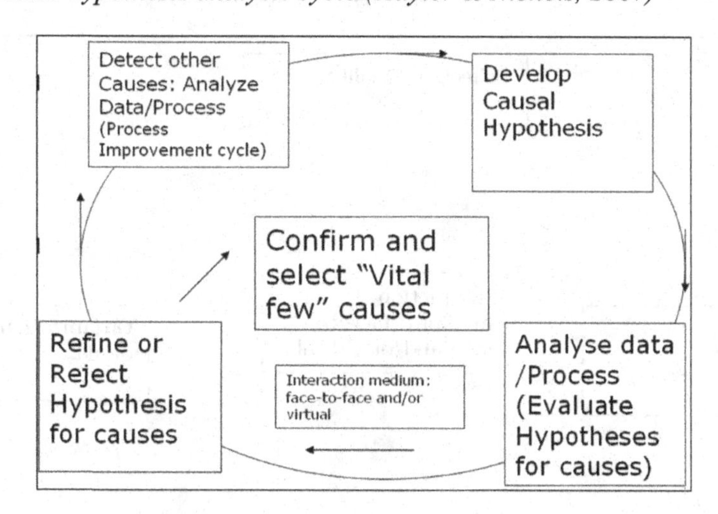

selected and confirmed for affecting the process in the process improvement cycle, which can be found by team members using both face-to-face and virtual interactions. Hence, root cause analysis is used to analyze the source of quality problems associated with quality improvement.

Similarly, in the "Task" stage of the CTIO Cycle, the Root Cause Hypothesis Analysis Cycle can be applied to detect the vital root causes affecting teamwork and virtual teamwork interactions showing a linkage of the conceptual cycles.

The Data Evolution Life Cycle

As described by Maier, Muegeli and Krejza (2007), a standard for information quality does not exist. The intuitive approach derives attributes for information quality based on personal experience (Wang, Reddy, & Kon, 1995; Miller, 1996; Redman, 1996; English, 1999), the empirical approach quantitatively illustrates the data consumer's point of view in the service sector about quality dimensions important to their tasks (Wang & Strong, 1996; Helfert, Zellner, & Sousa, 2002); and, the theoretical approach proposing quality dimensions that build upon established theory (Ballou & Pazer, 1985; Te'eni, 1993; Wand & Wang, 1996; Liu & Chi, 2002; as cited by Maier, Muegeli, & Krejza, 2007).

The data evolution lifecycle, shown in Figure 5, is normally used as a theoretical basis characterising the typical sequence of data evolution stages to derive four data quality stages. These consist of: (1) data collection, (2) organization, (3) presentation; and, (4) application (Liu & Chi, 2002; as cited by Maier, Muegeli, & Krejza, 2007). Each of these stages is derived through a task and an interaction medium, which can be either face-to-face and/or virtual. The objective is to effectively use collected information to focus on data quality. At each stage in the data evolution lifecycle, typical root causes of poor data quality as well as specific measurement attributes and models are derived through face-to-face and/or virtual interactions of team members.

At Credit Suisse, this conceptual model is used for illustrating information quality issues with improvements in the customer investigational process (Maier, Muegeli, & Krejza, 2007). The process consists of two connected data evolution lifecycles. The first evolution lifecycle is the customer inquiry process. It is specified by team member interactions. The second data evolution lifecycle is initiated by the application of the customer inquiry data triggers. An illustration of this is: (1) An inquiry phase results in a new customer inquiry; and, (2) the reply phase produces a response to the customer inquiry.

Figure 5. The data evolution life cycle (Maier, Muegeli, & Krejza, 2007)

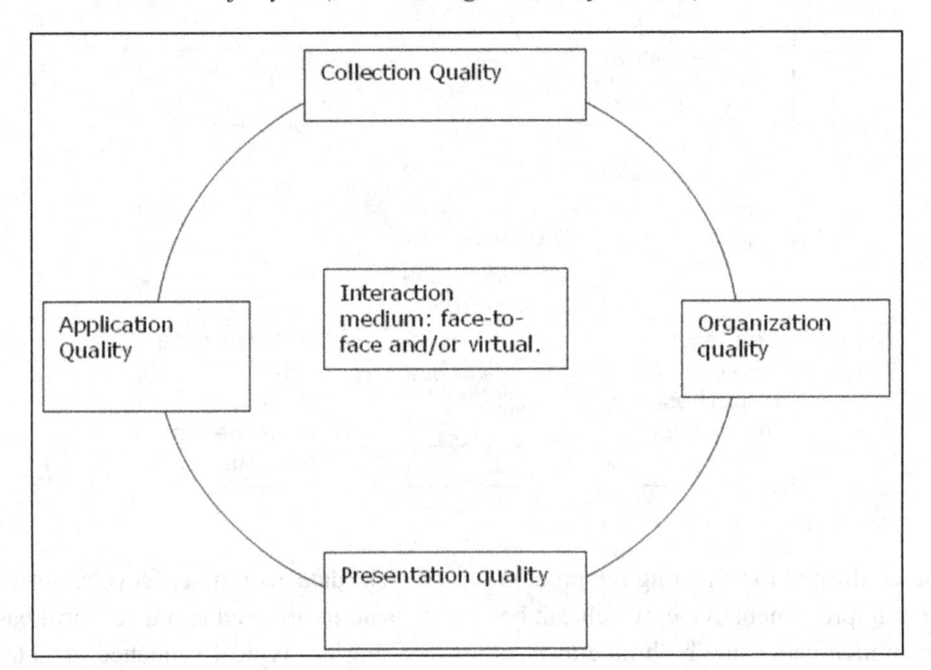

The CTIO Model implements a similar cycle in addressing customer issues and other concerns. The overall objective is the reliance on hybrid teams for resolution outcome. The data-oriented view of the customer investigation process relates mostly to data quality. There is nothing mentioned about team interaction in the data evolutionary life cycle. As such, it is unknown as to how a team utilizes the data generated by the use of the model in arriving at a valid outcome. The CTIO Model illustrates the interaction type being used in a CIT approach to effectively use quality data. The research conducted on CIT is important, as it has shown that using various medium of communication, either or both face-to-face and/or virtually, is important for the effective operations of the organization.

The Customer Investigation Process (CIP) is illustrated, as described by Maier, Muegeli and Krejza (2007) and carried out through banking staff interaction at Credit Suisse. The Credit Suisse internal investigation process initiates the inquiry asked by the inquirer, which may be external such as government or internal, to identify the appropriate receivers or consignees associated with the specific inquiry. These receivers will then start to identify the departments and people that might have relevant information as owners of information archives. The receiver will consolidate all the information obtained from respective departments. When there are no accurate results, the same investigation process will be done with other departments and the request will be repeated until sufficient information collected. The final information dossier with a summary is normally sent to the inquirer. This work is quite complex and time consuming for Credit Suisse.

Though there are many people from various departments, nothing is mentioned on how the team members from Credit Suisse interact as a team using the CIT approach in addressing the concern raise by the inquirer. The CTIO model does illustrate how team members interact routinely and interactively in addressing customer concerns to come to the right solutions.

Figure 6. The NEAT methodology data life cycle (Bobrowski, Marre & Yankelevich, 2002)

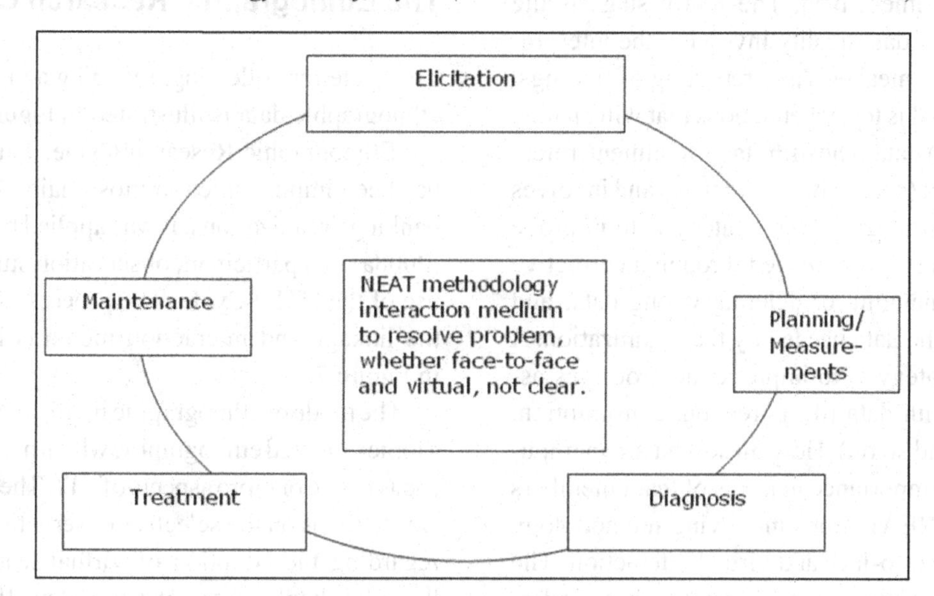

The NEAT Methodology Data Life Cycle

Many problems have emerged due to poor quality data collected through team member interaction, to which many software engineers can attest. To address these problems, the NEAT Methodology provides a systematic way in determining data quality for developing an improvement plan (Bobrowski, Marre & Yankelevich, 2002). The improvement plan constitutes both corrective and preventive actions in order to maintain the quality standards eventually met.

The NEAT Methodology is typically based on the Goal Question Metric (GQM) framework. The NEAT Methodology identifies the need for making a diagnosis to assess based on its output and provides the convenience of implementing a corrective improvement action on the data while maintaining focus improvement expectations. The interaction media used is also important, as the NEAT Methodology facilitates organizations for evaluating the investment for improving its data. The NEAT Methodology (shown in Figure 6) is typically presented as a theoretical model of the data life cycle; and also, as a practical approach to performing evaluations in controlled environments (Bobrowski, Marre & Yankelevich, 2002).

The NEAT Methodology guides data quality evaluation by the implementation of six stages: elicitation; planning; measurement; diagnosis; treatment; and maintenance. Each stage is composed of several tasks (Bobrowski, Marre & Yankelevich, 2002). Each task is carried out by team member interaction.

The first stage in the NEAT Methodology is elicitation, which is the acquisition of information on the organizational data life cycle. This includes the evaluation of the state of the data. The stage involves a precise description of the actual state (a snapshot) over different subsets of data with two levels of quality: the target (optimum) level and the minimum level required. The second stage, planning, is the development of a plan for assessing quality. It includes the use of an evaluation plan along with an explanation of how the evaluation plan is executed. The plan is updated with each task. The third stage, measurement, relates to the measurement of certain attributes using tools and techniques such as GQM tables, templates, and data analysis tools according to the measurement plan. The fourth stage is the diagnosis of data quality involving the interpretation of the measures and reporting of findings. Its main goal is to trigger actions that will change the status of data. The fifth stage, treatment, refers to treatment (corrective/preventive) and involves two main strategies. One strategy is to improve existing data accomplished through a corrective strategy, changing or deleting wrong data, and including the data needed by the organization. A second strategy is to improve the processes associated with data (data creation, consumption, updates, and so on). How these data are manipulated is of importance in terms of team members using the NEAT problem-solving methodology through face-to-face and virtual interaction. The latter, virtual interaction of team member, hinders incorporation of poor data into the system. The sixth stage, maintenance, provides a mechanism in maintaining quality once goals are met. This is strongly related to the treatment stage where data quality is monitored through time-based, systematic and periodic measurements, which in some cases are automated.

The NEAT Methodology is aligned with the CIT approach in terms of acquisition of information in the concern stage of the CTIO Cycle. The methodology performs certain phases that are part of the CTIO Model including planning, measurement and diagnosis. The interaction phase of the CTIO Cycle involves aspects of the NEAT Methodology treatment and maintenance phases where team interaction media offers critical support. The only difference found between the NEAT Methodology and the CTIO Cycle is the team interaction aspect. The CTIO Cycle uses both face-to-face and virtual teamworking. It is unclear which type of interaction media the NEAT Methodology uses as part of any of the six stages.

The Ethnographic Research Cycle

The cycle for collecting, recoding and analysing ethnographic data is illustrated in Figure 7 using the Ethnographic Research Cycle. This method has been implemented in most major Australian banking organizations. It was applied by Kissoon (2008a) in a participant observation study on the use of the CTIO Cycle. The process of showing the linkage and interaction media is illustrated in Figure 7.

The mode of ethnographic inquiry was used on a topic-oriented ethnography, which narrowed the focus to one or more aspects of CIT. The objective was to focus on the selective observation process regarding the adoption of virtual teamwork in the retail banking sector (Spradley, 1980). The process of narrowing participant observation was done by asking ethnographic questions related to the core elements of the CIT Model/CTIO Cycle. Throughout the observational process of nine months, several questions were asked (refer to Figure 7). The researcher analysed the responses to reach a reliable and valid conclusion about the effectiveness of the CIT approach. The Ethnographic Research Cycle may be considered linked to the CTIO Cycle, as there is a continual search through the participant observation study to diagnose and realize the CIT approach inclusive of face-to-face and virtual interactions. Both the Ethnographic Research Cycle and the CTIO Cycle have common, problem-solving characteristics, which have been used in the service sector.

Other Conceptual Cycle Linkages Using Face-to-Face and Virtual Interactions

The Continuous Improvement Teamwork, realised through the CTIO Cycle using both face-to-face and virtual team interaction, is linked to other conceptual models such as: the Benchmarking Cycle; the Kolb`s Cycle; the Reverse Logistics and Supply Chain Continual Cycle; the Action Re-

Figure 7. The Ethnographic Research Cycle, as Described by Spradley (1980) Used for The Observational Study. (The 5W-1H questioning skills technique has been introduced by the researcher to better focus on the selective observational aspects using both face-to-face and virtual interactions (Kissoon, 2008a)).

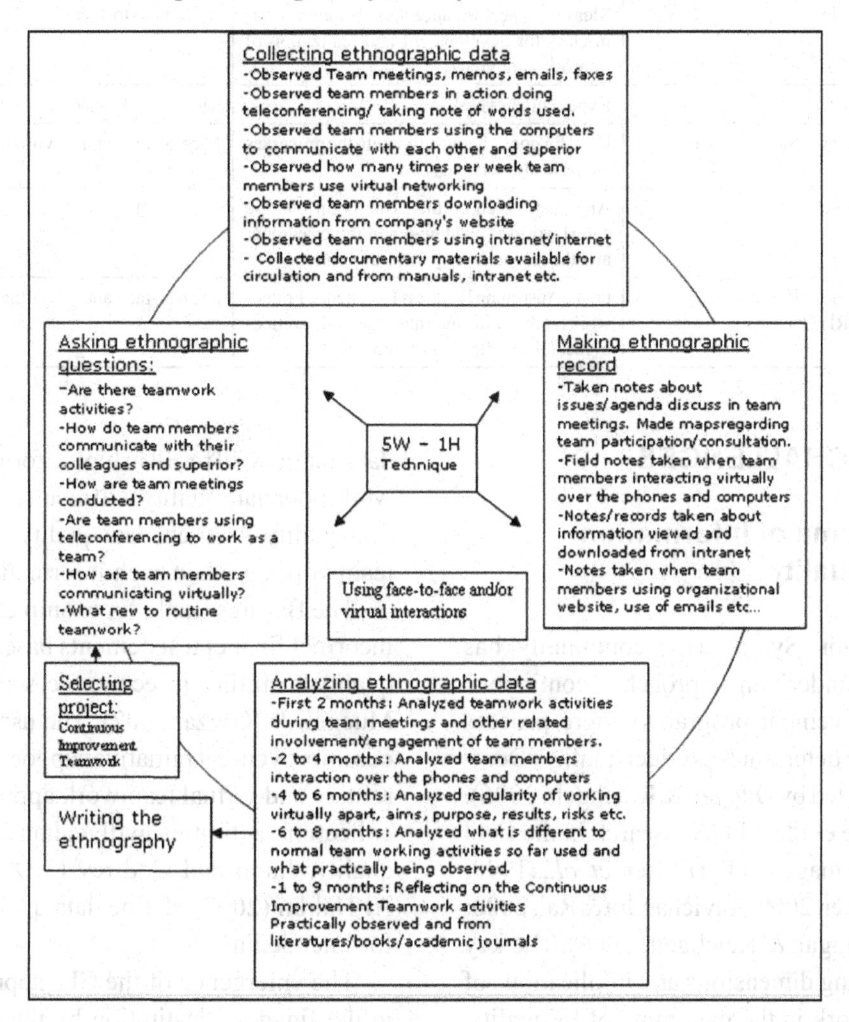

search Cycle (commonalities as a problem-solving approach) ; the Approach-Deployment-Results-Improvement(ADRI) Cycle of the Australian Business Excellence Framework. Table 1 presents a short description of each of these cycles.

The Information System Life Cycle, for example, consists of activities spanning the period between inspection and retirement of a system (Duggan & Reichgelt, 2006). The Information System Life Cycle searches for feasible influences in the planning of information during conceptualization, creation, consummation and consolidation. In the Information System Life Cycle, virtual teamwork may have a greater influence on project outcomes when compared to face-to-face, routine teamwork.

In the RADAR Cycle, as mentioned by Al-Zamany, Hoddell and Savage (2002), only the phases of assess, review and results have similarities to the CTIO Cycle. In these phases, both types of teamwork are deployed.

Table 1. Description of other performance cycles

Problem-solving cycle	Brief description of cycle	Interaction medium
Benchmarking Cycle	Measuring performance against that of best-in-class for development and realization of improvement goals.	More face-to-face
Kolb`s Cycle	Experiential cycle.	More face-to-face
Reverse Logistics and Supply Chain Continual Cycle	Used for goods being return into the processes for re-manufacturing.	Face-to-face and/or virtual
Action Research Cycle	Approach based on the assumption that the social world is changing, and the researcher and the research are part of this change.	Face-to-face
Approach-Deployment-Results-Improvement(ADRI) Cycle	Of the Australian Business Excellence Framework representing international best practices evident throughout the company.	Face-to-face and /or virtual

CURRENT CHALLENGES

The Panorama of Information Systems Quality

The Information System (IS) community has embraced the underlying approach of continuous process improvement programs where process quality largely determines product quality (Deming, 1986; as cited by Duggan & Reichgelt, 2006). This was more evident in IS research in the field of 'quality management' (Harter *et al.*, 1998; Khalifa & Verner, 2000; Ravichandra & Rai, 2000; as cited by Duggan & Reichgelt, 2006). The key issues, including dimensions and implications of virtual teamwork in the panorama of IS quality, are reviewed and discussed in relation to the CIT Model/ CTIO Cycle.

Virtual Teamwork for Data and Information Dimension

According to Al-Hakim (2007), there is a distinction between data and information. Data are items, activities, events, things, figures, numbers, and transactions that are classified and stored but not organised to express specific meaning. Information, on the other hand, is data that have been organized in a way that gives them meaning for the recipient. It is commonly known that high data quality will result in high information quality; while poor data quality will result in poor information quality. As such, data quality impacts virtual teamworking when such information is used.

The Barings Bank, for example, had provided incorrect financial statements based on poor data quality resulting in economic scandals (Maier, Muegeli & Krejza, 2007). The use of a continuous improvement initiative supported by effective routine and virtual teamwork approaches in data collection activities will enhance information quality. Juran and Godfrey (1999), as cited by Al-Hakim (2007), define data quality as "fitness for intended use".

The emergence of the CIT approach, as seen in the financial institution by the researcher, illustrates the consideration by employees involved in retail operations, decision-making and planning and service quality to effectively use data and information through normal face-to-face interaction and virtual teamworking with the intention of meeting and exceeding customer expectations.

Al-Hakim (2007) mentioned that information quality is multi-dimensional meaning that organisations must use multiple measures to evaluate the quality of data and information. Wand and Wang (1996), as cited by Al-Hakim (2007), identified two subcategories of information quality dimensions namely data-related and system-related. The CIT approach through the CTIO Cycle relates to these two subcategories in essential measures and dimensions. For example, the face-to-face teamwork may be data-related and virtual teamwork may be system-related, which will continue to change as new technologies emerge.

Virtual Teamwork for Information Quality (IQ)

As mentioned by Koronios (2006), cited by AL-Hakim (2007), issues of information quality (IQ) in team member interaction is becoming very important in the modern organization in making decisions very quickly and in order to gain information superiority and competitive advantage. The concept of Information Quality has been brought to management's attention because of the widespread, successive waves of technology-driven innovations in information and communication technologies (ICT). Enabling technologies bring widespread connectivity, real-time access and large volumes of data and information for banking staff to use in daily interaction domains.

It is also becoming a customer requirement that organizations provide quality information (English and Perez, 2003; as cited by Al-Hakim, 2007). As a result, Australian financial institutions do not want to run the risk of legislation that requires quality information be provided by employees to internal and external customers. Furthermore, many financial institutions have found that the right approach is resolving information quality problems using a continuous, interaction process, instead of a single-phase interaction process, is the right approach. It may be that one solution may lead to new problems and employees at all levels have to work together in solving the information quality problems. This is where the CIT approach comes into play, as it is a continuous process of handling information quality through team interaction in resolving problems relevant to the organization. In the CIT approach, team members use both face-to-face and virtual interactions to handle information quality to alleviate potential problems and for increased performance.

Al-Hakim (2007) provides an example. On December 9, 2005, brokers at the Mizuho Securities tried to sell 610,000 shares at 1 yen (0.8 US cents) each. A "typing error", which is an information quality problem, introduced chaos into the Japan trading market. As a result of such chaos, Mizuho Securities incurred a loss of 27 billion yen or US $21 billion. The example illustrates that human interaction through virtual means, must provide for data input that is free-of-error (degree to which information is correct). It must also promote interpretability and objectivity of the information quality dimension. In the CIT approach, both face-to-face and virtual team interactions promote information quality thus limiting simple errors (e g., typing mistakes). The more interaction employees have in an organization, such as collaborative error checking and problem solving, the better chance for improved efficiencies and higher levels of performance.

Continuous Improvement for Information Quality Process Control

For customer satisfaction to be realised, organizations should consider continuous process improvement as a key management practice (Evans & Lindsay, 2005; as cited by Al-Hakim 2007). Effective team interaction, face-to-face or virtual, promotes the implementation of process improvements through collaborative problem-solving and information sharing. As mentioned by Al-Hakim (2007), process and people are no longer enough for achieving the required output without continuously improving upon procedures, policies and regulations. This is aligned with the concept of Continuous Improvement as a derivative of TQM that controls the conversion process. Information quality dimensions need to be based on the same principles of Total Quality for Service, which as listed by Evans and Lindsay (2005, p.18), are as follows: (1) Focus on customers and stakeholders; (2) consider participation and teamwork by everyone in the organization; and, (3) have a process focus with continuous improvement and learning.

The CIT illustrates a more focused and in-depth approach that is in alignment with Total Quality for Service. Both address key issues directly related to interaction and participation through face-to-face and virtual teamwork and through the use of the Continuous Improvement principles. Virtual teamwork, for achieving effective and efficient quality information, needs to be further researched. Much of the current research has focused on face-to-face teamwork and Continuous Improvement. The knowledge gap in Quality Information is the notion of integrating both face-to-face and virtual teamwork along with Continuous Improvement for effectively achieving results and improving performance.

The Continuous Improvement in an environment of Information Quality System, and as described by Al-Hakim (2007), illustrates the mechanism of the Information Quality System Process. Emphasis is placed on the importance of leveraging process and people using information quality (IQ) dimensions and information orientation. AS mentioned by Al-Hakim (2007), the term people refer to staff and users and the process means the operational activities being done effectively through interactions to achieve better outcome in service organizations. The process alone will not result in well achieved performance intended from the process without the involvement of the employees (Al-Hakim, 2007). The system and communication means for the employees to interact together are also important. The Information Quality dimensions and the Information Orientation form the control dimension of the IT system process. To have an Information Quality System, both the IQ dimensions and Information Orientation are important. The Information Dimensions relate to issues that are important to customers while the Information Orientation refers to an organization`s capability to effectively manage and use information.

The Information Dimensions and Information Orientation, in relation to the CTIO Cycle, focuses on virtual and routine interactions among employees in the banking sector such that they can work effectively as a team. As part of the Information Quality System, the Information Dimensions and Information Orientation are crucial for team members to interact virtually wit the objective of achieving quality customer service. Thus, the interaction evidenced by the CIT Model/CTIO Cycle shows the hybrid approach of using both face-to-face teamwork and e-teamwork in a continuous improvement environment as a viable means of achieving high performance. In the banking sector, the interaction of team members in meeting the customer's needs is very important in terms of using, maintaining, and promoting quality information. This is important in the banking sector given high information quality offers a competitive advantage. Customers are becoming very demanding seeking for efficient service and information quality through the process of Continuous Improvement.

Information is defined as data having been organised in a manner that gives meaning for the recipient (Turban *et al.*, 2005; as cited by Al-Hakim, 2007). For customers in the banking sector, "meaning to the recipient" can be interpreted as accurate, error-free, reliable, effective and efficient information in processing transactions and addressing concerns. To provide such information quality, people working in the banking sector need to work together using both face-to-face and virtual interaction media in a team environment. This requires the proper use of interactive systems and emerging technologies. It also requires adopting the concepts of TQM through the Continuous Improvement approach emphasizing quality customer service.

Information Quality Using Virtual Networks in Service Sector

As mentioned by Melkas (2007), networking and virtualization networks are calling for new ways of information quality utilizing well-being technology, which team members use to interaction using problem-solving approaches. There has been a growing use of virtual communication in many organizations. The real goal of information quality is to enhance customer satisfaction (English, 1999; Huang, Lee, & Wang, 1999; as cited by Melkas, 2007). Networking and virtualization is not the only concern for companies, but also for public organizations, cooperatives and non-governmental organizations, which are forming networks (Melkas, 2007).

A study done by Melkas (2007) has identified both weaknesses and strengths in network collaboration affecting management of information quality. There has been a rapid increase in networking and virtualization within many organizations. Many researchers in recent years have also studied virtual organizations, virtual enterprises and virtual teams (Duarte & Tennant, 2001; Handy, 1995; Holton, 2001; van Hout & Bekkers, 2000; Kotorov, 2001; Lipnack & Stamps, 1997; Miles & Snow, 1992; Putnam, 2001; Rouse, 1999; van der Smagt, 2000; Voss, 1996; as described by Melkas, 2007). Nevertheless, very little research has been devoted to organizational requirements placed of the effective utilization of hybrid team with CI. Both face-to-face and virtual teamwork would be used for improved operationalization of the retail banking service sector taking into account an increase in information technology. The quality of service provided to customers can be improved by the use of hybrid teams with CI.

As mentioned by Melkas (2007), the way the employee interprets the customer's message and transfers the information forward to the collaboration network may have a major impact on service quality. Thus, the interaction of the staff with other team members in the banking sector using information technology may have an impact on the quality of information and service provided to the customer. This is what led the researcher to investigate the CIT approach to study teamwork in the retail banking sector of a major Australian banking organization.

The research done by Melkas (2007) demonstrated that information quality planning by organization or network has been possible. This research had been a first attempt to study quality in the branch of safety telephone services in Finland and that the results of the information quality analysis can also be utilized in individual organization's quality management systems. An information quality analysis could form one element of a general quality assessment at the networked collaboration level. The valuable interview data have contributed to the basis for action and scientific recommendations. The results of this investigation were incorporated into the general quality recommendations that were recently formulated for the whole branch safety telephone services in Finland. This research has opened up two new insights into the directions of analysis and management of information quality and service networks based on virtualization.

CONCLUSION

With deregulation, reforms, competition, globalisation, enabling technologies, virtualisation and network technology, virtual communications is becoming as important as face-to-face communication. The mode of interaction of team members in a continuous improvement approach is important when quality is initiated in an organization. Similarly the major Australian bank which had undergone the study for justification of the CIT approach had embarked on a quality program in 2004.

The present research has investigated the key components and measures as illustrated in Appendix 1 that are enhancing the interaction in the CIT approach. Using the CTIO Model within the existing technological infrastructure, results in small improvements in response time from team members interacting in a virtual environment. The end result is significant benefits to customers in the banking sector. The various other linkages to process improvement and data quality approaches, as presented in this chapter may offer additional benefits in the implementation of the CIT approach and achieving continual improvement in response time. Thus, the CTIO Cycle framed from the other conceptual cycles and validated in the participant observation study (Kissoon, 2008a; Kissoon, 2008b) demonstrated the important of using a hybrid team for real-time interaction thus achieving performance efficiencies.

This chapter has shown a practical and theoretical linkage of the CTIO Model with other conceptual cycles using face-to-face and/or virtual interaction media. Face-to-face interactions related to routine teamwork and virtual interactions related to virtual teamwork are being integrated with Continuous Improvement initiatives for the banking sector to perform better. Without teamwork and virtual teamwork, all the practical project work would not have obtained productive performance in the same period of time. Data quality and information quality are key issues for the success of achieving productivity gain and performance in team member daily interaction medium.

REFERENCES

Al-Hakim, L. (2007). Information quality function deployment. In Al-Hakim, L. (Ed.), *Challenges of Managing Information Quality in Service Organizations* (pp. 26–50). Hershey, PA: IGI Global.

Al-Zamany, Y. Hoddell. S. E. J., & Savage, B. M. (2002). Self assessment and obstacles to their implementation in Yemen, TQM and change management. In S.K.M. Ho & J. Dalrymple (Ed.), *Proceedings of the 7th International Conference on ISO 9000 and TQM*. RMIT University, Melbourne, Australia.

Arvan, A. (1988). Those fabulous Japanese banks. *Bankers Monthly, 105*(1), 29–35.

Ballou, D. P., & Pazer, H. L. (1995). Modelling data and process quality in multi-input, multi-output information systems. *Management Science, 31*(2), 150–162. doi:10.1287/mnsc.31.2.150

Bobrowski, M., Marre, M., & Yankelevich, D. (2002). A Neat Approach for Data Quality Assessment. In Piattini, Calero & Genero (Ed.), Information and Database Quality (pp. 135-162), Hershey,PA: IGI Global.

Duarte, D. L., & Tennant Snyder, N. (2001). *Mastering virtual teams: Strategies, tools, and techniques that succeed*. San Francisco, CA: Jossey-Bass.

Duggan, E. W., & Reichgelt, H. (2006). *Measuring information systems delivery quality*. Hershey, PA: IGI Global.

English, L. P. (1999). *Improving data warehouse and business information quality: Methods for reducing costs and increasing profits*. New York: Wiley.

Evans, J. R., & Lindsay, W. M. (2005). *The management and control of quality* (6th ed.). Ohio: Thomson/South-Western.

Evans., J. & Dean, J. (2003). *Total quality management, Organization and strategy*. Ohio: Thomson/ South-Western.

Handy, C. (1995, May-June). Trust and the virtual organization. *Harvard Business Review, 73*(3), 40–50.

Hayler, R., & Nichols, D. M. (2007). *Six Sigma for financial services, how leading companies are driving results with Lean, Six Sigma, and process improvement, profiles from global leaders including AMERICAN EXPRESS, BANK OF AMERICA, WACHOVIA, and LLOYDS TSB*. New York: McGraw-Hill.

Helfert, M., Zellner, G., & Sousa, C. (2002). Data quality problems and proactive data quality management in data-warehouse-systems. In *Proceedings of BIT-World 2002*, Guyaquil, Ecuador.

Holton, J. A. (2001). Building trust and collaboration in a virtual team. *Team Performance Management*, *7*(3/4), 36–47. doi:10.1108/13527590110395621

Huang, K.-T., Lee, Y. W., & Wang, R. Y. (1999). *Quality information and knowledge*. Upper Saddle River, NJ: Prentice Hall PTR.

Hutley, P. S. B., & Russell, P. A. (3rd ed.). (2005). An introduction to the Financial Services Reform Act, 2001. Australia: LexisNexis Butterworths.

Kissoon, S. V. (2008a). Toward the conceptual model of Continuous Improvement Teamwork: A participant observation study. In F. Zhao (Ed.), Information Technology Entrepreneurship and Innovation (250-276). Hershey, PA: IGI Global.

Kissoon, S. V. (2007). Continuous improvement teamwork in the Australian banking sector. In *Proceedings of the 5th ANZAM and 1st Asian Pacific Operations Management Symposium 2007*. RMIT University, Melbourne, Australia.

Kissoon, S. V. (2008b). Ethnographic research cycle to evidence the Continuous Improvement Teamwork model. *Qualitative Research Journal*, *8*(2), 134–136. doi:10.3316/QRJ0802134

Kotorov, R. (2001). Virtual organization: Conceptual analysis of the limits of its decentralization. *Knowledge and Process Management*, *8*(1), 55–62. doi:10.1002/kpm.93

Lipnack, J., & Stamps, J. (1997). *Virtual teams: Reaching across space, time and organizations with technology*. New York: John Wiley & Sons.

Liu, L., & Chi, L. N. (2002). Evolutional data quality: A theory-specific view. In *Proceedings of 7th International Conference on Information Quality (ICIQ 2002),* Cambridge, MA.

Maier, D., Muegeli, T., & Krejza, A. (2007). Customer investigation process at Credit Suisse: Meeting the rising demands of regulators. In Al-Hakim (Ed.), *Challenges of managing Information Quality in service Organizations* (pp. 52-76). Hershey, PA: IGI Global.

Melkas, H. (2007). Analyzing information quality in virtual networks of the services sector with qualitative interview data. In Al-Hakim, L. (Ed.), *Challenges of managing Information Quality in service Organizations* (pp. 187–212). Hershey, PA: IGI Global.

Miles, R. E., & Snow, C. C. (1992). Causes of failure in network organizations. *California Management Review*, *34*(4), 53–72.

Miller, H. (1996). The multiple dimensions of information quality. *Information Systems Management*, *13*(2), 79–82. doi:10.1080/10580539608906992

Pande, P. S., Newman, R. B., & Cavanagh, R. R. (2000). *The Six Sigma way, how GE, Motorola, and other top companies are honing the performance*. New York: McGraw-Hill.

Putnam, L. (2001). March/April). Distance teamwork: The realities of collaborating with virtual colleagues. *Online*, *25*(2), 54–57.

Rouse, W. B. (1999). Connectivity, creativity, and chaos: Challenges of loosely-structured organizations. *Information & Knowledge Systems Management*, *1*, 117–131.

Spradley, J. P. (1980). *Participant observation*. USA: Thomson Learning Academic Resource Centre. Te`eni, D. (1993). Behavioural aspects of data production and their impact on data quality. *Journal of Database Management*, *4*(2), 30–38.

STA-Singapore Technologies Automobile. (1996). *Business improvement Hhandbook* (5th ed.). Singapore: Singapore Technologies.

van der Smagt, T. (2000). Enhancing virtual teams: Social relations and communication technology. *Industrial Management + Data Systems, 100*(4), 148-156.

Van Hout, E. J. Th., & Bekkers, V. J. J. M. (2000). Patterns of virtual organization: The case of the National Clearinghouse for Geographic Information. *Information Infrastructure and Policy, 6*, 197–207.

Voss, H. (1996, July/August). Virtual organizations: The future is now. *Strategy and Leadership*, 12–16. doi:10.1108/eb054559

Wand, Y., & Wang, R. Y. (1996). Anchoring data quality dimensions in ontological foundations. *Communications of the ACM, 39*(11), 86–95. doi:10.1145/240455.240479

Wang, R. Y., Reddy, M. P., & Kon, H. B. (1995). Toward quality data: An attribute-based approach. *Decision Support Systems, 13*(3-4), 349–372. doi:10.1016/0167-9236(93)E0050-N

Wang, R. Y., & Strong, D. M. (1996). Beyond accuracy: What data quality means to data consumers. *Journal of Management Information Systems, 12*(4), 5–33.

ADDITIONAL READING

Avkiran, N. (1997). Models of retail performance for bank branches: predicting the level of key business drivers. *International Journal of Bank Marketing, 15*(6). doi:10.1108/02652329710184451

Gujarati, D. N. (1998). *Basic econometrics*. New York: McGraw-Hill Higher Education.

Kissoon, S. (2008). Ethnographic research cycle to evidence the Continuous Improvement Teamwork model. *Qualitative Research Journal, 8*(2), 134–136. doi:10.3316/QRJ0802134

Kock, N. (2005). *Business process improvement through e-collaboration: Knowledge sharing through the use of virtual groups*. Hershey, PA: IGI Global.

Maier, D., Muegeli, T., & Krejza, A. (2007). Customer investigation process at Credit Suisse: Meeting the rising demands of regulators. In Al-Hakim (Ed.), *Challenges of Managing Information Quality in Service Organizations* (52-76). Hershey, PA: Idea group Publishing.

Tagliaferri, L. E. (1982). As quality circles fade, a bank tries top-down teamwork. *American Bankers Association. ABA Banking Journal, 74*(7), 98.

APPENDIX A

Results for the paired units obtained for both FF and VI variables following face-to-face interviews with 29 retail banking managers.

Variable	Results only for coding done on both variables
Face-to-face Interaction (FF)-Dependent variable (Y)	Continuous reinforcement (CR) = 19 change management (CM) = 22 six-sigma (SS) = 21, audio/video/T.V sets (AV) = 25 visual communication (VC) = 27 voice mail (VM) = 24 brainstorming (BR) = 22 computer assisted work (CA) = 2 training (TR) = 28 coaching/mentoring (CT) =16 continuous support (CS) =21 team convergence (TC) = 19
Virtual Interaction (VI)-Independent Variable (X)	Synchronous conferencing (SC) = 16 asynchronous conferencing (AC) = 14 audio conferencing (AU) = 19 computer conferencing (CC) = 24 e-learning (EL) = 25 electronic meeting (EM) = 15 search engine (SE) = 20 virtual group networking (VG) = 19 computer-mediated communication (CO) = 24 interactive multimedia communication (IM) = 15 WWW communication (WW) = 19 virtual working environment (VW) = 11

After obtaining these quantitative data, the bivariate analysis can be conducted to measure the relationship between face-to-face and virtual interactions variables. For example CR = 19 paired with SC= 21, CM = 22 paired with AC= 14 and follows the same order of paired-matching patterns up to TC= 19 paired with VW=11.

The contents of Appendix 1 show two types of data for face-to-face communication and virtual communication of team members working as problem-solving teams. The data represents the measures for the face-to-face interaction and virtual interaction variables paired together from the participant observation, as illustrated in Figure 7. For instance, computer assisted work was paired with virtual group networking in the participant observation study. Computer assisted work can be considered as peer-to-peer network, which is much like a company run by a decentralised management philosophy in which computers on the network used by team members on their desks communicate with each other as equals. Virtual group networking is similar to a company run by centralised management using a server computer where decisions are made centrally, which is more reliable and facilitates backup.

Section 3
Innovations in Information and Communication Technologies and Software Systems

Chapter 10
Collaboration, Innovation, and Value Creation:
The Case of Wikimedia's Emergence as the Center for Collaborative Content

Divakaran Liginlal
Carnegie Mellon University, USA

Lara Khansa
Virginia Polytechnic and State University, USA

Jeffrey P. Landry
University of South Alabama, USA

ABSTRACT

This chapter describes the entrepreneurial vision and business model of Wikimedia, particularly the successes and challenges of its innovations, the wiki and Wikipedia. The case study first traces the history of how Wikimedia was founded, as such providing a rich descriptive background, using information obtained from scholarly news sources and websites. This historical overview is followed by a description of Wikimedia's business model, including the sources of capital and flows of revenues. The business model is then compared and contrasted to other Internet business models such as Knol, Google's open encyclopedia. This is followed by a discussion of a balanced scorecard to analyze how the wiki business model generates value. Finally, the case explores the use of Wikipedia from a societal and ethical perspective and provides an illustrative example of its use for collaborative work in a funded academic research project.

BACKGROUND

Wikipedia, the online encyclopedia, is one of the top ten most visited websites in the world and a truly innovative product concept (Wikimedia Foundation, 2009a). The Wikipedia product is one of a growing number of innovative uses of the World Wide Web known as Web 2.0 (O'Reilly 2005). Despite the format of its name, Web 2.0 does not denote a specific version of a particular product. The 2.0 means, in a general sense, a break from the past uses of the World Wide Web and the dawning of a new era—a

DOI: 10.4018/978-1-61520-609-4.ch010

second wave, if you will—of Web applications. This second wave of the Web is characterized by such applications as social networking, blogs, and video-sharing, as well as wikis like Wikipedia. The values inherent in Web 2.0 include collaboration, creativity, shared ownership, a sense of community, and participation.

Just as innovative is Wikipedia's originator, the pioneering Wikimedia Foundation, established in 2003 by founder Jimmy Wales. This chapter describes the entrepreneurial vision and business model of Wikimedia and the success and challenges of its Web 2.0 innovations, the wiki and Wikipedia. The case study traces the history of how Wikimedia was founded, as such providing a rich descriptive background, using information obtained from scholarly news sources and websites, including the Wikimedia website itself. This historical overview is followed by a description of Wikimedia's business model, including the sources of capital and flows of revenues. The business model is then compared and contrasted to other Internet business models such as Knol, Google's open encyclopedia, and Britannica's Webshare offering. A business value perspective is used to analyze how the wiki model generates value for businesses.

Wikipedia is truly unique in its collaborative philosophy. Its innovativeness is both the secret to its success and the biggest challenge to its credibility. The case explores this concept, its controversy, and the associated ramifications to society, along with an illustrative example of its use for collaborative work in a funded academic research project.

THE WIKI MODEL: INNOVATION THROUGH COLLABORATION

The Wikimedia Foundation is a nonprofit organization dedicated to encouraging the growth, development, and distribution of free, multilingual content, and to providing the full content of wiki-based projects to the public free of charge. It developed the Wiki—the Hawaiian word for "quick" and the acronym "what I know is" to be easily and quickly learned and used (Wikipedia, 2009a; Survey: New Media, 2006; Kirschner, 2006). It enables a group or larger community of participants to post, edit, and structure a variety of information and knowledge on the Web. Rather than using an editorial staff of researchers and writers, Wikipedia allows its content to be driven by a global network of communities of volunteer contributors. The organization, based in San Francisco, California, is managed by just over 20 paid staff. They maintain the technical infrastructure, software, and servers that allow millions of people to collaborate for knowledge creation.

A key characteristic of entrepreneurship is that it is a risky endeavor, with as many potential successes as pitfalls. Jimmy Wales' success with Wikimedia followed an earlier failure in an online collaborative project from which he learned important lessons. Wales' success with Wikimedia, therefore, is a case of getting it right the *second* time. The Wikimedia Foundation operates some of the largest collaboratively edited reference projects in the world. All contributions are licensed as open source, meaning that their content may be freely used, edited, copied and redistributed, subject to the restrictions of that license. At the creation level, the foundation's objective is to provide the community with freely-licensed tools for participation and collaboration. To ensure worldwide, unrestricted dissemination of knowledge, the foundation does not enter into exclusive partnerships with regards to access to their content or use of their trademarks. Although US-based, the organization is basically international in its nature, managed by a diverse group of trustees, staff members, and volunteers with emphasis on transparency. Being community-based, its fundamental mission is to support community-led collaborative projects, listening and taking into account the voice of these communities in all important decisions.

WHAT ARE WIKIS AND WIKI ENGINES?

A wiki enables documents to be written and edited collaboratively using an ordinary web browser. A single page in a wiki website is referred to as a "wiki page", while the entire collection of pages, which are usually well interconnected through hyperlinks is "the wiki". In simple words, a wiki is nothing but a database for creating, browsing, and searching through information. A wiki fosters strong and meaningful associations between the content of different pages by making the process of creation of the link quite intuitive. Further, the key objective of a wiki is to involve a casual visitor not just in perusing the content but in meaningful participation and 'co-creation.' A strength and at the same time a weakness of a wiki is the ease with which documents can be created and modified with no review before modifications are accepted. This openness of wikis often results in abuse and could very well spell the demise of open 'wikis.' However, private wikis could be created with some built-in authentication for participation, sometimes both for editing and browsing.

The server-side software that makes a wiki system work is called a wiki engine. There are numerous implementations of wiki engines. The common manifestation of a wiki engine is an application server running on one or more web servers. Every user is empowered with the ability to create a comprehensive knowledge base on a relevant topic of interest to the user's community. The content itself is maintained in a file system and changes are tracked through a database management system. Each member of the community has the ability to not only search and read the collaborative content, but also to add and edit articles. It requires only very basic programming skills for using and maintaining a wiki. MediaWiki is the most prominent of the wiki engines. It is a free software package licensed under the GNU General Public License (GPL). MediaWiki supports a variety of languages, styles, multimedia and other extensions, content indexing, and tracking features. There are literally hundreds of other wiki engines, all implementing the same concept but in a variety of feature-rich ways using different programming languages. Another category of wikis are the PersonalWikis or DesktopWikis, which are not intended for collaborative work but for organizing personal information. Also, a MobileWiki, as the name implies, is an extension of web-based wikis optimized for mobile devices such as BlackBerry or iPhone.

Facilitating Innovation, Collaboration, and Value Creation

Wikis provide effective means of embedding the collective knowledge of a group. As argued in the Wisdom of Crowds (Surowiecki, 2004), "under the right circumstances, groups are remarkably intelligent, and are often smarter than the smartest people in them." The underlying principle of collaborative innovation is accentuated by the fundamental features of wikis such as their self-organizing nature, democratization, and leadership by merit. For a specific collaborative work context anybody can be sent an invitation to participate in a wiki, and the collaborative process has the potential to proceed in a seamless manner. Each individual's contribution is evaluated based on its merit, with no specific status or power structure. The wiki model thus exhibits true democratic characteristics where each individual accepts responsibility and leadership is by merit. The communities thus formed are self-organizing with individuals tasked with their own editing and improvements and fostering creative contributions of content and ideas. It has now been clearly demonstrated that wikis empower the individual by coaxing them to shed the trappings of their individual silos and participate in co-creating activities, for instance jointly creating a better product. In fact, history provides us with many examples of 'innovation through collaboration,' including the open source revolution heralded by such products as Linux.

The most common examples of Wikimedia projects are Wikipedia, Wikimedia Commons, Wikiversity, Wiktionary, Wikinews, and Wikibooks. Of these, the best known Wikimedia project is Wikipedia, founded in 2001 by Jimmy Wales and Larry Sanger. The name Wikipedia combines the words "wiki" and "encyclopedia." In the words of Wales, the objective is to create a world in which every single person on the planet is given free access to the sum of all human knowledge. Wikipedia has emerged as a prominent source of current and multifarious information. The word 'wiki' itself yields nearly 400,000 hits on Google.

Wikipedia's long-term success may depend on the effectiveness of its communities of volunteer contributors. As Wikimedia's executive director, Sue Gardner, has put it, "Wikipedia has proven that mass volunteer collaboration can create real social value: people will participate in large numbers, out of sheer goodwill and helpfulness, to develop great educational material that other people want and use." Still, critics (Schneider, 2007) argue that the volunteer model will bring out not only those with noble intentions, but those that are self-serving as well. Three factors that define the effectiveness of online communities of contributors include the following: (i) the built-in capability to discern fraudulent or false information and correct it; (ii) the ability to block such sources of data through effective monitoring; and (iii) the ability to quickly compile authoritative and reliable information. Indeed, if there are editorial conflicts or a large number of corrections, the communities are likely to disintegrate or dissolve. Unlike many other online communities that require identity verification using credit cards, Wikipedia does not require the establishment of genuine identities. Often the requirement of identity verification through, say, a credit card will turn off many likely participants and also raise privacy concerns.

The Wikimedia Commons was founded in 2004 as a free media repository of educational media content such as images, sound, and video clips. The primary objective of setting up this central repository was the desire to reduce duplication of effort across the Wikimedia projects and languages, as the same file had to be uploaded to many different wikis separately. Wikiversity is devoted to learning sources, learning projects, and research for use at all levels, types, and styles of education. It is intended to be an open repository of educational resources contributed by teachers, students, and researchers thus helping foster collaborative learning communities. Wikibooks was established with the aim of building a free library of educational textbooks that anyone can edit. Wikinews embodies collaborative journalism, where in the words of Jimmy Wales, 'each story is written as a news story as opposed to an encyclopedia article.' Wikinews permits original contributions in the form of original reports or interviews. All of these collaborative projects underline the common theme of 'innovation through collaboration.'

BUSINESS MODELS

Analyzing Wikimedia's Business Model

Wikimedia is a non-profit organization which does not accept any advertising on its sites. Its financial sustainability is dependent mostly (80%) on gifts such as donations, grants, and sponsorships for its operation. In order to ensure that the organization stays free of influence in the way it operates, they strictly follow the donation policy. They reserve the right to refuse donations that could generate constraints and try to multiply the diversity of revenue sources. Independent fund raising serves as another source of revenues. Recently, Wikipedia launched a fund raiser to help them tide over the hard times of the economic recession. Over $6 million was raised from about 125,000 donors. The French Wikipedia created a poster service that makes images available in hard copy. The profits from this service were donated to Wikimedia.

The wiki open source model emphasizes use value while standard business models emphasize end value. Thus, the key objective is to maximize use value and the most important strategic issue is the question of how to harness the input of many users to improve use value. This use value can be transformed into economic value through hidden revenue-generating models. Niki Scevak, the founder of Homethinking (Scevak, 2007) recently proposed such a model. The keyword traffic data of Wikipedia, which would presumably be both deep enough (being one of the top 10 visited websites) and informative enough (articles read, and referring keywords) is extremely valuable for marketing companies. Such an approach would ensure privacy protection, while keeping Wikipedia clean from visible advertising.

Based on a modified version of Weill and Vitale's model schematics (Weill and Vitale, 2001), we depict the flows of information and revenues within Wikipedia's business model. Figure 1 shows the firm of interest, in our case Wikimedia, as a knowledge integrator facilitating the aggregation of content from any contributing source and its presentation in a meaningful format. Information flows from these content contributors to the firm and from there to the consumers, who may themselves be contributors. Both contributors and consumers serve as sources of revenue, albeit in the form of donations. Further, Wikimedia also receives free hosting services, which is depicted in the form of information flow in both directions and no money flow between the firm and the service provider. Also captured is a plausible revenue source that works in the spirit of the Wikimedia foundation, i.e. shirking the use of ads along with content. The figure shows a hidden revenue generating model in the form of flow of money from allies in exchange for keyword traffic data.

A look at Wikimedia's financial statements over the past 4 years reveals a considerable increase in cash and cash equivalents (from around $10,000 in 2004 to around $3 million in 2008, an increase of 300 times). Wikimedia relies on donations and gifts of up to $100,000 in cash or publicly traded securities as their revenue model. Donations of more than $100,000 are subject to restrictions (Wikimedia Foundation, 2009b).

Although the financial statements confirm that Wikimedia has been quite successful in sustaining themselves over the years, the fundamental question is whether this model is viable in the long run. At some point of time the foundation will be forced to examine other revenue options such as advertising (Google), licensing Wikimedia's technology, hosting other wiki's, or collecting membership fees. Given the fact that websites such as Wikipedia are among the most visited on the Internet, advertising constitutes a lucrative source of income. There is also the possibility that despite the high traffic on sites sponsored by the foundation, there is no guarantee that Wikimedia can bring in hoards of advertisers who would like to splash their wares next to content over which they have no control. The problem is exacerbated by the fact that the content is open and editable by anyone. Quite plausibly, some innovative business models are required to make Wikimedia viable in the long run.

Developing such business models would incur more costs. Currently, Wikimedia's balance sheets for the past four years indicate that their current liabilities are quite low (less than 12% of current assets in 2008 and less than 10% in 2006). Many argue that Wikimedia's model would die if it fades away from its grass roots and movement of being independent from big money. In fact, having subscribed to a knowledge diffusion model, Wikimedia could be branded as 'diabolical' if it were to succumb to commercial pressures.

Competing Business Models

It is worthwhile examining competing models such as Britannica's Webshare and Google's Knol. Encyclopedia Britannica is a classical example of how the advent of online business models has disrupted traditional business models. Comscore

Figure 1. Business model for Wikimedia

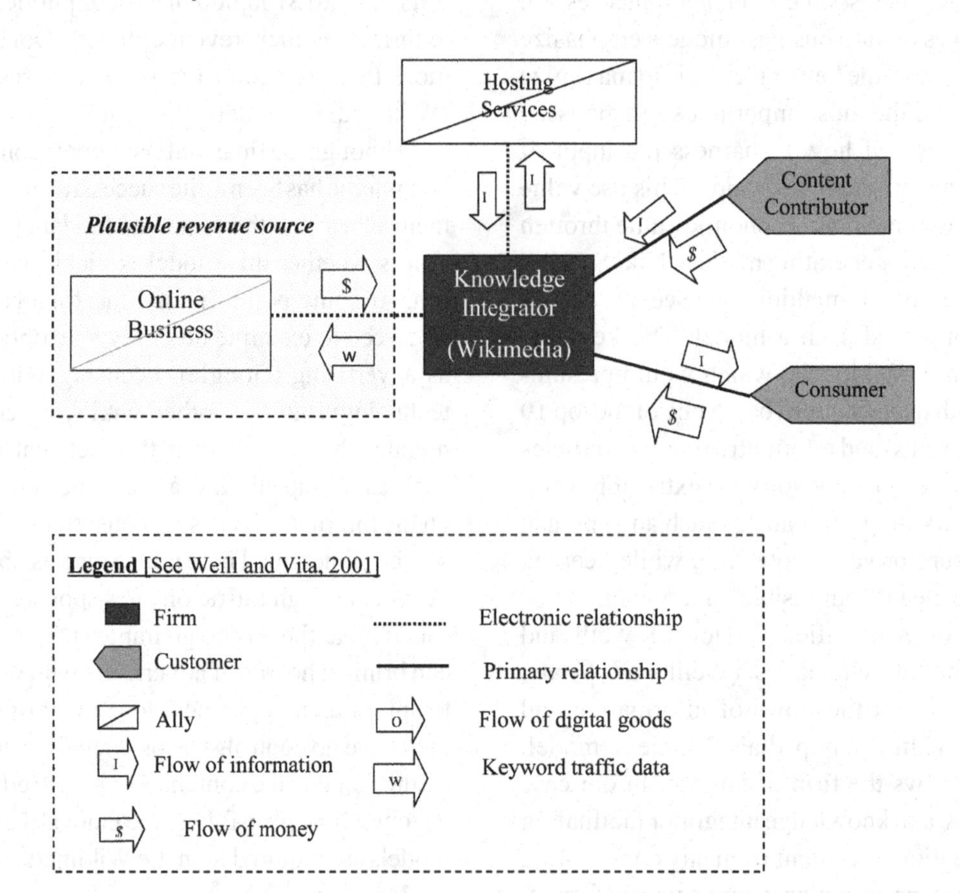

estimates that for every page viewed on Britannica.com, 184 pages are viewed on Wikipedia (Arrington, 2008). The answer Britannica has come up with is Webshare. Any web publisher, i.e., blogger, webmaster, or writer, who publishes with some regularity on the Internet, can now access Britannica for free. The web publisher signs up and provides relevant information such as site URL and a description of the content to Britannica. Britannica reviews it and grants permission to link the publisher's site to a full version of related articles. Thus people clicking the link can read that specific article but do not have access to other parts of the Britannica site. Britannica's future is under threat and they need such innovative business models to survive.

The term 'Knol,' which Google defines as a "unit of knowledge" refers to both the project and an article in the project (Wikipedia, 2009b). Knols are authoritative articles about specific topics, written by people who know about those subjects. Knols include strong community tools, which allow for many modes of interaction between readers and authors. The contributor can create (and own) new Knols, and there can be multiple articles on the same topic, each written by a different author. Because multiple articles can have the same title, readers find a topic by searching, rather than just by looking up the article's title. The authors have an option to allow their Knols to be edited by the public, to make them editable only to co-authors or to make them closed entirely. At the discretion of the author, a Knol may include ads from Google's AdSense. AdSense is an advertisement application run by Google, which will enable text and image advertisements on the

Figure 2. Contrasting business model – Knol

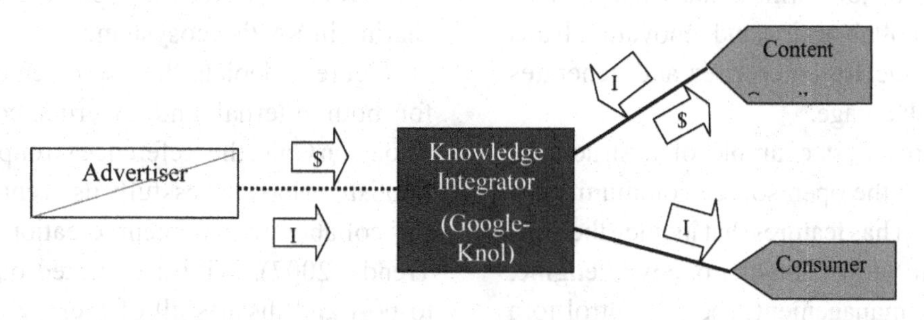

pages created by the author. These advertisements are administered by Google and generate revenues on the basis of the number of times a reader clicks on the ad or the number of times the ad appears on the page. If an author chooses to include ads, Google provides the author with a revenue share from the proceeds of those ad placements. Figure 2 portrays the modified business model for Knol. Money flows from advertisers to the firm of interest, in this case Google. The contributors of pages with ads enabled, in turn, receive part of the proceeds for their contributions.

Business Models Based on Wikis

Open collaboration models have become crucial to survival for modern businesses, especially small and medium sized companies. Even for larger companies innovation at a fast pace implies getting connected globally with the most creative minds. Ascend of wikis in the so called "Enterprise 2.0" provides a potent tool to power such open collaboration and innovation.

There are two different kinds of wikis - publishing wikis and structured wikis (TWIKI.NET, 2007). Publishing wikis, an example of which is the Mediawiki, are the most common form of wikis. They are focused on publishing content and allowing users to publish text, images, or video. Structured wikis, in addition to what publishing wikis do, allow the user to build applications designed for enterprise use. An example of a structured wiki is the enterprise wiki, which allows for information to be organized in individual workspaces based on a functional organization such as departments or teams. Unlike Internet-based wikis such as Wikipedia, enterprise wikis are not intended for open access. Instead they often implement access control through login requirements and page-level or within-page permissions assigned to users. Further such access control may be provisioned from the enterprise directory. Within the organization, enterprise wikis may be used for project management, coordinating meeting agendas and minutes, and creating and maintaining documentations, reports, and knowledge bases. Being crucial to success in daily job functions, the creation and maintenance of such wikis should necessarily be the result of a well-orchestrated adoption strategy that places the wiki at the center of the core activities of a team.

Wikis create business value through internal collaboration, content management, and the creation of portals, such as Blackboard systems and meeting notes. Further, they facilitate knowledge management in the form of self-help, glossaries, handbooks, organizational memory, frequently asked questions, and better software development processes. They lead to the creation of communities that are cross-functional and with shared interests. Wikis are useful not only within organizations, i.e., among employees and personnel, but, more importantly outside of the organization, among strategic partners to promote effective collaboration creating significant business value and making good use of internal and external ideas leading to success. In short, wiki-based technologies can potentially integrate an organization and facilitate new forms of knowledge work. Wiki-enabled collaboration and innovation leads to business value for enterprises and generates competitive advantage.

TWiki is the best example of a structured wiki created by the open source community for enterprise use. It has features that include the ability to run multiple wikis under one wiki engine, advanced link management, access control to a web or to a set of pages, integration with enterprise directory infrastructure, audit trails, and a plug-in architecture supported by a rich set of existing plug-ins such as tracking action items, calendars with highlighted events, database, RSS news feed, and presentation converter (TWIKI.NET, 2009). The basic wiki page is supplemented by simple application elements such as forms and database queries, including the possibility of using individual wiki pages as databases. TWikis have been adopted by several Fortune 500 companies for a variety of innovative applications including SOX compliance, bug tracking, customer relationship management (CRM), and enterprise blogging. Many have successfully integrated the TWiki infrastructure with key business systems for these applications.

Wikis are replacing intranets on a large scale. At IBM Research, Lotus Notes has been ignored in favor of wikis which enable open access, contribution, and discussion of research. Google uses wiki technologies to enhance internal communication and knowledge sharing, particularly for collaborative authoring such as specifications and documentation generation and idea brainstorming. Many software companies facilitate internal collaboration with standalone tools based on wikis. For example, Chordiant uses wikis for project management and IBM uses wikis to develop its IP manifesto. The model can be used to engage with the ecosystem, i.e., a tool of open collaboration, well integrated in the ecosystem. Examples of open collaboration integrated into the ecosystem are DevCentral used by the technical users' community for shared development and the SAP Developer Network (SDN) for knowledge sharing in SAP's ecosystem.

Figure 3 depicts the use of enterprise wikis for both internal and external collaboration. Thomson Gale, the Reference Group of Thomson Publishing has successfully used enterprise wikis for collaborative content creation (Publishing Trends, 2007). Wiki use started out as a place to post and discuss all of their standard forms, white papers, meeting minutes, and ideas. It was the center for communication. This use has dramatically cut down on the amount of time it takes to circulate, edit and collaborate on documents. With Microsoft Word-based editing each person in a product development team had to complete everything, make sure the document was perfect, and then send it to a second person. Then they'd go through everything, track changes, and send it to the third person and so on, it would take 2-3 weeks. Now, with the wiki, someone posts, and immediately everyone in the group can respond and manipulate one document. They can get an RFP out in 2-3 days now. By better understanding the technology, Thomson employees are thinking about new products in light of new business models, and making the Kuhnian shift from print to digital (Publishing Trends, 2007).

A wiki is not only used for controlled access for authors. There are even certain parts of the site that are shared with vendors as well. As the business model in Figure 3 depicts, wikis provide the common platform for authors, editorial and production staff, editorial partners, and content syndicators to collaborate. Authors serve as the source of digital products, in this case books and other reading materials, for which they are paid by Thomson. The editorial and production staff, which may be geographically dispersed, in multiple units, and often external to the firm interact with the authors through the common platform enabled through wikis. Digital products flow from Thomson to the customer, who in turn pays for these products. Information in the form of product support flows from Thomson to the

Figure 3. Thomson Publishing's use of wiki for collaborative content creation

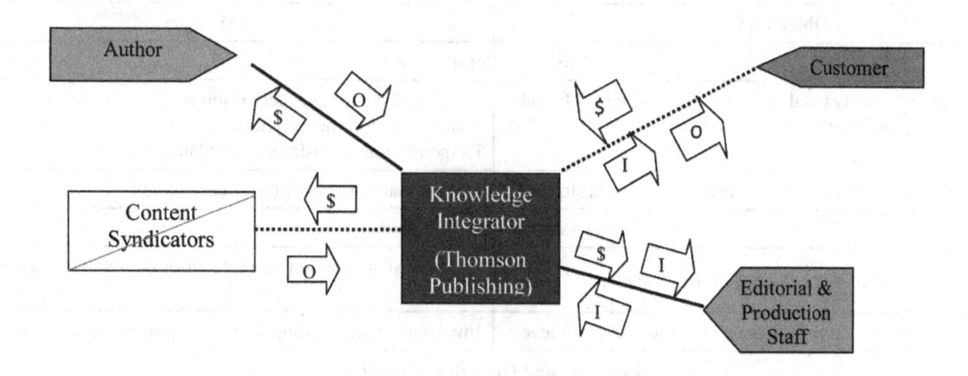

customer who may also contribute to building the support sites in the form of pedagogical materials and other educational materials besides wish lists and product selection suggestions.

THE WIKI SCORECARD

We next discuss how the use of wikis by enterprises can be evaluated using a balanced scorecard. Wikis permit a higher Return on Investment (ROI) by reducing development and maintenance cost. Further, they solve accessibility issues with external partners and thus reduce communication costs. Wikis also benefit shareholders by improving internal business processes, encouraging learning and growth within organizations, helping organizations better serve the customer, and increasing organizations' competitive advantage. The Balanced Scorecard approach, developed by Kaplan and Norton (1992), has been widely used to evaluate information technology and its investments. Using this method, Kaplan and Norton (1992, 1993, 1996) were able to evaluate an organization's projects from four perspectives, namely the customer's perspective, the perspective of internal business processes, the growth and learning perspectives, and the financial perspective.

We use Kaplan and Norton's Balanced Scorecard approach to demonstrate how wikis can drive business performance by supporting the vision and strategy of an organization through improving the four different facets discussed earlier, as shown in Table 1.

1. Internal Business Processes. Wikis allow capturing and leveraging knowledge within organizations by facilitating collaboration. All employees can contribute to organizational planning so employees feel empowered, involved, and satisfied, which increases their productivity. The simplicity of wikis, i.e. the ability to edit a wiki page in Word, encourages non-technology savvy employees to add their ideas and inputs. Wikis also encourage transparency by exposing how employees are thinking and can be used to make management aware of employees' contributions to the organization. The wiki model when implemented in an organization allows employees to identify themselves and their contributions. Employees can gain good reputation capital and are able to establish their credibility in the organization. Wikis help organizations migrate into a virtual flatter structure, which has proved very valuable in the business world. The virtual flatness of an organization is more valuable than an actual flatness because it keeps decision making focused at a high level, while virtually empowering employees to actively contribute. The wiki business model allows

Table 1. Applying the balanced scorecard perspective to uncover the business value of wikis

Objectives	Measure
Customer Perspective	
The organization's ability to address customers' complaints and increase customer retention	Amount of complaints per customer Amount of customer inquiries Response time to customer complaints
The organization's ability to compete and attract new customers	Market share (number of new customers)
Internal Processes Perspective	
Elimination of hierarchies in the organization	Amount of interaction among upper management and other employees
Increased collaboration and empowering of employees at all levels	Involvement and participation of employees at all levels
Learning and Growth Perspective	
Facilitating employee training	Employee satisfaction
An inexpensive tool for continuous learning	Employee performance / reduced errors
Financial Perspective	
Streamlining internal business processes	Transaction cost reduction
Producing more innovative products faster	Better product quality
Retaining current customers and attracting new ones	Increased market share/ competitiveness / profits

openness and fairness because it ensures that innovative employees get credit for their ideas and that exploitive middle managers don't take undeserved credit for their subordinates' efforts.

2. Learning and growth. Wikis enable information dissemination within organizations and stimulate organizational learning. Wikis allow businesses to use forums for inexpensive shared learning, as a continuous education for existing employees and to improve the on-boarding of new employees. The wiki model offers good feedback mechanisms whereby more experienced employees in the organization can revise the contributions of less experienced employees and suggest changes or revert to a prior state. Also, the wiki model encourages individual learning as employees are recognized for their valuable contributions. The wiki model is a progressive model that builds upon a shared vision in the organization. More importantly, the wiki model allows establishing and improving shared vision in an organization. Wikis are mostly valuable when organizations are dispersed across geographical locations because they allow distant contributors to participate, thus ensuring that information is widely relevant, valid, and unbounded.
3. Customers. Wikis allow customer knowledge to be used for effective marketing decision making. Customers are able to voice their preferences and concerns through wikis. For example, tech support in organizations can use wikis to give faster and higher quality help, and convey consistent answers and responses to customers, while keeping up with their demands. As such, wikis can also be an effective tool to create competitive advantage and attract new customers.
4. Financial. Internally, Wikis improve an organization's productivity by streamlining its internal business processes through reducing bureaucracy and eliminating hierarchies. Wikis also constitute an inexpensive tool to ensure employee training and continuous education. As such, wikis can be used to increase productivity at low cost. Externally,

organizations using wikis can get products to market faster and with more agility by using innovation of many more people across organizational groups and across more regions of the world than would have been possible with earlier generations of workplace technology. Further, wikis can be used to quickly respond to customers' complaints, thus encouraging their retention. They can also be used to compete with rival firms, thus attracting new customers and increasing market share.

Some companies are applying the wiki model beyond the boundaries of the organization by benefiting from outsiders to improve their business operations. For example, Wikinomics is a tool that allows firms to solicit mass collaboration from professional volunteers to improve their businesses. According to Tapscott and Williams (2008), Wikinomics reaches beyond outsourcing and relies on "prosumers", which the authors define as "savvy consumers working with companies to help create and innovate -using "ideagoras", i.e. online forums where ideas and solutions change hands."

Many challenges persist in the face of fully exploiting the benefits of the wiki business model, mostly because it is hard to keep track of the ownership of intellectual property. Reliable systems should be put in place to ensure that identity, intellectual property, and security are well-taken care of. The following section discussed the risks and legal issues associated with wikis.

RISKS AND LEGAL ISSUES

One of the significant issues associated with the content hosted by Wikipedia is that of ensuring the accuracy and authenticity of information, as well as the exclusion of personal opinion. The submission process is open to anyone. In a similar vein, anyone can start an 'Article for Deletion' review process if they believe the piece does not live up to their standards. The vetting process is moderated by a Wikipedia administrator who ultimately decides the fate of the article. Not all iterations of the wiki are successful. Recently the LA Times had to shut down its editorial wiki due to an onslaught of vandals posting irrelevant material. The most noteworthy case of inaccurate information posting was the biography of John Seigenthaler.

The Seigenthaler Case

Seigenthaler, a former journalist, was an assistant to Attorney General Robert Kennedy in the 1960's. He was implicated by the Wikipedia article as being involved in the assassinations of both Kennedy brothers after living in the Soviet Union—none of which is true. The Wikipedia article remained unchecked for more than four months. Automated software copied the false biography to two other popular Web reference sites (Answers.com and References.com), as part of the normal practices at those sites (Seigenthaler, 2005). Eventually, the material was removed from all three sites, but Seigenthaler had great difficulty discovering the identity of his accuser. Wikipedia at the time had no way of identifying authors of its content, and relied on the community to quickly check and correct mistakes.

When Seigenthaler checked with an Internet Service Provider (ISP) to investigate, he was met with legal resistance, as federal law protects the privacy of the ISP and the identity of authors, even if defamatory. Only a lawsuit would force BellSouth, the ISP in this case, to reveal the identity of the defamatory author. Eventually, the perpetrator was discovered as Brian Chase, a manager at a Nashville, Tennessee courier firm. Chase admitted his wrongdoing as a hoax and apologized (Survey: New Media, 2006).

This and other incidents (Breitkopf, 2007) forced Wikipedia to rethink its policies. Wales, in an interview with Business Week (Helm, 2005), stated that Wikipedia should be used only for "background" information and not directly cited as an authoritative source. Wikipedia began enforcing its stated policies by publishing a lengthy set of guidelines for biographies of living persons, for example (Wikipedia 2009c). These guidelines emphasize its pillars of verifiability, neutrality, and avoiding original research, although its access policies remain quite loose. Creators of original content (pages) are required to register at the site, and certain prominent pages are locked, but unregistered users are still allowed to perform edits on existing pages with IP address histories as a check (Phillips, 2005; Schneider, 2007; Pressley and McCallum, 2008). When Wikipedia started getting pulled in the direction of greater control and accountability, the actions of its editorial community have been criticized as being conflicted, emphasizing more deletes and less transparency and inclusiveness (Schneider, 2007).

A Note About Crowdsourcing

Crowdsourcing is a neologism for the act of taking a task traditionally performed by an employee or contractor, and outsourcing it to an undefined, generally large group of people or community in the form of an open call (Wikipedia, 2009d). Wikis are touted as a common platform for such crowdsourcing. A crowdsourced project may often cost more for businesses than conventional outsourced projects. The lack of financial incentive, inadequate number of participants, poor quality, lack of motivation, cultural and global issues such as language, and above all the burden of managing such large scale projects are negative aspects of using wikis in such contexts.

Wikipedia: An Ethical Analysis

Any entrepreneurial venture involving innovative uses of computers and the Web will be at the forefront of computer ethics. Why? Because as Moor (1985) pointed out in his seminal work, innovative uses of technology usually precede a full ethical understanding of its implications. The trend is that technology appears first and ethics follow, with cyber law lagging behind. Rapid proliferation of participative Web media, including sites like Wikipedia, other Web 2.0 media, and the WWW in general, have created ethical dilemmas in practice. A pair of ethical theories is used to illustrate ethical dilemmas for Wikipedia: act utilitarianism and social contract theory (Quinn, 2005).

According to the ethical philosophy of act utilitarianism, it is the value of the consequences of the particular act that counts when determining whether the act is right. Act utilitarianism requires one to calculate the total net benefit from a course of action being evaluated. In the case of Wikipedia, the benefits include widespread—through the WWW—access to valuable, rapidly evolving encyclopedic information. The harm, meanwhile, is exemplified in cases such as Seigenthaler's. From an act utilitarian perspective, one can argue that the great deal of good that comes from the use value of information on Wikipedia far outweighs the harm that results from a few isolated abuses. That is, the total net benefit to society justifies the continued use of Wikipedia's current model.

On the other hand, one of the biggest drawbacks to using act utilitarianism as a moral lens is difficulty in calculating "utility" and the moral problems that result. Calculating "net societal benefit" in the aggregate means that individual outcomes are cancelled out. The possible disastrous consequences to someone who is potentially libeled by Wikipedia, as could have happened to Seigenthaler, are offset in act utilitarianism, by the benefits calculated elsewhere. Act utilitarianism alone provides an incomplete ethical analysis of the situation. Individual rights issues need to be considered, so a perspective like social contract theory is useful.

According to social contract theory, consent is the basis of law. Morality consists in the set of rules governing how people are to treat one another; a set of rules that every member of society must accept for mutual benefit. Rights are rules agreed to be accepted as part of a social contract. In the case of Wikipedia, there is a clear conflict between the right to freedom of expression and an individual's right to be protected against harmful speech.

On the one hand, a free, participative, and uncensored Wikipedia advances the principles of freedom of expression, opening up avenues of participation for many people, enabling more ideas, competing ideas, opportunities for modification and correction of content. On the other hand, the lack of control and accountability are a threat to individual rights, making it difficult to detect the source of inaccurate content, and putting individuals at greater risk of harm.

An open question as a result of the Seigenthaler case is whether Wikipedia would or should be legally liable had Seigenthaler brought a lawsuit. Wikipedia self-corrected by adding a degree of transparency by requiring contributors to register, but have they gone far enough? What will happen with future web-based innovations? Will entrepreneurs simply test the waters as before, and as Moor predicted, push ahead with new technologies while the world struggles to understand the ethical costs and benefits, or will they learn the lessons of Wikipedia and take greater social responsibility in their initial offerings?

LESSONS LEARNED FROM A WIKI PROJECT

One of the co-authors of this case study is involved in a federally-funded research project using a wiki as a collaborative tool for the project team, comprised of faculty, students, consultants, and legal counsel as members. The wiki is proving to be a versatile tool for the project team, enabling project management tasks to collaborative content-creation. The wiki is the engine that runs sites like Wikipedia. It is an open-source, downloadable utility for quickly creating a collaborative site with the look-and-feel of Wikipedia. It can be set up in less than an hour on a Web server with either a public URL or else requiring users to log in.

A common use of a wiki is for project management tasks. The team uses its wiki for the following project data:

- the solicitation (request for proposal)
- proposal
- project management plan
- timesheets
- status reports
- product-related content

The product-related content includes collaboratively-created content. This includes an annotated bibliography created and used by the team and being considered for public consumption. There are links to scholarly articles, reports, news stories, and activist web sites related to the project goals. Each bibliography entry has a complete citation and annotation, including article summary, evaluation of its usefulness to the project, and a keywords list. Students were trained to write entries and given assignments through the wiki itself, using the category tags, which were updated as students were assigned articles and as they completed them.

The use of the wiki has had a noticeable effect on patterns of collaboration. The wiki is clearly impacting faculty-student relationships, including the interplay between trust and control. With the wiki, not everything has to be reviewed before posting. There is trust that anyone can upload when they feel it's ready. Other members of the team freely volunteer to edit, add, and format the content. On Wikipedia, automated Wiki-bots do a lot of the formatting (Kirschner, 2006).

On our project wiki, it's usually student volunteers. Yet, there is a push-pull; and still, some resistance to totally letting go. Lessons learned include:

- wiki soft security (Rasmusson and Jansson, 1996) principles means review after posting, not before
- posting is itself a kind of control
- category tags enable assignments to be made
- history page is a check against social loafing
- bad fixes can be reversed through edit history

The culture of the wiki meant that we could post "unfinished" work, and expect that others would review and edit as they saw fit, but there was also the pressure to publish with better quality, because of the "public" nature of the deliverable. Both forces were felt to be at work, pulling authors in opposite directions. Because user names get posted when pages are edited, the history buffer acts as an additional control mechanism against social loafing. The history buffer is also reversible, so if an individual makes a poor change or accidentally deletes content, it is nevertheless recoverable.

The pace of the work flowed through the ability of the wiki to serve as a self-control mechanism and a medium for immediate uploading. The wiki reduced and can eliminate review bottlenecks. Presently, there are explicit sections of the wiki for students to post intermediate deliverables so that a project audit might explicitly link reported hours worked in a time sheet to actual deliverables posted.

Another issue is the controversy over making the wiki site public. The federally-funded project has several stakeholders with a vested interest in the content of the site. These stakeholders include local, state, and federal government officials, the major political parties, private government contractors, and political activists. The content on the site would almost certainly be viewed by the various stakeholders as supporting or hurting their cause, and could generate controversy. The choice facing the team, and its primary government sponsor, is either public transparency, a desirable trait in an open democracy, or confidential deliverable.

CONCLUSION

Based on the principles of collaborative innovation, self-organization, democratization, and leadership by merit, wikis can generate tremendous value for businesses. The wiki open source model harnesses users' inputs to generate use value, rather than end value. We analyzed the particular case of Wikimedia, a non-profit organization that sustains itself on donations, grants, and sponsorships, but not advertising. We studied Wikimedia's business model using a modified version of Weill and Vitale's model schematics, which are useful in visualizing the flows of information, resources, and revenues among Wikimedia's contributors and consumers. We then compared and contrasted Wikimedia's business models with other business models, i.e. Britannica's Webshare and Google's Knol. Britannica's Webshare allows regular Internet publishers to access Britannica for free; others have to subscribe to be granted access. While Britannica is struggling to survive, Google's Knol seems to be thriving. Knols include strong community tools, and allow users to generate revenues through Google's AdSense. Other than Wikimedia, we also discussed enterprise wikis, which were successfully implemented by Thomson Gale for collaborative content creation. Thomson pays authors for their books and other reading materials and makes use of editorial and production staff that are geographically scattered.

To evaluate the wiki business model, we used the Balanced Scorecard approach by Kaplan and Norton. Internally, wikis improve internal business processes and encourage learning and growth within organizations. As important is the ability of wikis to increase customer satisfaction, reduce costs, and boost profits. The wiki business model helps enforce policies and important procedures for effective collaboration. It ensures accountability and encourages employees' learning and contribution.

While the mass communication feature of wikis can eliminate the boundaries of organizations and generate infinite resources, it creates many technical and legal challenges. We presented instances of wiki vandalism, namely the Seigenthaler case, and presented an ethical analysis based on the utilitarianism and the social contract theories. We raised several issues and presented some lessons learned from our experience with a federally-funded research project using a wiki as a collaborative tool for a project team. While wikis could become a way of life, their successful implementation necessitates a sturdy security infrastructure to ensure that contributors' identities are tracked, intruders and wrongdoers are kept out, and that intellectual property is safeguarded to avoid unnecessary legal problems.

REFERENCES

Arrington, M. (2008). Encyclopedia Britannica now free for bloggers. *TechCrunch*. Retrieved Aug 3, 2009 from http://www.techcrunch.com/2008/04/18/encyclopedia-britannica-now-free-for-bloggers/

Breitkopf, A. (2007). Diebold machine was used to alter Wikipedia entry. *American Banker, 172*(159), 8–8.

Helm, B. (2005, December 14). Wikipedia: a work in progress. *Business Week Online*. Retrieved April 1, 2009, from http://www.businessweek.com/technology/content/dec2005/tc20051214_441708.htm.

Kaplan, R. S., & Norton, D. P. (1992). The balanced scorecard – Measures that drive performance. *Harvard Business Review, 70*(1), 71–80.

Kaplan, R. S., & Norton, D. P. (1993). Putting the balanced scorecard to work. *Harvard Business Review, 71*(5), 134–142.

Kaplan, R. S., & Norton, D. P. (1996). Using the balanced scorecard as a strategic management system. *Harvard Business Review, 74*(1), 75–85.

Kirschner, A. (2006, June). What I know is. *PublishingTrends.Com*. Marketing Partners International. Retrieved March 31, 2009, from http://www.publishingtrends.com/copy/06/0606/0606WhatIKnowIsWiki.html.

Moor, J. H. (1985). What is computer ethics? *Metaphilosophy, 16*(4), 266–275. doi:10.1111/j.1467-9973.1985.tb00173.x

O'Reilly, T. (2005). What is Web 2.0: Design patterns and business models for the next generation of software? *O'Reilly Media*. Retrieved March 13, 2009, from http://www.oreillynet.com/pub/a/oreilly/tim/news/2005/09/20/what-is-web-20.html.

Phillips, K. (2005, December 5). Live from... transcript: interview with John Seigenthaler and Jimmy Wales. *CNN*. Retrieved April 1, 2009, from http://transcripts.cnn.com/TRANSCRIPTS/0512/05/lol.02.html.

Pressley, L., & McCallum, C. J. (2008, September/October). Putting the library in Wikipedia. *Online*. Retrieved July 15, 2009, from http://www.onlinemag.net

Publishing Trends. (2007). Thomson teaches tech through TWikis. Retrieved April 25, 2009, from http://www.publishingtrends.com/copy/07/0702/0702Thomson.html.

Quinn, M. J. (2005). Ethics for the information age. Boston, MA: Pearson Addison Wesley.

Rao, L. (2009). The French come calling for Wikipedia. Retrieved April 25, 2009, from http://www.techcrunch.com/tag/wikipedia/

Rasmusson, L., & Jansson, S. (1996). Simulated social control for secure internet commerce. In M. C. (Ed.), *Proceedings of the 1996 New Security Paradigms Workshop*. Retrieved April 1, 2009, from http://en.wikipedia.org/wiki/Soft_security.

Scevak, N. (2007, March 12). Wikipedia as a proxy for keyword analysis. *Bronte Media*. Retrieved April 26, 2009, from http://brontemedia.com/2007/03/12/wikipaedia-as-a-proxy-for-keyword-analysis/.

Schneider, K. G. (2007, September). Wikipedia's awkward adolescence. *Cio.com*. Retrieved July 16, 2009, from http://www.cio.com/article/141650/Wikipedia_s_Awkward_Adolescence.

Seigenthaler, J. (2005, November 29). A false Wikipedia "biography". *USA Today*. Retrieved April 1, 2009, from http://www.usatoday.com/news/opinion/editorials/2005-11-29-wikipedia-edit_x.htm.

Surowiecki, J. (2004). The wisdom of crowds: why the many are smarter than the few and how collective wisdom shapes business, economies, societies and nations. London: Little, Brown.

Survey: New Media (2006, April 20). The wiki principle. *Economist*. Retrieved March 31, 2009, from http://www.economist.com/surveys/displaystory.cfm?story_id=6794228.

Tapscott, D., & Williams, A. (2008). Wikinomics: How mass collaboration changes everything. *Journal of Information Technology & Politics*, *5*(2), 259–262. doi:10.1080/19331680802294487

TWIKI. (2009). TWiki tutorial. Retrieved April 25, 2009, from http://twiki.org/cgi-bin/view/TWiki.TWikiTutorial.

TWIKI.NET. (2007). White paper: Bringing wikis to the enterprise. Retrieved April 25, 2009, from http://www.twiki.net/.

Weill, P., & Vitale, M. (2001). From place to space: migrating to e-business models. Cambridge, MA: Harvard Business School Press.

Wikimedia Foundation. (2009a). *Home*. Retrieved April 22, 2009, from http://wikimediafoundation.org.

Wikimedia Foundation. (2009b). *Gift policy*. Retrieved April 25, 2009, from http://wikimediafoundation.org/wiki/Gift_policy.

Wikipedia (2009a). Wiki. *Knol*. Retrieved April 25, 2009, from http://en.wikipedia.org/wiki/Knol.

Wikipedia (2009b). Wiki. *Wikipedia: The Free Encyclopedia*. Retrieved April 1, 2009, from http://en.wikipedia.org/wiki/Wiki.

Wikipedia (2009c). Wikipedia: Biographies of living persons. *Wikipedia: The Free Encyclopedia*. Retrieved July 16, 2009, from http://en.wikipedia.org/wiki/Wikipedia:Biographies_of_living_persons.

Wikipedia (2009d). Wiki. *Crowdsourcing*. Retrieved 10 July, 2009 from http://en.wikipedia.org/wiki/Crowdsourcing.

Chapter 11
Social Networking for Distance Caregiving and Aging in Place:
A Case on Web 2.0 Technologies for Virtual Support

S. Ann Becker
Florida Institute of Technology, USA

ABSTRACT

This case examines the business development process for launching a social network targeting older adult caregivers many of whom have chronic health conditions. An older adult becomes a member of a social network called iShare-With-U.biz to monitor online one or more health conditions. He or she invites family and friends to join a private network for support in distance caregiving, staying connected in personal health management, and socializing using common social networking features. Web site design is discussed in terms of usability by older adults. Health Web sites and social networks are assessed in terms of usage by age group. Options for revenue generation are identified when taking into account free and fee-based Web site membership. The case concludes with a discussion of challenges facing online startups given rapid changes in technology, minimal barriers to market entry, and a near saturation point for Web sites with social networking capabilities.

BACKGROUND

An explosion of user-generated content is reshaping the media landscape, shattering the status quo and creating new opportunities for marketers (eMarketer.com, 2007).

*i*Share-With-U.biz [1] is a social networking Web site co-created by a college professor and a healthcare professional with the intention of bringing together friends, family, and communities in the common pursuit of good health, quality of life, and aging in place. *i*Share-With-U.biz offers proprietary features for self monitoring, distant caregiving, and data aggregation with the objective of supporting every member in a virtual network of support.

*i*Share-With-U.biz was initially the brainchild of Professor Katherine Adams for staying in touch with aging parents geographically distant from her. Dr. Adams' father suffers from chronic pain

DOI: 10.4018/978-1-61520-609-4.ch011

associated with aging. Her mother plays a caregiving role often needing family support to deal with health and daily living issues. Both parents are in their early eighties, are first time users of computers, have high school educations, and have few typing skills.

Dr. Adams asked her friend and colleague, Dr. Elizabeth Schmidt, to become a collaborator in the development of a social network Web site targeting aging caregivers. Dr. Schmidt lectures and writes articles on health and wellness inclusive of healthy aging. Dr. Schmidt is a firm believer in family involvement with aging parents to maintain quality of life and promote living at home (often referred to as "aging in place") for as long as possible. Dr. Schmidt agreed, so a small business called *i*Share-With-U, LLC, was formed.

Dr. Adams conducts research in related fields including human computer interaction and assistive technology to support caregivers and loved ones aging in place. Dr. Adams found that older adults, typically characterized as sixty years plus, are increasingly getting online to socialize and search for health information. Dr. Adam's parents had not previously used a computer, and they are excited about the prospect of staying connected to their daughter living thousands of miles away.

*i*Share-With-U.biz provides a simple capability of dynamically tracking data at personal, family, and group levels. It supports a personal journal customized by the user to track one or more health-related issues. These include: digestion and sleep quality; behaviors and moods; stress and energy levels; blood pressure and pulse; vision, taste, and hearing changes; exercise and social interaction, and others. *i*Share-With-U.biz applications support "on-the-go" usage through both mobile and Web interfaces.

The user has total control over personal data gathered and shared in his or her Web space in both raw and compiled formats. A novel feature is proposed using personal data and external data sources (e.g., weather) to generate reports showing data correlations. This and other features add personal value to the online experience and promote daily use of the Web site by members in a network of support.

Critical alerts, set by the user to appear in his or her Web space, highlight in real-time personal data that exceeds specified limits, external data that may impact personal health, and insights into journal entries or lack of them. The user may share alerts and in turn receive alerts from invited members. This feature promotes higher levels of site usage by engaging members and strengthening their online relationships.

An exciting innovation for *i*Share-With-U.biz is a proposed community forum feature whereby a user has the capability of pooling data with other community members in identifying group health trends. This engages users at a level beyond family and friends with the potential for large membership growth in both domestic and international markets.

Web Design for Older Adults

Dr. Adams has an opportunity to apply lessons learned in Web design research for older adult users taking into account normal aging factors. Figure 1(a) illustrates a data entry component of the Web site whereby a user enters sleep quality information. Most design components only require a button to make an entry. For sleep quality, pulse, blood pressure, blood sugar, and a few others, the user types a data value into the box. Figure 1(b) shows the capability of deleting an entry when a mistake is made. The user clicks on the checkbox and "Delete Selected" button. A message box is displayed to validate the intention to delete data. Then, the data value is removed. If the user has trouble with data entry or deletions, a member in his or her support network can provide assistance.

Because of usability recommendations by Nielsen (1999), Schneiderman (1998), and other experts, Web sites in general have become more user-friendly. Many Web sites meet the online needs of younger adult users. But, some do not

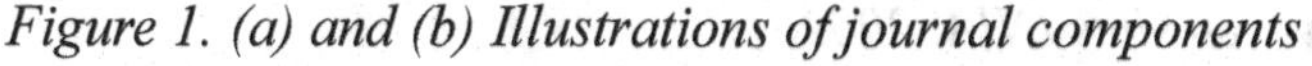

Figure 1. (a) and (b) Illustrations of journal components

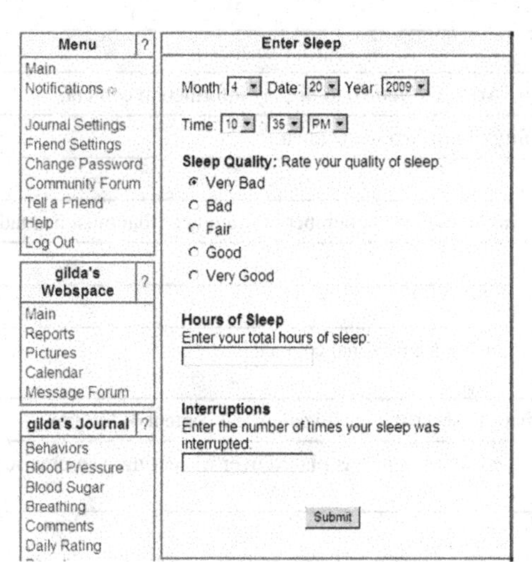

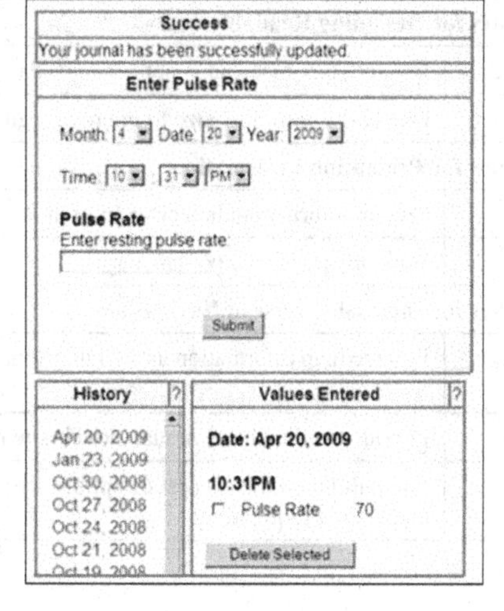

take into account usability needs of older adults in terms of vision, cognition, and motor skills associated with normal aging (Becker, 2004; Becker, 2005). These aging factors are further described.

Vision. The aging eye has a reduced ability to focus on close objects due to a reduction in the elasticity in the lens. There is a decline in visual acuity affecting the ability to see objects clearly. The lens of the eye yellows and thickens thus impacting color perception. There is decreased light sensitivity affecting adaptation to changes in light levels, and increased sensitivity to glare from light reflecting into the eye. Depth perception is reduced making it more difficult to judge the distance of an object (American Foundation for the Blind, 2009).

Together, these vision changes affect the use of a Web site in terms of legibility, reading speed, comprehension, navigation, searches, and visual distractions (Charness, 2001). For many older adult users, eye fatigue and strain occur even with corrective lenses, due to Web design factors including font size, font style, and font type, foreground and background colors, patterned background images, and animation, among others (Echt, 2002).

Cognition. Strong, Walker and Rogers (2001) identify problem solving, working memory, attention, and concept formation, as the cognitive issues influencing an older adult's use of Web features. Working memory entails temporarily holding and manipulating information while engaging in a variety of cognitive tasks (Baddeley, 1986; as cited in Strong, Walker, & Rogers, 2001, p.263). An older adult's performance on working memory tasks declines with age (Holt & Morrell, 2002), and he or she has a reduced ability to discern details in the presence of distracting information. As a result, complex navigation schemas, poorly designed search capabilities, and cluttered Web pages all negatively affect the older adult's online experience.

Motor Skills. Older adults have decreased motor coordination such that it becomes difficult moving and clicking a mouse, scrolling down a Web page (Hawthorne, 2000), and clicking on standard-size links (Ellis & Kurniawan, 2000). An older adult typically takes longer to complete a movement than younger adults (Chadwick-Dias, McNulty, & Tullis, 2003), and their movements tend to be less smooth and less coordinated (Seidler & Stelmach, 1996). Due to a reduction in fine motor skills, cursor positioning of the mouse is difficult

Table 1. Guidelines for making senior-friendly web sites (NIA, 2002)

Sample Guidelines for Designing Readable Text	
Sans serif typeface	Use font typeface that is not condensed (e.g., Arial, Helvetica) to display information content.
Large font size	Use 12-14 point font size to improve legibility of information content.
Sample Guidelines for Presenting Information	
Style	Present information in a clear and familiar way to reduce the number of inferences that must be made.
Simplicity	Write the text in simple language.
Sample Guidelines for Increasing Ease of Navigation	
Help and Contact Information	Provide help information as well as phone numbers for personal contact.
Site Map	Provide a hierarchical, visual model (site map) to show the organization and content of the site.
Menus	Use pull down menus (list of options displayed when mouse is placed over it) sparingly so precise mouse movement is not required.

for older adults, especially when interacting with small objects (Chaparro, Bohan, Fernandez, Choi, & Kattel, 1999; Walker, Millians, & Worden, 1996; Worden, Walker, Bharat, & Hudson, 1997).

Web sites may have usability impediments that make it difficult if not impossible for use by middle-age and older adults. Web sites may use an 8 point font size, for example, on low contrasting links appearing at the bottom of Web pages. Some use saturated colors for both foreground and background colors thus causing eye fatigue when using the site. Other Web sites use mouseover technology for triggering an event, such as a link changing color, images appearing, or menu lists expanded, when the user places the mouse over these objects. These and other potential design impediments make it difficult, if not impossible, for older adults to utilize the site fully.

User interface components for *i*Share-With-U.biz have been designed using the U.S. National Institute on Aging (NIA, 2002) guidelines. Dr. Adams knows that the NIA guidelines are based on scientific findings from research in aging, cognition, and human factors. (Morrell et al. (2004) provides a compilation of research findings.) Web design features for older adult users include sans serif font type, 12 point or larger font size, contrasting foreground and background colors, no patterned background images underlying text, and no mouseover technology. The NIA research-based guidelines, summarized in Table 1, provide a much-needed framework for addressing online design barriers facing older adult users.

Dr. Adams knows the importance of designing a Web site that takes into account aging factors. The objective is to promote readability, ease-of-use, and memory recall such that anyone can use the site with no usability impediments. She feels that if adults encounter usability barriers, then site usage will remain low.

Figure 2 shows a journal entry for Gilda who is an aging caregiver living at home. Gilda tracks personal health by daily monitoring of blood pressure, blood sugar, and breathing. Gilda tracks her emotions related to daily caregiving and the behaviors exhibited by the person receiving care. She posts comments to be read by family and friends in her network. Gilda has available to her other network features including picture album, shared calendar, message forum, and report generator. She can also participate in community forums thus connecting to other aging caregivers or those monitoring similar health conditions. Not shown is Gilda's network of friends and family each of whom has access to a specified set of features.

Figure 2. Journal entry by aging caregiver

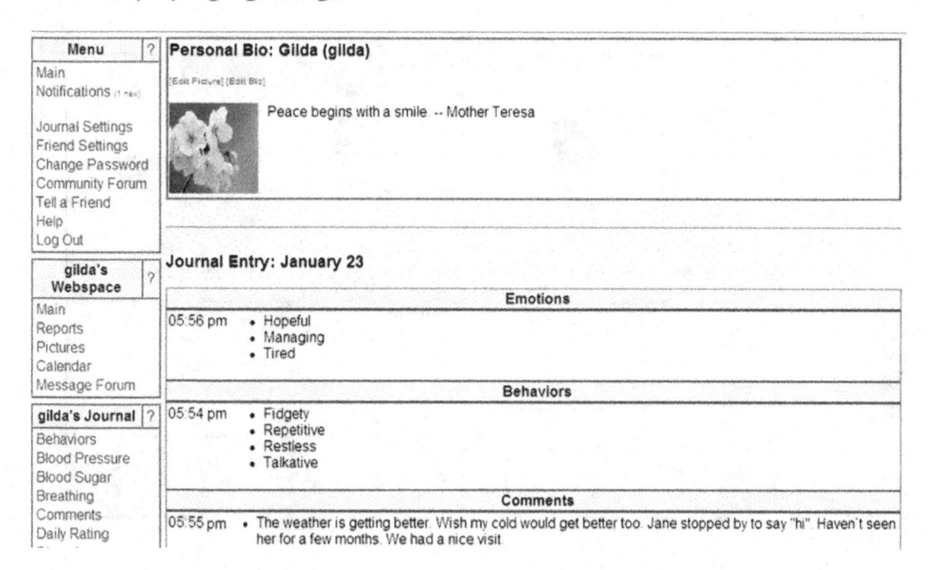

Gilda for a long time has wanted to send email to family and friends. She has tried email systems made popular by Web browsers; but became flustered when she forgot the steps of usage. The comments feature in *i*Share-With-U.biz has become a substitute for email in that Gilda posts messages for invited members. Several unedited messages are listed.

- *Bill is home. Still is tired and weak, but is getting better. Still coughs some.*
- *News from ice box Vermont. 28 below this morning. 20 below tonight at 7 o'clock. 30s this weekend will be wearing shorts. Bill is getting a <treatment> for his back. Pray that it will help heal the pain in his back. Love to all mom.*
- *Baked bread today. Haven't done that for a while. Turned out good. Bill's back is still sore. Happy spring, hope its going to warm up soon. Had rain today. Love mom and dad.*

Figure 3 shows the report feature whereby friends and family granted access can view personal health trends. This example shows Gilda's pulse rate for a specified time frame and presented in both histogram and time series formats. Not shown are other report options to help track multiple health trends over a period of time.

Demographics

Drs. Adams and Schmidt performed some additional research to find that life expectancy continues to be on the rise. The baby boomer generation has an adult turning 50 years old every seven seconds (Coughlin, 1999). They also found that older adults comprise about 12.4% of the U.S. population with about one in every eight Americans being in this age group (Administration on Aging, 2002). By 2030, this percentage will increase to 20% of the total U.S. population. The 85 years and older age group is the fastest growing (He et al., 2005).

Dr. Schmidt found that older adults in the United States are living longer with many aging in place without institutionalized care. It is reported that more than 55% of older adults live at home with their spouse, and the number of adults living without a spouse increases with age. About 50% of women aged 75 years and older live alone

Figure 3. Report generation capability of iShare-With-U.biz social networking site

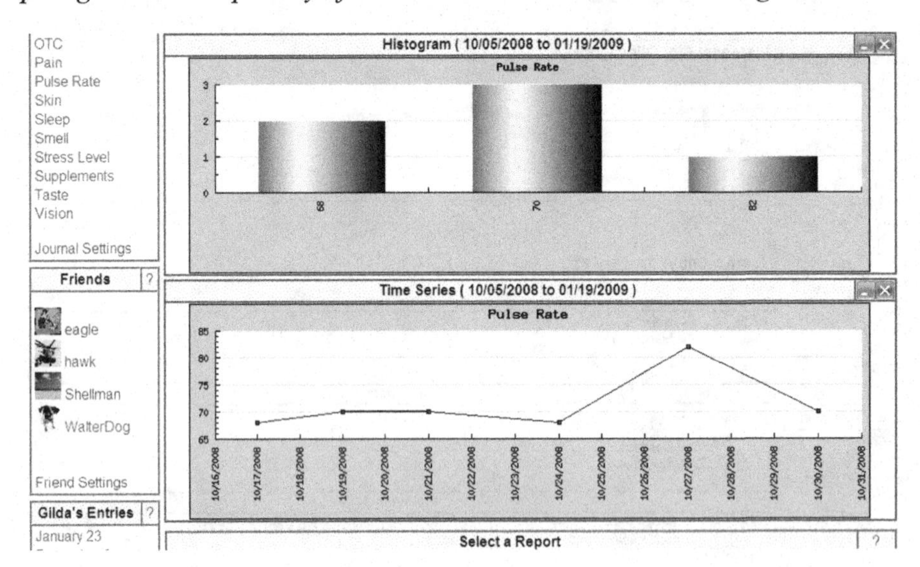

at home ((Fields & Casper, 2001). Dr. Schmidt also found that older adults prefer to age in place rather than in an assisted living facility (Riche & Mackay, 2005); many citing independence and social interaction as being critical to their well being (Hirsch et al., 2000). She also found that increasingly older adults require long-term care from unpaid or paid caregivers due to diseases associated with aging. About 7 million older adults over 65 years experience mobility or self-care limitations (Administration on Aging, 2002).

Because *i*Share-With-U.biz focuses on virtually connecting family and friends for healthy living, Drs Adam and Schmidt expanded their research on aging caregivers in the U.S. Fifty-five percent of the U.S. adult population is currently providing or has provided unpaid caregiving services to family or friends, and family caregivers perform 80% of all long-term care services (Opinion Research Corporation, 2006). Other societal trends compound the burden placed on unpaid caregivers. By 2030, a family will have on average about 2 children compared to 3 children in 1990 (Fields & Casper, 2001). Smaller family sizes along with geographically dispersed family members make it increasingly difficult to provide long-term care without some type of external support system.

Dr. Schmidt is concerned that the U.S. faces a critical challenge in dealing with an aging population that has unprecedented life expectancies. Emerging technologies offer the hope of allowing older adults to remain in their homes longer by empowering individuals to manage daily activities while dealing with chronic health conditions and age-related diseases. These technologies increasingly target a home environment whereby on a regular basis an individual can obtain assistance in performing daily living activities, stay connected to family and friends, manage medication, and have monitoring support for health-related changes.

The cited demographic changes highlight the need for innovative support systems for family members and their caregivers. For those with family and friends geographically dispersed, with busy schedules, or work constraints, *i*Share-With-U.biz.com and other such technologies, offer the means to monitor the health and well-being of aging caregivers and those in their care.

Table 2. Estimated health web site usage by age category

Percentages in Each Age Category						
	18 -24	25-34	35-44	45-54	55-64	65+
Drugs.com	7	20	16	19	21	18
Emedicine.com	11	24	22	19	14	10
Everydayhealth.com	5	12	15	20	24	22
Familydoctor.org	12	24	21	18	14	11
Healthline.com	11	17	19	19	17	15
Healthology.com	10	15	15	19	19	23
MayoClinic.com	8	18	17	18	19	21
Medhelp.org	11	26	22	18	13	10
MedicineNet.com	10	19	19	19	18	16
Prevention.com	11	15	18	20	18	16
Qualityhealth.com	6	16	19	21	22	16
Realage.com	5	10	13	22	25	25
RevolutionHealth.com	10	18	18	18	19	18
Rxlist.com	8	16	18	20	20	19
WebMD.com	10	23	19	18	17	14
(www.quantcast.com October, 2007).						

Opportunity and Strategy

Mr. Morgan is the Marketing Director of *i*Share-With-U.biz, LLC, responsible for developing a marketing plan. Mr. Morgan has several years of experience in online marketing. He worked with a Web hosting company for five years and an online retailer for three years. This is his first venture into the social networking arena; however, Mr. Morgan feels confident that his experience in online marketing will be a valuable asset.

Dr. Schmidt shared with Mr. Morgan her perception of a growing trend in adults using the Internet as a primary source for healthcare information. Dr. Schmidt noted that patients increasingly refer to content from WebMD.com, MayoClinic.com, and other Web sites in the self-diagnosis and self-treatment of diseases and chronic conditions.

Dr. Schmidt and Mr. Morgan studied the use of health Web sites categorized by adult age group. They compiled the data, shown in Table 2, for six age groups. They found that for many health Web sites there was a relatively high percentage of site visitors in the middle-age or older adult age groups. Forty percent of estimated users of MayoClinic.com site, for example, are 55 years plus. Almost 60% of estimated users of the Web site are adults middle-aged or older.

Next, Mr. Morgan compiled the data in Table 3 to study the popularity of social network sites. What he found is that MySpace.com, YouTube.com, Wikipedia.org, Craigslist.com, Photobucket.com, and Facebook.com have tens of millions of unique visitors per month.

Table 3 identifies the number of inbound links from external Web sites, as reported in an article posted by eBizMBA.com. Inbound links result in significant visitor traffic when users click through from the linking sites. The Google page rank is also shown in Table 3. This metric is important in determining the likelihood of a Web page appearing at the top of search results. Most social network sites listed in Table 3 will likely appear

Table 3. Search engine optimization statistics for social media web sites

Company Name	Inbound Links	Monthly Visitors in the U.S.	Google Page Rank
Myspace.com	92,285,805	72,505,214	8
Youtube.com	52,375,050	50,827,728	8
Wikipedia.org	75,539,411	42,177,402	8
Photobucket.com	144,789,477	26,424,436	8
Craigslist.com	2,218,413	22,434,271	8
Facebook.com	18,261,829	22,609,677	8
Flickr.com	51,779,871	21,045,679	8
Digg.com	118,353,522	18,443,249	8
IMDB.com	22,521,797	16,202,097	8
Typepad.com	39,468,446	6,134,284	8
Feedburner.com	100,244,672	5,381,882	8
Netscape.com	21,622,171	4,870,107	9
Topix.com	12,338,857	4,496,803	6
Xanga.com	8,229,423	3,713,235	7
Livejournal.com	33,451,748	3,389,292	8
Cafepress.com	39,894,526	3,225,855	8

Source of inbound links and Google page rank from eBizMBA (www.ebizmba.com/articles/user-generated-content.html). Source of monthly visitors in the U.S. is Compete.com (2007).

on the first page of the search results. Mr. Morgan is quick to point out that the data reflects an increasing use of the Internet in staying connected to others through a virtual environment.

Drs. Adams and Schmidt feel the time is right for taking advantage of the growing popularity of social networking sites. They both agree that being in the health niche of the social networking industry offers a competitive edge over the popular social network sites. It appears that no other social media site brings together family, friends, and invited members to share and aggregate personal data taking into account the usability needs of older adult users.

Mobile technology, with the introduction of smart phones, is emerging as a means of expanding the use of social media. *i*Share-With-U.biz will be exploiting this opportunity by offering both mobile and Web features to promote "on-the-go" connectivity to members in a support network. The integration of mobile and Web technologies is viewed as critical for *i*Share-With-U.biz to become an industry leader in the social media arena.

Drs. Adams and Schmidt and Mr. Morgan have come to the same conclusion that young, middle-aged, and older adults, comprising two-thirds of the U.S. population, can all benefit from the use of *i*Share-With-U.biz. Younger adults are tuned into proactive health management. Middle-aged and older adults are focused on managing chronic health conditions and are often involved in caregiving for one or more loved ones with chronic conditions. Older adults are focused on maintaining quality of life and aging in place. These trends reflect significant market potential for social networking that promote better living for all members.

SETTING THE STAGE

*i*Share-With-U.biz proposes to take advantage of the social networking marketplace through the use of mobile and Web technologies to track personal data related to everyday life and to engage others in viewing it. Mr. Morgan put together the following scenarios using "Joe" and a network of his family and friends to illustrate the social network's potential.

- Scenario. *A member selects from a range of health and daily living factors as part of a personal tracking system.* Joe sets up his account to track nutrition, exercise, blood pressure, and blood sugar in managing his diabetes. Joe makes data entries several times a day that are summarized in various reporting formats. He adds a new one to start tracking pulse rate.
- Scenario. *A member monitors his or her health and the health of members in the network, performs caregiving activities for distant members, and supports "aging in place" for members living at home with health issues.* Joe invites his Mom to join *i*Share-With-U.biz. Joe monitors Mom's data entries. Joe also tracks Dad's progress, a sufferer of chronic disease, as Mom enters observed behaviors as part of her data entries. Web alerts notify Joe when Mom posts that Dad is having a bad day or that she is feeling overwhelmed.
- Scenario. *A virtual network of support is broadened by the inclusion of linked in professionals making connections with potential customers.* Joe is searching for a therapist specializing in medical massage. Joe searches the professional connections component of *i*Share-With-U.biz to find a specialist that meets his busy schedule.
- Scenario.*A member may "pool" personal data with other members in online communities.* Joe joins an online community for Floridians suffering from chronic arthritis. His data, stripped of personally identifiable information, is aggregated with other members to identify data relationships with weather patterns in southern regions of the U.S.

Market Analysis

Mr. Morgan compiled data, presented in Table 4, identifying site usage by age group. The data showed a significant disparity between younger and older adults getting online to use social network sites. Mr. Morgan used Tables 1 and 3 to compare Web site usage of health sites and social networks. He found that a significant percentage of younger adults use both social networking and healthcare sites; whereas, middle-age and older adults are more apt to use health Web sites. While a small percentage of older adults use social networking sites, more than double the percentages are using healthcare Web sites.

To determine the potential of the targeted market, Mr. Thomas Morgan used the U.S. Census data to compile population estimates for targeted market segments. His data are shown in Table 5 for years 2000 and 2006.

Mr. Morgan aggregated the population data into three age groups: 25 to 44 years, 45 to 64 years, and 65 to 75 years plus. In total, these segments comprise about 65% of the total U.S. population. Each is below described in terms of population size and health management opportunities. Included are percentages of use for health Web sites (Table 2) and social network sites (Table 4).

- **Proactive health management (25 to 44 years of age).** User segment comprises 30% and 28% of U.S. population, respectively; for 2000 and 2006 census estimates. Focus is on proactive health management and staving off the onset of chronic health conditions. Approximately 37% visit healthcare sites and 40% visit social networking sites.
- **Quality of life and better living (45 to 64 years of age).** User segment comprises 22% and 26% of U.S. population, respectively;

Table 4. Estimated social network site usage by age category

Percentages in Each Age Category						
	18 -24	25-34	35-44	45-54	55-64	65+
Myspace.com	38	21	17	13	7.7	3
YouTube.com	25	18	19	18	11	10
Wikipedia.org	23	21	20	19	11	7
Photobucket.com	41	19	16	13	7	3
Craigslist.com	20	31	22	14	9.9	4
Facebook.com	56	11	10	11	8.8	3
Flickr.com	17	22	20	19	13	9
Digg.com	20	20	19	19	12	9
IMBD.com	19	23	20	20	13	7
Typepad.com	14	23	20	19	13	9
Feedburner.com	11	21	21	20	15	12
Netscape.com	16	18	17	20	16	11
Topix.com	6	11	15	21	23	24
Xanga.com	45	18	14	12	7	3
Cafepress.com	14	23	21	20	13	7
Linkedin.com	9	28	24	19	13	6
Livejournal.com	34	24	13	16	8	4
Friendster.com	25	42	14	10	7	3

(www.quantcast.com, posted October, 2007).
The data shows estimated breakdown of population use by age category for each Web site. These estimations are based on Quantcast.com data for interpreted patterns of Internet use.

Table 5. Percent of U.S. population by age group (U.S. Census actual and estimations)

Population	25-44 yrs	45-54 yrs	55-64 yrs	65-74 yrs	75 plus
2000 Estimation of the U.S. Population in Percentages					
Population Numbers	85,040,251	37,677,952	24,274,684	18,390,986	16,600,767
% of Population Rounded	30	13	9	7	6
% of Population Aggregate	**30**	**22**		**13**	
2006 Estimation of the U.S. Population in Percentages					
Population Numbers	84,082,929	43,378,174	31,586,683	18,916,844	18,343,508
% of Population Rounded	28	15	11	6	6
% of Population Aggregate	**28**	**26**		**12**	

U.S. Census Bureau (www.census.gov). Annual estimates of the U.S. population by age groups. (April 1, 2000 to July 1, 2006 (NC-EST2006-01)). Release date 5-17-2007. Note that the percentages do not add to 100% because persons younger than 25 are not included in the data.

for 2000 and 2006 census estimates. Focus is on managing chronic health conditions and staving off the onset of additional ones to maintain quality of life and living well. Approximately 38% visit healthcare sites and 29% visit social networking sites.

Table 6. Percent of U.S. population with top ten diseases

Condition:	Percent 25-44 yrs	Percent 45-54 yrs	Percent 55-64 yrs	Percent 65-74 yrs	Percent 75 plus
Arthritis	9.4	23.4	37.7	46.5	54.0
Hypertension	9.0	24.0	39.8	49.4	54.8
Heart disease	4.6	9.4	17.8	27.1	36.8
Coronary heart disease	1.1	4.3	10.7	18.5	25.1
Cancer, all	2.3	5.9	11.1	18.6	23.7
Diabetes	2.5	7.3	13.5	18.2	15.6
Heart attack	0.4	2.1	6.1	10.1	13.3
Stroke	0.5	1.4	3.6	6.8	11.9
Chronic obstructive pulmonary disease	3.4	5.3	7.4	9.8	9.4
Asthma	6.3	6.9	7.3	7.5	6.1

*Rounded to nearest whole number.
Center for Disease Control, Self-reported Chronic Conditions Among Adults (2003-05).
Source: http://209.217.72.34/HDAA/TableViewer/tableView.aspx?ReportId=221

- **Aging in place and life longevity (65 years plus).** User segment comprises 13% and 12% of U.S. population, respectively; for 2000 and 2006 census estimates. Focus is on managing multiple chronic health conditions while dealing with normal aging factors to prolong life and age in place. Approximately 16% visit healthcare sites and 7% visit social networking sites.

Mr. Morgan also analyzed data about diseases and chronic conditions across U.S. population using the same five age groups as in Table 4. The data, presented in Table 6, shows the percentage of the U.S. population with one or more diseases increases with age. The one exception is asthma, which has a consistent population base across all age groups. The data show that for those 65 years and older about 50% have arthritis or hypertension and almost one-third have heart disease. It also shows that almost one fourth of the middle-aged adult population in the U.S. suffer from arthritis or hypertension.

Mr. Morgan shared this information with Drs. Adams and Schmidt as a market need for monitoring health conditions and distance caregiving. Everyone agreed that there is a sufficiently large market in each of the three groups for high growth in site usage and strategic revenue generation.

Revenue Projections

The owners identified four revenue streams that offer long-term sustainability and growth. These include online advertising, sponsored advertising and affiliation programs, fee for services, and commercial sales of data lists. Each revenue stream is part of a two year roll-out of the social network. Each is briefly described in terms of revenue potential.

Advertising (Ad) Revenues

Click rate is the number of clicks on an ad as a percentage of the number of times that the ad was downloaded with the Web page. The cost-per-clickthrough (CPC) is the amount paid by the online vendor for each click by a user on an ad. Tables 7, 8, and 9 each show a revenue stream based on a click rate and CPC. The data in the three tables are based on several assumptions: site visits almost double each month during the first

Table 7. Projected ad revenues using $.03 CPC with 1% and 5% click rates

Month	Unique Visitors	Number of Visits/Month	Pages Per Visit	1% Click Rate		5% Click Rate	
				$.03/CPC Ad Revenue		$.03/CPC Ad Revenue	
Apr-09	5,000	60,000	8	0.0024	144	0.012	720
May-09	10,000	120,000	8	0.0024	288	0.012	1,440
Jun-09	20,000	240,000	8	0.0024	576	0.012	2,880
Jul-09	40,000	480,000	8	0.0024	1,152	0.012	5,760
Aug-09	80,000	960,000	8	0.0024	2,304	0.012	11,520
Sep-09	140,000	1,680,000	8	0.0024	4,032	0.012	20,160
Oct-09	200,000	2,400,000	8	0.0024	5,760	0.012	28,800
Nov-09	260,000	3,120,000	8	0.0024	7,488	0.012	37,440
Dec-09	320,000	3,840,000	8	0.0024	9,216	0.012	46,080
TOTAL for 2009:					**$30,960**		**$154,800**
Qtr1-10	1,080,000	12,960,000	8	0.0024	31,104	0.012	155,520
Qtr2-10	1,560,000	18,720,000	8	0.0024	44,928	0.012	224,640
Qtr3-10	1,720,000	20,640,000	8	0.0024	49,536	0.012	247,680
Qtr4-10	1,800,000	21,600,000	8	0.0024	51,840	0.012	259,200
TOTAL for 2010:					**$177,408**		**$887,040**

Table 8. Projected ad revenues using $.05 CPC with 1% and 5% click rates

Month	Unique Visitors	Number of Visits/ Month	Pages Per Visit	1% Click Rate		5% Click Rate	
				$.05/CPC Ad Revenue		$.05/CPC Ad Revenue	
Apr-09	5,000	60,000	8	0.004	240	0.02	1,200
May-09	10,000	120,000	8	0.004	480	0.02	2,400
Jun-09	20,000	240,000	8	0.004	960	0.02	4,800
Jul-09	40,000	480,000	8	0.004	1,920	0.02	9,600
Aug-09	80,000	960,000	8	0.004	3,840	0.02	19,200
Sep-09	140,000	1,680,000	8	0.004	6,720	0.02	33,600
Oct-09	200,000	2,400,000	8	0.004	9,600	0.02	48,000
Nov-09	260,000	3,120,000	8	0.004	12,480	0.02	62,400
Dec-09	320,000	3,840,000	8	0.004	15,360	0.02	76,800
Total for 2009:					**$51,600**		**$258,000**
Qtr1 10	1,080,000	12,960,000	8	0.004	51,840	0.02	259,200
Qtr2 10	1,560,000	18,720,000	8	0.004	74,880	0.02	374,400
Qtr3 10	1,720,000	20,640,000	8	0.004	82,560	0.02	412,800
Qtr4 10	1,800,000	21,600,000	8	0.004	86,400	0.02	432,000
Total for 2010:					**$295,680**		**$1,478,400**

six months of operation; a user visits the Web site on average 12 times per month; and a user visits on average 8 pages per visit. Dr. Adams wants to project a range of revenues for cash flow purposes; hence, there are three separate tables showing scenarios of low to high ad revenues.

Table 9. Ad revenues using $.10 CPC with 1% and 5% click rates

Month	Unique Visitors	Number of Visits/Month	Pages Per Visit	1% Click Rate		5% Click Rate	
				$.10/CPC Ad Revenue		$.10/CPC Ad Revenue	
Apr-09	5,000	60,000	8	0.008	480	0.04	2,400
May-09	10,000	120,000	8	0.008	960	0.04	4,800
Jun-09	20,000	240,000	8	0.008	1,920	0.04	9,600
Jul-09	40,000	480,000	8	0.008	3,840	0.04	19,200
Aug-09	80,000	960,000	8	0.008	7,680	0.04	38,400
Sep-09	140,000	1,680,000	8	0.008	13,440	0.04	67,200
Oct-09	200,000	2,400,000	8	0.008	19,200	0.04	96,000
Nov-09	260,000	3,120,000	8	0.008	24,960	0.04	124,800
Dec-09	320,000	3,840,000	8	0.008	30,720	0.04	153,600
Total for 2009:					**$103,200**		**$516,000**
Qtr1 10	1,080,000	12,960,000	8	0.008	103,680	0.04	518,400
Qtr2 10	1,560,000	18,720,000	8	0.008	149,760	0.04	748,800
Qtr3 10	1,720,000	20,640,000	8	0.008	165,120	0.04	825,600
Qtr4 10	1,800,000	21,600,000	8	0.008	172,800	0.04	864,000
Total for 2010:					**$591,360**		**$2,956,800**

The ad revenues for 2009, as shown in Table 7, are projected to range from $30,960 for a 1% click rate to $154,800 for a 5% click rate using a CPC of $.03. The ad revenues for 2010 are projected to range from $177,408 for a 1% click rate to $887,040 for a 5% click rate having a CPC of $.03. A higher click rate is used in 2010 projections as it is anticipated that there will be strategic learning about ad placement.

Tables 8 and 9 show the potential for growth in ad revenues based on aggressive marketing techniques to increase site traffic. Banner advertising commonly is cited at a 1% click rate; however, this can be increased through targeted advertising with achievable click rates of 5% (Irwin, 2007).

The ad revenues for 2009, as shown in Table 8, are projected to range from $51,600 for a 1% click rate to $258,000 for a 5% click rate using a CPC of $.03. The ad revenues for 2010 are projected to range from $295,680 for a 1% click rate to $1,478,400 for a 5% click rate having a CPC of $.03. A higher click rate is used in 2010 to reflect ongoing learning about strategic ad placement.

The ad revenues for 2009, as shown in Table 9, are projected to range from $103,200 for a 1% click rate to $516,000 for a 5% click rate both using a CPC of $.03. The ad revenues for 2010 are projected to range from $591360 for a 1% click rate to $2,956,800 for a 5% click rate both having a CPC of $.03. A higher click rate is used in 2010 as a result of strategic learning about ad placement.

Mr. Morgan feels confident that the ad revenues in Table 8 are attainable by working with Google Analytics and other tools to strategically place ads on the Web site. He feels that Table 7 ad revenues are too low given market niche opportunities. He plans on monitoring on a weekly basis age and health appropriate ads associated with targeted user groups. Mr. Morgan will rely on analytics tools to further customize ads per site visit.

Sponsored Advertising and Affiliation Programs

Dr. Schmidt will use her contacts in pharmaceuticals, health and medical supplies, general retailers,

Table 10. Sponsored advertising revenue

Month and Year	$500/month	$1,000/month	$2,500/month	$5,000/month
Apr-09	$0	$0	$0	$0
May-09	$500	$0	$0	$0
Jun-09	$500	$0	$0	$0
Jul-09	$500	$1,000	$0	$0
Aug-09	$500	$1,000	$0	$0
Sep-09	$1,000	$1,000	$0	$0
Oct-09	$1,000	$2,000	$2,500	$0
Nov-09	$1,000	$2,000	$2,500	$0
Dec-09	$1,000	$2,000	$2,500	$0
Total by category 2009:	$6,000	$9,000	$7,500	$0
Total for 2009:				**$22,500**
Qtr 1	$6,000	$6,000	$7,500	$0
Qtr 2	$6,000	$9,000	$7,500	$0
Qtr 3	$6,000	$9,000	$7,500	$5,000
Qtr 4	$6,000	$9,000	$10,000	$15,000
Total by category 2010:	$24,000	$33,000	$32,500	$20,000
Total for 2010:				**$109,500**

and insurance companies as potential targets for sponsorships and affiliation programs. Table 10 shows sponsorship revenue generated through the end of 2009 and for each of four quarters in 2010. The projected sponsorship revenue for 2009 is $22,500. The projected sponsorship revenue for 2010 is $109,500. The higher sponsorship rates reflect campaigns that would focus on a company throughout the site using variations in media formats. The ad revenue generated in 2009 is significantly higher than 2010 reflecting increased site traffic and brand recognition in the online marketplace. At the onset of launching the Web site, there is no high-level sponsorship advertising. This reflects the marketing effort that is needed to promote sponsorship advertising sales.

Dr. Adams found some interesting references in calculating revenue generated by sponsored advertising. She found that a Facebook.com sponsored group package costs $300,000 for three months and it has 150 sponsors participating in this program for a revenue base of over $90 million a year ((Internet Outsider, 2007). Eons.com, a social networking site for fifty year plus adults, recommended on its Web site sponsored advertising using a mix of banner and text ads with a minimum cost of $10,000.

Report Generator Fee for Services

Dr. Adams thinks a "user fee for service" will be popular in providing access to sophisticated report capabilities for personal data tracking. Dr. Schmidt is less optimistic that site visitors would pay for reports especially if simple ones are provided at no cost. Dr. Schmidt is concerned about the market perception that any usage of social networks is free. Dr. Adams recommends a market strategy of offering the service for free for an introductory time period.

Dr. Adams recommends that the Web site offer for a $2.95 monthly fee a report feature with sophisticated data correlations. Table 11 shows four scenarios of revenue generation for the nine

Table 11. Fee for service revenue estimations (report generator)

Month & Year	Unique Visitors	One Percent of Unique Visitors	Two Percent of Unique Visitors	Five Percent of Unique Visitors	Ten Percent of Unique Visitors
4/2009	5,000	$0	$0	$0	$0
5/2009	10,000	$0	$0	$0	$0
6/2009	20,000	$0	$0	$0	$0
7/2009	40,000	$1,180	$2,360	$5,900	$11,800
8/2009	80,000	$2,360	$4,720	$11,800	$23,600
9/2009	140,000	$4,130	$8,260	$20,650	$41,300
10/2009	200,000	$5,900	$11,800	$29,500	$59,000
11/2009	260,000	$7,670	$15,340	$38,350	$76,700
12/2009	320,000	$9,440	$18,880	$47,200	$94,400
Total for 2009:		**$30,680**	**$61,360**	**$153,400**	**$306,800**
Qtr1 2010	1,080,000	$31,860	$63,720	$159,300	$318,600
Qtr2 2010	1,560,000	$46,020	$92,040	$230,100	$460,200
Qtr3 2010	1,720,000	$50,740	$101,480	$253,700	$507,400
Qtr4 2010	1,800,000	$53,100	$106,200	$265,500	$531,000
Total for 2010:		**$181,720**	**$363,440**	**$908,600**	**$1,817,200**

month period of 2009 and the four quarters of 2010. Each scenario is based on a certain percentage of site visitors using the paid service. Dr. Adams points out that even if only 1% of site visitors pay for using the advanced report feature, the revenue generated ranges from $30,680 in 2009 to $181,720 in 2010. If 10% of projected site visitors used this feature, the revenue generated could near $2 million in 2010.

Professional Connections Fee for Services

Dr. Adams feels strongly that many health professionals will use a "fee-for service" in posting information to end users of the Web site. Once again, Dr. Schmidt is somewhat skeptical of a paid service, but she is willing to add the functionality for a trial time period. The professional connections feature is similar to one provided by a popular social networking site called LinkedIn.com. A physical therapist, for example, could post specialty areas, contact information, location, and hours of operation that would appear as an ad or a link on selected site pages (e.g., user's journal entry page for tracking pain level.)

Table 12 shows revenue projections for the nine month period of 2009 and the four quarters of 2010. The first few months show an introductory offer of free services for three months. The projected revenues for 2009 and 2010 are shown for two different monthly fees. The lower fee of $4.95 per month is projected in 2009 to generate $14,108 and the higher fee of $11.95 per month is projected in 2009 to generate $34,058. With strategic marketing efforts, it is anticipated that 2010 will generate significantly higher revenue at either rates. The lower fee of $4.95 per month is projected in 2010 to generate $92,813 and the higher fee of $11.95 per month is projected in 2010 to generate $224,063.

Data Lists

Dr. Schmidt identified the revenue potential of mining data for health trends. The data entered

Table 12. Fee for service revenue estimations (professional connections)

Month and Year	Professional Connections	$4.95/mth	$11.95/mth
4/2009	50	$0	$0
5/2009	75	$0	$0
6/2009	100	$0	$0
7/2009	350	$1,733	$4,183
8/2009	400	$1,980	$4,780
9/2009	450	$2,228	$5,378
10/2009	500	$2,475	$5,975
11/2009	550	$2,723	$6,573
12/2009	600	$2,970	$7,170
Total for 2009:		**$14,108**	**$34,058**
Qtr1 10	1,000	$12,375	$29,875
Qtr2 10	1,500	$19,058	$46,008
Qtr 3 10	2,000	$26,978	$65,128
Qtr 4 10	2,500	$34,403	$83,053
Total for 2010:		**$92,813**	**$224,063**

by *i*Share-With-U.biz users will be stripped of personally identifiable data. It then will be made available to companies searching for health-related information. Table 13 shows projected revenues. There are fewer projected revenues for 2009, as data will have to be compiled. The projected revenues for both years are based on $.07 per record sold. The data are based on industry rates (e.g., data in 1000 increments sells for $70 per thousand customer records).

Revenue Generation Market Strategies

Significant site traffic is critical in achieving revenue projections for both years. Mr. Morgan proposed that several marketing efforts be implemented to achieve an acceptable level of site traffic.

Table 13. Data aggregation

Data Records for 2009	$0.07/Visitor Record
300,000	$21,000
Data Records for 2010	**$0.07/Visitor Record**
1,500,000	$105,000

- **Email Lists and Email Alerts.** Use email and alerts for notifications about online communities, invited membership, special offers, new features, and company events.
- **Publicity.** Tap into social media sources for getting the word out about proprietary features. Issue press releases and pursue active engagement with organizations both regional and national levels for broader dissemination. Develop YouTube.com video explaining major features of the Web site.
- **Awards Program**. Develop award programs for design of innovative features to promote the use of *i*Share-With-U.biz. Similar award programs have been successful at promoting the development of mobile applications.
- **Free Services.** Announce limited offer for free access to fee for service features.

Target regional areas for a timed roll-out of professional connections.

- **Affiliate for Trade and Sponsorships.** Identify community and industry partners to trade links or share features on each other's Web sites. Identify social networking and health sites to post sponsored ads.
- **Web Analytics**. Use software tools and technologies to track data about bounce rates, hits, page views, requests, visits, inbound links, and sessions to strategically increase site traffic.
- **Search Engine and Web Page Optimization**. Optimize Web design to strategically place ads to increase click rates. Use software tools and technologies to analyze site statistics in pursuit of a higher Web site ranking.

Industry Outlook and Trends

The company anticipates that several trends in the social networking industry will contribute to the success of *i*Share-With-U.biz in the global marketplace. These trends include:

- Global memberships in social networking sites are expected to reach 230 million and global revenue is expected to reach $965 million by the end of 2007 (Pierce-Grove, 2007).
- North America accounts for only 25% of today's global social networking market offering the potential for significant growth in the international marketplace.
- Internet advertising revenues in the U.S. totaled nearly $10 billion for first six months of 2007, a 26.4% increase from the previous year (IAB, 2007).

Mr. Morgan put together advertising sales information, shown in Table 14, highlighting growth potential in the U.S. for online advertising sales. The table shows opportunities in sales revenue beyond what has been projected for the first two years. An example opportunity is the use of email to increase repeat visits to the Web site, generate new members, and sell strategically placed products and services.

Product Development

Drs. Adam and Schmidt have hired a database developer and a programmer to add enhancements and additional features to the Web site. Their work plan includes four versions of *i*Share-With-U.

Table 14. Advertising sales in the U.S dollars rounded to millions

Advertising Media	Year 1: First Six Months	Year 2: First Six Months
Search	40% ($3,164)	41% ($4,097)
Classifieds	20% ($1,582)	17% ($1,699)
Referrals/Lead Generation	7%($592)	8% ($799)
E-mail	2% ($158)	2% ($200)
Display-related:	31% ($2,413)	32%($3,198)
– Rich Media (Includes Video)	6% ($475)	8%($799)
– Ad Banners / Display Ads	21% ($1,622)	21%($2,099)
– Sponsorships	4% ($316)	3% ($300)
– Slotting Fees	<1% (<1$)	<1% (<$1)
Data for online advertising revenues from Web sites, commercial online services, free e-mail providers, and all other companies selling online advertising (IAB, 2007).		

biz. Each version is to be released in sequential order during a 15 month period. Each version is considered a milestone integrating mobile, database, and Web technologies in the development of proprietary software components.

Version 1 - Originally designed by Dr. Adams with assistance from local consultants. The nonproprietary features, shown in Table 15, are a major component of this version. Features include many of those associated with white label platforms, such as personal journal, personal data tracking, basic Web alerts, smart phone connectivity, shared calendars, photo album, and family blogs. Version 1 focuses on common social network features that many users will expect to be available upon product launch. Version 1 is released with no fee for service components.

Version 2 - Builds upon the previous release offering personalized exercise and simple nutrition components, enhanced report features, group blogs, file sharing, and a simple version of professional connections. Mobile applications offer transparent "on-the-go" connectivity to features on the social network site.

Table 15. Nonproprietary features

Feature	Version 1	Version 2	Version 3	Version 4
Profiles	X			
Profile Customization	X			
Personal History	X			
Member Portraits, Avatars	X			
Relationship Types	X			
Groups	X			
Photos	X			
Music/Audio		X		
Videos		X		
Forums/Message Boards	X			
Blogging	X			
Create a Group	X			
Group Blogging	X	X		
Messaging	X			
System-wide Messaging		X		
RSS Feeds	X			
Search			X	X
Widgets	X	X		
Tagging	X			
Events/Calendar	X			
File Sharing	X	X		
Member Updates	X			
Email Notification	X	X		
Settings	X			
Instant Messaging	X	X		
Invitations	X			

Version 3 - Personalization for personal health management is added along with data mining capabilities associated with the nutrition component. The report generator is enhanced with data correlation features, real-time Web alerts, and expanded mobile features. The site is also enhanced with aggregation of online community data. The user will add his or her health data to a community group for tracking health conditions. Fee for service components are fully implemented.

Version 4 - Enhanced mobile solutions are released that promote "on-the-go" usage to engage the user multiple times each day. Customizable report features are released to further engage members of online communities. A Spanish version is released to promote further expansion into both domestic and international markets.

Table 16 shows the proprietary components to be developed during versions 2 through 4. Dr. Adams is working with an attorney to secure patent disclosures as early as possible. The intention is to introduce barriers to market entry such that features in Table 16 cannot be duplicated on competitor Web sites. The sophistication of these features, such as statistical correlation of health data presented in a readable format, will promote fee for service revenue generation. Dr. Adams feels that by offering advanced features there is a significant opportunity for *i*Share-With-U.biz to fill a market niche in distance caregiving.

*i*Share-With-U.biz will continue to expand operations in the third and fourth years both domestically and internationally. It will pursue an international presence in South America, Europe, and the Asian-pacific areas by offering localized versions of the Web site. The penetration of these markets will result in a moderate increase in operational expenses to integrate local data sources, localize site features, and offer native language support in multiple languages.

CURRENT CHALLENGES FACING THE ORGANIZATION

Drs. Adam and Schmidt, with input by Mr. Morgan, have completed a Strengths, Weaknesses, Opportunities, and Threats (SWOT) analysis as part of the business planning process. They feel the SWOT analysis is necessary for identifying major challenges especially during the initial launch of the Web site. Their results are briefly described below.

Table 16. Proprietary and advanced features

Feature Description V1 V2 V3 V4					
Personalized Exercise and Nutrition Components	Proprietary software, external data, enhanced mobile design.		X	X	
Simple Reports	Personal data tracked	X	X		
Correlated Data Reports	Proprietary software, statistical manipulation, alerts, warnings		X	X	
Customizable Nutrition	Proprietary software for personal tracking of nutrition and diet.		X	X	
Professional Connections	Proprietary software.		X	X	
Advanced Alert Mechanisms, Enhanced Mobile features	Proprietary software for personal and group alerts. Enhanced mobile feature set.			X	X
Customizable by Condition	Proprietary software.				X
Group Reports	Proprietary software, statistical analysis.				X
Multilanguage Support	Design initiated for localized versions.				X

Strengths

- Web usability factors well understood by Dr. Schmidt who is the designer of the site.
- Proprietary software prototyped using Microsoft Mobile operating system and SQL Server 2005.
- Programming expertise in C#, PHP, and SQL readily available through local university talent.
- Marketing expertise associated with large projects. Mr. Morgan has been the marketing director for several Internet-based companies.

Weaknesses

- Executive management expertise is currently not part of the organization structure.
- Lack of experience in affiliate and sponsorship revenue generation.
- No brand name recognition.

Opportunities

- Increasing use of health sites and social networks by all age groups.
- Increased involvement by family and friends in distant caregiving activities.
- Increased internationalization of social networking sites for global reach.
- Increased demand for software tools and technologies that promote health and wellness.

Threats

- Technological innovation making social networking features obsolete.
- Crowded market makes it difficult for new social networking sites to establish a market base.
- Cultural impediments on sharing health information with others in a social setting.
- Low usage of social networks by middle-aged and older adults
- Low interest by Angel investors in social networks due to minimal barriers to market entry.

A major concern by the company is launching Version 1 with minimal barriers to market entry. The initial version makes use of Web 2.0 technologies used by many social networking sites offering basic features. Drs. Adams and Schmidt haven't reached agreement on whether to delay the initial launch until after Version 2 is completed. This delay would allow the development team to introduce several proprietary features.

Dr. Adams and the development staff have expressed concern about technology innovation. They feel it is important to focus on emerging technologies, which should be integrated into the initial release of software applications. It would be important, from their perspectives, to integrate mobile device usage. Dr. Adams points to the explosive usage of the microblogging site, Twitter.com, by cell phone users around the globe.

Dr. Schmidt has expressed a concern about whether older adults will use the Web site without some type of training support. She conducted an informal survey of older patients as to whether they would monitor personal health and wellness. The answers were mixed. Several patients expressed concern about privacy and security of data given it is available in a shared environment. Others expressed the notion that, "ignorance is bliss". They did not want reminders of health areas that needed improvement. Younger adult patients were excited about the opportunity to be proactive about their health particularly in weight and nutrition management. Several patients were distant from aging family members, so they expressed interest in distance caregiving.

Dr. Adams recently noticed that several health Web sites added nonproprietary features similar to what is available on Version 1 of *i*Share-With-U.biz. She is concerned that it may be difficult to compete with health sites if too many social network features are added. The popular health sites have millions of site visitors many of whom return regularly to obtain health information.

After discussing this issue with the company's staff, Dr. Adams quickly checked out health Web sites to determine the number and type of social network and health monitoring features made available to end users. The results of her informal validation are presented in Appendix A. All of these Web sites support group blog and email alert features. Most support RRS feed, community forum, and personal profile features. Some of the health sites offer features similar to *i*Share-With-U.biz for personal data tracking of blood pressure, blood sugar, and pulse, among others. Several Web sites generate simple reports based on data entered over a period of time (e.g., systolic and diastolic data for blood pressure), though none appeared to be very sophisticated.

Another challenge is to initiate revenue generation upon release of Version 1. Mr. Morgan identified a risk of having much lower click rates given the Web site is targeting older adults. His concern is associated with ease-of-use such that an older adult doesn't get "lost" on the Internet when clicking on an ad. Dr. Adams admits there hasn't been much research on online advertising and the impact on usability by older adults.

Dr. Schmidt has initially approached several colleagues in the health profession about Angel investment opportunities. So far, no one has expressed interest in becoming an investor. She will continue to foster relationships in the health care arena to get the word out about the newly launched site. Dr. Adams compiled a list of venture capitalists and investments in popular social network sites. Appendix B shows the results of her research efforts. Dr. Adams will seek venture capital investments using this list given the amount of funding made available to popular social network sites.

Immediately following the completion of Version 1, the company issued an electronic press release, posted a YouTube.com video, held a social event in the local community, and issued announcements on several listservs. Mr. Morgan has developed a marketing strategy that targets older adult organizations such as AARP. He plans on rolling it out in the next two weeks. Meanwhile, the development team is working on several proprietary features to be released to the general public as "Version 1.1".

REFERENCES

Administration on Aging. (2002). A profile of older Americans. *Administration on Aging, U.S. Department of Health and Human Services*. Retrieved May 1, 2009, from http://www.aoa.gov/prof/Statistics/profile/2002profile.pdf

American Foundation for the Blind. (2009). *What is normal vision? The aging eye- normal changes and their symptoms*. Retrieved May 2, 2009, from www.afb.org/seniorsite.asp?SectionID=63&DocumentID=3194.

Becker, S. A. (2004). A Study of Web Usability for Older Adults Seeking Online Health Resources. *ACM Transactions on Human-Computer Interaction*, *11*(4), 387–406. doi:10.1145/1035575.1035578

Becker, S. A. (2005). Web Accessibility and Critical Issues Facing Older Adult Users. In M. Khosrow-Pour (Ed.), Encyclopedia of Information Science and Technology (pp. 3036-3041). Hershey, PA: IGI Global. Reprinted 2008, Encyclopedia of Information Science and Technology (2nd ed.) (pp. 4041-4046).

Chadwick-Dias, A., Mcnulty, M., & Tullis, T. (2003). Web usability and age: How design changes can improve performance. *2003 Conference on Universal Usability*, Vancouver, British Columbia, Canada, 30-36. Chaparro, A., Bohan, M., Fernandez, J. E., Choi, S., D., & Kattel, B. (199). The impact of age on computer input device use: Psychophysical and physiological measures. *International Journal of Industrial Ergonomics*, *24*, 503–513.

Charness, N. (2001). Aging and communication: Human factors issues. In N. Charness, D. C. Park, & B. A. Sabel (Eds.), Communication, Technology, and Aging: Opportunities and Challenges for the Future (pp. 1-29). New York: Springer Publishing Company.

Coughlin, J. F. (1999). Technology need of aging boomers. *Issues in Science and Technology*, (Fall): 53–60.

Echt, K. V. (2002). Designing Web-based health information for older adults: Visual considerations and design directives. In R.W. Morrell (Ed.), Older Adults, Health Information, and the World Wide Web (pp. 61-88). Mahwah, NJ: Lawrence Erlbaum Associates, Inc.

Ellis, R. D., & Kurniawan, S. H. (2000). Increasing the usability of online information for older users: A case study in participatory design. *International Journal of Human-Computer Interaction*, *12*(2), 263–276. doi:10.1207/S15327590IJHC1202_6

eMarketer.com. (2007). User-Generated Content: Will Web 2.0 Pay Its Way? *eMarketer.com*. Retrieved May 2, 2009, from http://www.emarketer.com/Reports/All/Emarketer_2000421.aspx?src=report2_home

Fields, J., & Casper, L. M. (2001). America's families and living arrangement population characteristics. P20-537. U.S. Bureau of the Census, U.S. Department of Commerce, Washington, D.C.

Hawthorne, D. (2000). Possible implication of aging for interface designers. *Interacting with Computers*, *12*(5), 507–528. doi:10.1016/S0953-5438(99)00021-1

He, W., Sengupta, M., Velkoff, V. A., & DeBarros, K. A. (2005). 65+ in the United States: 2005. Special report issued by the U.S. Department of Health and Human Services and the U.S. Department of Commerce, Washington, DC.

Hirsch, T., Forlizzi, J., Hyder, E., Goetz, J., Kurtz, C., & Stroback, J. (2000). The ELDer project: Social, emotional, and environmental factors in the design of eldercare technologies. In *Proceedings of CM Conference on Universal Usability*, Arlington, Virginia (pp. 72-79).

Holt, B. J., & Morrell, R. W. (2002). Guidelines for Web site design for older adults: The ultimate influence of cognitive factors. In R. W. Morrell (Ed.), Older Adults, Health Information, and the World Wide Web (pp. 109-132). Mahwah, NJ: Lawrence Erlbaum Associates, Inc.

IAB. (2007). IAB Internet Advertising Revenue Report. Industry survey conducted by PricewaterhouseCoopers and Sponsored by the Interactive Advertising Bureau (IAB). Retrieved from http://www.iab.net/media/file/IAB_PwC-2007Q2.pdf.

Internet Outsider. (2007). *Time to update those Facebook revenue estimates*. Retrieved May 2, 2009, from www.internetoutsider.com/2007/07/time-to-update-.html.

Irwin, T. (2007). Web site click-through rates soar with human touch. *MediaPost Publications*. Retrieved May 2, 2009, from http://www.mediapost.com/publications/index.cfm?fa=Articles.showArticle&art_aid=65206.

Morrell, R. W., Dailey, S. R., Feldman, C., Mayhorn, C. B., Echt, K. V., Holt, B. J., & Podany, K. I. (2004). Older adults and information technology: A compendium of scientific research and Web site accessibility guidelines. National Institute on Aging, Bethesda, MD.

National Institute on Aging. (2002). *Making your Web site senior-friendly: A checklist.* Retrieved May 2, 2009, from www.nlm.nih.gov/pubs/staffpubs/od/ocpl/agingchecklist.html.

Nielsen, J. (1999). Designing Web Usability: The Art of Simplicity. Indianapolis, IN: New Riders Publishing.

Opinion Research Corporation. (2006). Attitudes and beliefs about caregiving in the U.S. Findings of a national opinion survey. *Johnson and Johnson.* Retrieved May 1, 2009, from www.strengthforcaring.com/util/press/research/index.html.

Pierce-Grove, R. (2007). *The Future of Social Networking: understanding market strategic and technological developments.* Datamonitor.com.

Riche, Y., & Mackay, W. (2005). Peercare: Challenging the monitoring approach to care for the elderly. In *Proceedings of the British Human Computer Interaction 2005 Workshop on HCI and the Older Population.* Edinburgh, United Kingdom.

Schneiderman, B. (1998). Designing the User Interface: Strategies for Effective Human-Computer Interaction (3rd Ed.). Boston, MA: Addison-Wesley.

Seidler, R., & Stelmach, G. (1996). Motor control. Encyclopedia of Gerontology: Age, Aging, and the Aged. San Diego, CA: Academic Press.

Strong, A. J., Walker, N., & Rogers, W. A. (2001). Searching the World Wide Web: Can older adults get what they need? In W.A. Rogers & A.D. Fisk (Eds.), Human Factors Interventions for the Health Care of Older Adults (pp. 255-269). Mahwah, NJ: Lawrence Erlbaum Associates, Inc.

Walker, N., Millians, J., & Worden, A. (1996). Mouse accelerations and performance of older computer users. In *Proceedings of the Human Factors and Ergonomics Society 40th Annual Meeting,* Santa Monica, CA (pp. 151-154).

Worden, A., Walker, N., Bharat, K., & Hudson, S. (1997). Making computers easier for older adults to use: Area cursors and sticky icons. In *Proceedings of the SIGCHI conference on Human factors in computing systems,* Atlanta, GA (pp. 266-271).

ENDNOTE

1 The Web site, company, owner, and employee names have been changed. The *i*Share-With-U.biz is a fictitious name used only in this case.

APPENDIX A: FEATURE COMPARISON OF TOP HEALTH SITES AND /SHARE-WITH-U.BIZ(TABLE 17)

Table 17.

Feature	iShareWithU	Revolution Health	Prevention	Everyday Health	WebMD	MedHelp	Medicine Net	Mayo Clinic
Built-in Personal Tracking								
a. Diet	X		X	X				
b. Moods	X	X	X					
c. Sleep	X							
d. Exercise	X	X	X					
e. Weight	X	X	X	X				
f. Stress	X							
g. Fatigue	X							
h. Herbal/Vitamins	X							
i. OTC Meds	X							
j. Behaviors	X							
k. Vitals	X	X	X	X				
l. Sleep Quality	X							
Subtotal:	100%	60%	50%	30%	0%	0%	0%	0%
Customized Personal Tracking								
a. Weight	X	X	X	X				
b. Nutrition	X		X	X				
c. Personal Reports	X	X	X	X				
d. Correlated Data External/Personal	X							
e) Automated Alerts - Personal Data	X							
f. Email Notifications	X							
g. Mobile Interface	X							
Subtotal:	100%	29%	43%	43%	0%	0%	0%	0%

Feature	iShareWithU	Revolution Health	Prevention	Everyday Health	WebMD	MedHelp	Medicine Net	Mayo Clinic
Social Network Tracking								
a. Share Data	X	X						
b. Customized Alerts - Distant Caregiving	X							
c. Aggregated Data Shared with Groups	X	X						
Subtotal:	100%	67%	30%	30%	0%	0%	0%	0%
Social Network Features								
a. Alerts to Members	X	X						
b. Photos, Files	X					X		
c. Forums/ Chat	X	X	X	X	X	X		
d. Widgets	X	X	X	X	X		X	
e. Tagging	X				X	X		
f. Personal Profile	X	X	X	X	X	X		X
g. Groups	X	X	X	X	X	X	X	
h. RSS feeds	X	X	X	X	X		X	X
i. Personal Blogs	X	X				X		
j. Group Blogs	X	X	X	X	X	X	X	X
k. Email Alerts	X	X	X	X	X	X	X	X
Subtotal:	100%	82%	64%	64%	73%	73%	46%	36%
Total Percentages	100%	52%	45%	39%	24%	24%	15%	12%
Web site analysis performed in December, 2007. Familydoctor.org uses the RevolutionHealth tracker toolset. Menshealth.com, realage.com, emedicine.com, healthology.com, rxlist.com, drugs.com had 10% or fewer features. Qualityhealth.com had fewer than 10% features with a posting that several social networking features were coming soon.								

APPENDIX B VENTURE CAPITAL AND SOCIAL NETWORKING SITES

Table 18 is a compilation of press releases, Web postings, and online articles that identify venture capital investments, acquisitions, and partnerships.

Table 18.

Venture Capital Funding	Company	Amount	Owner	Key Partners
Sequoia Capital www.youtube.com/press_room_entry?entry=jwIToyFs2Lc	YouTube	11.5 M	Google	None
Trinity Ventures www.photobucket.com	Photobucket	10.5 M (Series B)	Fox Inter-active Media	Unknown
Accel Partners (13 M), Greylock Partners (25 M) www.accel.com/news/news_one_up.php?news_id=1	Facebook	13 M 25 M	Independent	Microsoft
Greylock Partners http://publications.mediapost.com/index.cfm?fuseaction=Articles.showArticleHomePage&art_aid=37077	Digg	3 M	Independent	Microsoft
Mayfield Fund www.prnewswire.com/cgi-bin/stories.pl?ACCT=104&STORY= /www/story/02-08-2006/0004277319&EDATE=	Tagged	7 M	Independent	Jangl.com
Mayfield Fund, founding investor Idealab www.mayfield.com/newsarticles/SnapFundingRelease071805.pdf	Snap	10 M	Perfect Market Tech., Inc.	QuoteMedia, Picsearch
Index Ventures, Accel Partners www.techcrunch.com/2006/08/13/netvibes-secures-a-15million-investment/	Netvibes	15 M	Independent	MIVA Inc.
Benchmark Capital www.americanventuremagazine.com/news.php?newsid=1093	Bebo	15 M	Independent	Yahoo, Google, French Telecom
Benchmark Capital, Accel Partners Highland Capital Partners and DAG Ventures (Series C) www.archival.tv/2006/07/05/accel-benchmark-invest-15-million-in-metacafe/	Metacafe	15 M 30 M (Series C)	Independent (speculation Yahoo)	MTV Net-works
Silicon Valley-based VantagePoint Venture Partners, Point Judith Capital, Transcosmos Investments http://multiply.com/info/press/seriesb	Multiply	16.6 M (Series B)	Independent	ABS-CBNi
August Capital (Series B), Focus Ventures, August Capital, and Intel Capital (Series C) www.sixapart.com/about/press/2006/03/six_apart_close_1.html	Typepad	10 M (Series B) 12 M (Series C)	Independent	Part of Sixapart.com including LiveJournal.com
Portage Ventures (Series A), Mobius Venture Capital, Union Square Ventures (Series B) www.techcrunch.com/2007/05/23/100-million-payday-for-feedburner-this-deal-is-confirmed/	Feedburner	1 M (Series A) 9 M (Series B)	Google (100 M)	None

Chapter 12
Case "Mobile-INTEGRAL"

L-F Pau
Copenhagen Business School, Denmark & Rotterdam School of Management, The Netherlands

ABSTRACT

The case "Integral" is about how a multinational company specializing in machinery goods uses high technology in its field support and mandated safety solutions to migrate its customer relationships into partnerships of growing scope and with new revenue streams. The key technologies are in-situ equipment monitoring and wireless communications. The key management ingredients are top management's understanding and respect for operational issues. The history of the case also illustrates the importance of the strategic choice of the in-house vs. in-sourced nature of the needed technical expertise, and of a gradual deployment compatible with the fast technology evolution.

BACKGROUND OF THE CASE

Multinational industrial company, essentially family controlled, with roots in one part of Europe, and main sites in all of Europe, USA, South America, Canada, Japan (joint venture), Australia, China and most Asian countries. The size of the company is large with over 30,000 employees.

The company Integral produces mostly machinery goods, and related services, based on mechanical, electrical, and hydraulic engineering. Their customers are physical premise owners inclusive of building management companies, public buildings, and individual houses.

DOI: 10.4018/978-1-61520-609-4.ch012

Integral is an almost one hundred years old European company of which the majority is family owned. Today, Integral has over 30,000 employees worldwide (duplicate of sentence above). Its sales are in excess of 4 Billion Euros (approximately 5.2 Billion USD) and Integral has a wide customer base in excess of 200,000 clients. It supplies primarily mechanical and electromechanical machinery sold to owners of physical premises. Integral's machinery is widely deployed in many public, private and industry physical sites with regulated high safety and operational requirements. The regulators mandate that end users always can get help in an emergency or machinery failure situation. Customers mandate a very small downtime to allow end users to be on those sites.

Integral's roots and traditions are in industrial goods, largely linked to the construction and public works sector. The company culture is one of high trust to employees. Its strict enforcement of a strategy to give a performance edge to its customers is achieved by creating the best user experience with innovative solutions. The corporate culture, inclusive of human resource management and processes, focuses on enablement of operational excellence and cost competitiveness.

On the finance front, long-term value and customer relationships far outweigh short-term reactivity to financial results. This is especially important given the customers are all in cyclical industries.

SETTING THE STAGE

The overall management culture at Integral is one of delegation with consensus building. The roots for this are craftsmanship and the trust mindset by the owning family, far from hierarchical structures of many mechanical industries. However, there have been cases in some recent national subsidiaries where the hierarchical mindset prevails. They are slowly adapting to the corporate culture focused on improving cohesion.

Integral has a centrally-driven technology management process. As such, the company is continuously looking for innovations in the process, resource utilization, social behaviors as they relate to their customers, and advances in technology. In addition, Integral encourages "novelties" in technology when they lead to reduced cost of ownership from the end user perspective. Selected Integral customers are indirectly involved in the technology management process and much of corporate communications are targeting operational people rather than top executives in client firms, the financial community, or the general public.

Over the past 15 years, Integral had divested itself from a range of products families or generations, in order on one hand to focus on a very innovative new product line introduced only 10 years ago, and also on a very lucrative service business dealing with the installation and maintenance of all its product lines. Another major new product line has been introduced over the last two years.

This case will refer to some widespread technical abbreviations.

- B2B: Business-to-business
- CDMA: Code division multiple access
- D-AMPS: Digital advanced mobile phone system
- ERP: Enterprise resource planning
- GPRS: Global packet radio system
- GSM: Groupe système mobile (also called "2G")
- HSCSD: High speed circuit switched data (also called "4G")
- IT: information technology
- "Leased line": facility and contract granting a user exclusive usage of a fixed communication channel
- LTE: Long term evolution (also called "4G")
- Mobitex: Proprietary mobile packet data communications technology
- NMT: Nordic Mobile technology
- SLA: Service level agreement
- SS7: Signaling system 7
- UMTS: Universal mobile technology standards (also called "3G")
- VPN: Virtual private network

Brief Overview of Wireless Technology

Wireless communications technology allows humans or machines to exchange voice, data and multimedia. Users of this technology typically access such communications services via access terminals with a user interface (such as mobile phones, smartphones, portable computers), or dedicated stand-alone terminals embedded in other systems. The facilities and resources required to run communications services are owned and operated by communications operators, specialized companies or organizations, and other types of entities, subject to transmission licenses granted by national regulators. Large organizations may build their own communications facilities, and lease communications operators' backbone facilities. These organizations then enable communications exchange mostly internal to these organizations or with selected users. They also establish virtual private networks (VPN) or they develop customized wireless applications (Gao, Shin, Mei & Su, 2006).

Taking the case of wireless field support with field technicians as users, applications have been developed this way in the industry and over the past 20 years. Though this is the case, there are extremely few academic references about the management aspects thereof (see e.g. Barnes (2004) and Pesonen, Rossi & Tuunainen (2004)).

Almost all communication networks use a command protocol suite, also called signaling system SS7, to enable a connection to be set up, to continue, and to terminate. There exists many standards and competing wireless transmission technologies, all controlled by a command protocol suite, with very diverse transmission performance levels, although worldwide the dominant ones are the GSM, CDMA, UMTS and soon LTE standards. These standards have normally limited lifetimes due to technological progress, so that e.g. older technologies like NMT, D-AMPS have almost faded away, proprietary technologies like Mobitex are entrenched in their application niches, while new ones such as LTE are being put in service from 2008 (Lindmark, 2002). Some are incompatible or service integration is cumbersome, despite the now dominant role of the evolving unified technology family made of GSM (also called 2G), UMTS (also called 3G), HSCSD, LTE (also called 4G). Transmission technologies furthermore rely on standardized protocols, with are very detailed interaction process specifications, of which the number and complexity is very high; GPRS is one such protocol family for sending packet data over wireless networks, while Mobitex is another one, and the Mobile Internet protocols in the future may yield a migration path. The performance levels of all technologies and protocol suites is very different in terms of geographical area (coverage) reachable by these wireless technologies, of voice quality, of data transmission speeds, service diversity, as well as of security (Gao, Shin, Mei & Su,2006).

Finally, a word on costs; public communication operators for fixed telephony as well as wireless communications, charge subscription fees plus usage fees which are service dependent; most offer only pre-packaged "bundles" with catalog prices for up to a maximum usage per month, while some allow big users to negotiate a special tariff bundle for precisely the type of traffic and usage they require.

CASE

In the fall of 2002, Integral acquired another conglomerate in the industry sector of equipments for logistics. This sector is represented by a large number of diverse brands and products. The newly acquired entity is called PortB, which has a customer base for its products that does not overlap much with Integral's customer base. Both companies share a very critical dependency on some key mechanical /electromechanical technologies; and they also have their highest revenue margins in installation and maintenance of such technologies.

The two companies also share some similar external pressures such as providing very high availability for the products serviced by the installation and maintenance organizations. In addition, Integral must comply with very high safety requirements for all of its products, and in case of life being put at risk, dispatch times must be under 1 h in city environments and 2 h in the countryside. This is the regulated delay between a safety related alarm and having someone on the spot to solve the problem or verify the false alarm. Both companies have clients operating on very tight delivery plans often with guaranteed delivery times. So, they must ensure a very high degree of product availability and very short intervention times even in the roughest physical environments.

In the past, Integral and the acquired PortB used to operate with service contracts to their customers, with varying intervention times, and varying schemes for preventive maintenance and inspection. They supplied to customers their own or third party spares and consumables.

Over the past 40 years, Integral alongside its competitors developed a concept of remote emergency calls from end users over fixed telephone lines that are sent to emergency call centers manned 24 hours and seven days a week. This technology had evolved into on-demand flexible distance monitoring supported by the dispatch of mobile human intervention resources. Though the Integral products and its services "worked" well, over time the accrued engineering, project handling, interaction with customers, and interaction with other contractors grew too large in terms of increased expenditures and resources. In addition, indoor installation costs for the fixed telephone lines solution were quite high as well as being very time consuming and difficult to reconfigure when physical premises were changed.

Whether there was access to fixed telephone lines on or near the customer premises was almost never an issue. Regulators would not impose the same safety requirements (nor grant the same safety and intervention delay levels) when no reasonable access was available to fixed telecommunications. When fixed telephone lines were accessible, it was very costly over time to maintain dedicated subscriptions for these fixed distance monitoring solutions. Also, the case-by-case dispatching of support staff reduced scheduling flexibility of resources and thus significantly reduced response time. In fact, case-by-case dispatching often lead to intervention technicians having to drive or fly back to the emergency centre for debriefings, instructions, and get spare supplies before their next dispatch. In the 1980's, there had been many attempts in such then-highly centralized information handling organization to design planning and dispatching tools for minimizing both transportation costs and personnel employment constraints (e.g., schedules, mandatory rest periods).

In 1994, largely under the influence of the Nordic mobility culture and the dependability of NMT wireless phones, the interaction between Integral and a Technical University in a Scandinavian country on one hand, and the interaction between a local subsidiary of Integral and business consultants on the other hand, lead to the realization that the centralized field support command scheme did not make sense. A concept emerged to evaluate the use of mobile voice and data from wireless enabled terminals mounted inside Integral's products, thus avoiding in some cases (those where wireless coverage was sufficient) the primary dependency on specialized fixed telephone lines. (For further discussion of this concept, refer to Pesonen, Rossi, & Tuunaimen (2004)).

Over time, as wireless operators "spread their wings", wireless communications would allow geographical coverage even where no fixed telephone lines were accessible. Some early wireless technologies allowed for more cost-efficient use of communication systems, as transmission did not have to be continuous. Also, the service technicians could maintain mobile voice contact to the field support information centre, avoiding their use of separate fixed telephony lines and subscriptions. By roaming and call transfers, the field support information centre did not have to be manned necessarily from one location.

Early experiments in completely different industry sectors, such as agriculture (see Wang, Zhang, & Wang (2006) for a recent review), inspired a product development whereby mobile data from sensors on or inside Integral's products may be sent to the field support information centre to feed computer-based maintenance and diagnostic software, and can thereafter be dispatched in raw or processed form to the service technicians on the move. Furthermore, by integrating the design, manufacturing and operational know-how with the sensor, data collection and wireless communications capabilities, better product maintainability was achieved at product design level using new performance monitoring and failure diagnosis techniques (Pau, 1989). Integral pioneered this concept starting in 1994; however, most competitors only discovered its advantages much later (from about 2005). As recent as 2008, competitors still prioritized "traditional" fixed telephone lines whenever possible.

Business consultants were used by Integral to identify technology solutions for competitive advantage. The consultants' critical remarks highlighted Integral's technical solution as being too narrow a view of the potential of the proposed field support integrated information and communication technologies. The consultants pointed out that by having a communications operator allow Integral to get access to network traffic measurements and call originating data (via SS7), Integral could offer its clients detailed equipment service usage statements. In essence, service contracts, inclusive of terms and features, could be customizable based on a customer's needs. Furthermore, this would allow third party facilities management companies to become Integral affiliates to whom field support and operational information centers could be licensed. It also provided for a revenue stream in terms of selling or licensing monitoring software with upgrade proposals for products in the field.

These benefits were not "new motivations" driven by a push by mobile technology innovation, but a natural evolution from the way of doing business while exploiting mobile technology. An outcome of the technical capabilities of wireless technology was an incremental advantage over competitors not using this technology of achieving longer, more sustainable and often more profitable business relations. Given the potential of wireless technology, explained to them by the Technical University, it was never a problem to have Integral's executive management and board of directors' study, prototype and carry out field trials of wireless technology capability. Integral's management identified a fundamental, competitive advantage in the use of wireless technology with field support as the entry point. As such, the company did not want competitors or suppliers to fully understand the business consequences. The potential for reduced costs and increased revenue sources offered a competitive advantage in a highly competitive low-margin industry.

Integral could now change the traditional nature and "philosophy" of field support contracts into key service contracts linked, *or not*, to their own product offering while meeting safety and maintenance goals. Integral's business units were quickly asked to take on the maintenance and support of products supplied by both the company itself and competing product suppliers. The business units could take over the operational handling of these products from end customers receiving a service provisioning rent or a usage-based fee. Integral won the trust of customers to sometimes bundle the field support with the product sales. The customer was not even required to maintain ownership of products thus allowing Integral over time to manage an installed base resulting in both economies of scale and product portfolio streamlining. Customer contacts and contracts changed completely with dramatic increases in revenues generated. Such an evolution is in line with the most strategic evolutions of technology-driven changes associated with e-business (Chaffey, 2004).

The recent acquisition of the new affiliate PortB, which had been looking for a strong parent company, offered Integral operational synergies. PortB chose Integral for two reasons, one of which was this advanced field support capability and the second being the related installed base. PortB revamped its business model to enable progressive synergies associated with combining Integral's field support tasks with its own.

History of Technology Adoption

The wireless capability innovated by Integral has gone through several generations since the first prototype in 1994 and supporting trials in Italy, Singapore, and a Nordic country.

In the early phases of Integral's technology innovation, the mobile/wireless capability was installed side by side with the fixed telephone line connected terminals attached to Integral's products and made available to service technicians not able to access phone booths due to their zone of operation. The wireless boxes initially only handled voice. In two of the trial countries, about 30 wireless data enabled Mobitex terminals were installed with autonomous operations sending sensor measurements about Integral's product while in service.

Later, from 1998 to 1999, and in collaboration with a key wireless terminal supplier and the above mentioned Technical University, GSM and D-AMPS modules were instituted as a configuration option for new Integral equipment installations. The equipment end users would still call by voice from a microphone in case of an emergency or problem, but they would not know that the call was sent wirelessly given the surrounding noise environment of the equipment operations hid a lower voice quality. The field support information centers were essentially transcribing manually the user's voice messages, and transcribed formatted messages were thereafter routed semi-automatically. However, there were major problems. These included high tariffs by the mobile operators and sometimes low levels of support by the same for the wireless connections. They also included endless hassles about accessing so-called SS7 signaling data for the control of fixed and mobile voice communications.

In 1999, there were also major problems with some large suppliers and integrators involved in the innovative field support system development. Despite the strict non-disclosure and clear technical specifications given to them, these technical subcontractors over-engineered the solution. They had hoped in the reengineering process, to develop, at Integral's expense, some generic remote support/maintenance solution or product of their own. Some of the greediest were information technology (IT) "outsourcing" companies claiming special competencies or products for embedded software to be used in diverse markets. These commercial parties and in also the mobile operators, despite long lasting relations, wanted to position themselves as part of a Virtual Private Network (VPN). Their overall goal was to add themselves as required intermediaries to offer under their own brands the field support capabilities of Integral. Internally, from some local Integral subsidiaries, strong criticism came back to executive management as communications operator costs and time based tariffs were unjustifiably high (20% to 25% of in-field support costs) in view of the allowable transmission budget share in any field support service offering.

In 2000, Integral's management contacted an honest supplier along with the original influential consultants to develop a simple, understandable deployment strategy. In a nutshell, they recommended to initially use only the wireless data Mobitex standard where tariffs were essentially data volume based and for Integral to focus on major markets. Next, deployment was left as an initiative for local subsidiaries based on profit targets. This was especially beneficial in geographical areas where service technician time and logistics were the most expensive (had a wireless business solution not been used). Finally, engineering development of the wireless capability was no longer to be outsourced, but more than 50% was to be carried out internally at Integral. The bulk of the balance of work effort was to be assigned to one key global supplier of wireless terminals with good technical and operational understanding of Integral's needs. In addition, outsourced work would by given to a small, dedicated one-task integration company with 20 employees.

The above strategy was implemented amidst all kinds of hyped pressures bestowed upon Integral from IT consultants and financial analysts. They failed to realize that Integral was far ahead of their aspirations having had real trials and a real business vision it did not reveal. Some of the external pressures came from wireless operators hoping for large volumes of wireless data traffic (compliant with 2.5 or 3G wireless standards GPRS and HSCSD) in their "empty" new networks shunned by individual mobile data users due to overblown tariffs.

Actors Involved and Their Roles

Though Integral's facilities and products are typically labeled as safety critical for both the general public and company client users, industrial safety is always the key concern. Industrial safety regulations and requirements, in terms of wireless capability, vary according to the risk aversion profile of administrations and regulators. In some countries or regions, the wireless terminals must operate adjacent to fixed lines solutions, even if the latter just collect dust thus representing a wasted cost.

Integral has gotten some understanding by telecommunications regulators, in terms of business needs and customers' needs; particularly in countries with a technical and operational understanding of wireless technology as opposed elsewhere to a more legalistic and policy role view in safeguarding the end users' safety. In either situation, further innovations in telecommunications technologies could be deployed and enableb low-cost monitoring of Integral's systems without paying huge communications usage fees. Integral has sometimes been unsuccessful in promoting the possibility for third parties (such as facilities management companies) to carry out at low cost safety critical monitoring of Integral's systems. This would be accomplished through accessing the telecommunications operator's infrastructure.

From the start, the relation with the cellular wireless communications operators (which offer GSM, CDMA or 3G services) has been stormy, whether dealing with technical features, tariffs, support, or business contracts. In comparison, the fewer in number wireless data Mobitex operators worldwide have shown respectful concern. They have helped Integral and its clients by setting tariffs under the "wounding limit"; these would be bulk wireless data prices representing over time typically less than 8-10% of the total field support costs

End users have always been very positive when they have received explanations of what was being deployed and about the operational aspects associated with the use of Integral's equipments. This was especially the case in all the countries or regions where industrial safety authorities gave their blessing to the use of Integral's technologies.

Integral's direct and indirect customers accepted the better service level offered, sometimes paying more or granting longer term contracts. Due to needs for stability in the management of physical premises and in view of the long lifecycle of the systems (15-30 years or more), none of these customers used shorter-termed competitive service level agreements as offered by the IT industry. Typical contract length of two years with milestones, allowed Integral to increase revenue and reduce risk.

With the exception of the one major partner, a communications terminals supplier with global coverage, Integral has learned a lesson not to rely on IT consultants or suppliers for solutions of key strategic, business and technical importance. To accommodate this policy, Integral set up an internal organization involving in total, only 55 employees worldwide; this organization is considered by Integral as "lean" and it is very dedicated thanks to the prospects of evolution and future challenges.

The external company with 20 specialists focuses on providing customized integration services and independent testing for Integral. This company is not owned by Integral but it has gotten visibility with some of Integral's customers. As such, they perform tasks for these customers that are not provided by Integral.

The role of the two to three academic instigators of the concept should be highlighted. These academicians have a worldwide reputation in failure diagnosis and maintenance and also by chance, in wireless software and mobile business. Unfortunately, many of their peers in academia do not recognize the value of this combination of talents. In industry, however, they are highly regarded as true business and technical innovators.

Last but not least, the Integral field support staff benefited directly and quickly from the adoption of wireless capabilities, with significant time and travel savings, better connectivity to backend systems, and other "quality of work" advantages. (Refer to Barnes (2004) for further discussion). Especially highlighted is the dramatic increase in job satisfaction and the reduced poaching risk by competitors.

RESULTS AND CRITICAL SUCCESS ASSESSMENT

Operational Results

In 2006, six years after Integral management changed their wireless field support initiative, the company has been able, with varying and ever increasing number of client sites, to validate:

- On average 25% gains in operational time available to its service technicians or efficiencies in solving client tasks.
- An 18% higher margin on the field service contracts where the wireless solution is offered.
- Systematic observations of high costs (subscriptions and usage) charged by mobile communication operators, with in average 25% higher mobile communication or fixed leased line costs in comparison to old fashioned fixed telephony lines. However, the migration to more adequate Mobitex and GPRS mobile data transmission subscription bundles has lead to a 35% cost reduction over fixed telephony lines when available, with in addition the much wider coverage offered by wireless communication. When no wireless coverage is available, leased fixed lines are used, and the costs have started to trickle down as well.
- High costs for software development in view of the deployment of the innovative concept in terminals/modules attached to equipment on customer premises. The highest costs are associated with field support information centre server resources where ERP (Enterprise Resource Planning) software in particular is rapidly becoming more costly than the wireless application software.
- An immediate impact from the use of the mobile enhanced field services on Integral's product strategy, and the identification of where this mobile capability should be deployed by default.
- The ramping up of the number of third party field support organizations serving also competitors products, and operating with Integral's field support technology
- A 30% reduction in the variance of dispatching time in case of emergency or failure when the mobile field service was deployed in the customer area.
- On average a 50% reduction in false alarms as more diverse sensor data could be processed in almost real time, to show often bad usage or handling of products by end users.

Table 1. Values of success factors

Success Factors	(Positive, Neutral, Negative)
• Operationally minded top management	Positive
• Acceptance of local initiatives	Positive
• Very low key disclosure and high confidentiality	Positive
• Mobile/wireless business solution as an evolution from a mandated regulated technical facility	Positive
• Progressive and localized introduction	Positive
• Mapping of the mobile field support capability into operational customer cash flow and relationship terms	Positive
• Success in merger and acquisitions with fast synergies	Positive
• Trust some multidisciplinary expert individuals with simple operational mindsets	Positive
• Focus, and simplify all the time, while increase dependability of solution for end users, customers and regulators	Positive

Critical Success Assessment

The list below summarizes success and failure factors as selected and reviewed by Integral's top management, associated with Integral's field support innovations (See Table 1)

Management success factors:

- Operationally minded top management.
- Acceptance of local initiatives.
- Very low key disclosure and high confidentiality on the technical development and trials.

Management failure factors:

- Errors by Integral suppliers, who were inclined to steal an idea, grow 'fat' on Integral with outsourcing contracts.
- Attempts at product replication by suppliers/subcontractors in deriving a generic product to replace Integral's requirements.
- Negative attitude towards Integral by business management of wireless cellular operators.

Business success factors:

- Mobile/wireless business solution seen as an evolution from a mandated regulated technical facility (fixed telephone line safety feature).
- Progressive and localized introduction.
- Mapping of the mobile field support capability into operational customer cash flow and relationship terms.
- Success in merger and acquisitions with fast synergies (PortB).

Business failure factors:

- Risks associated with high fixed costs.
- Risks associated with early stage complexity associated with the development of customer premises services and products.

Technical success factors:

- Trust in experts with multidisciplinary backgrounds supplemented with "keep it simple" operational mindsets.
- Focus kept on a solution to satisfy the needs of a range of constituents including end users, customers, and regulators, simplifying it all the time and increasing dependability.
- The engineering of the wireless field support system from the beginning on packet data, enabling a smooth migration from

Mobitex to GPRS and now to Mobile Internet.

Technical failure factors:

- Too long external development times by large outsourcing integrators.
- Risks of getting too exposed to "trendy" technical solutions.

CURRENT CHALLENGES

Operational Challenges

For Integral, the progresses made in field support have been the key reasons why it can handle decreasing orders for new installations during depressed economic climates. From 2000 to 2008, operating income as a percentage of sales has grown consistently from 7% to 12%, while total sales have increased between 5% and 18%. The main business challenge is whether the operating profit level reached represents a ceiling. Another challenge is how to sustain competitive advantage. Competitors will soon start to adopt similar approaches to Integral in field support and maintenance.

One management camp inside Integral wants to focus on quality and maintainability of the systems as deployed, and to reduce the frequency of physical interventions. Another camp wants to capitalize better on the field support system by opening it up to other suppliers. A third is focused on system monitoring and the overall energy efficiency of installed systems. This camp aims at Integral becoming the only overall energy efficient supplier, reusing the same wireless monitoring capability, enabling in this way at attaining a market niche for refurbishing the installed base with the potential to pass on benefits from buyers to customers.

Corporate Image

A corporate debate has emerged in terms of how to classify Integral's achievements in technology innovation from a business model perspective with the objective of catering to a higher management style profile for the company. This case "Mobile Integral" does not fit into the conventional e-Business model classification (with B2B, "click and mortar", e-Tailer, Personalized portals, Auctions, Group buying, etc.), but is a case in how usage-led technology innovation leads to entrepreneurial successes and pitfalls.

The fundamental reasons are:

a) Most e-Business models apply to physical or virtual goods (such as physical or content), and not to revenues, costs and enhanced business relations from operational services like here field support.
b) e-Business models focus primarily on the value-added transactions and chains. They typically do not focus on: implications and design changes to the goods needing the operational services; internal productivity gains (time, costs); and pricing changes such as bundling wireless enhanced field support service with other services including equipment rent.
c) e-Business models externalize the required communication network(s) as pure content transport facilities and assume that all customers have access to them. Wireless communication technology has some fundamental attributes such as: ubiquity, no need for local cabling infrastructure, personalization, and eventual localization. Each of these impacts the acceptance of the business model, and each of these attributes has temporal characteristics.

REFERENCES

Barnes, S. (2004). Wireless Support for Mobile Distributed Work: Taxonomy and Examples. In *Proc. 37th Hawaii International Conference on System Sciences* (pp 2-10), Big Island, Hawaii. Retrieved January 10, 2009, from http://csdl.computer.org/comp/proceedings/hicss/2004/2056/03/205630078a.pdf.

Chaffey, D. (2004). E-business and e-commerce management (2nd ed.). Harlow, Essex: Pearson Education Ltd.

Gao, J. Z., Shim, S., Mei, H., & Su, X. (2006). Engineering wireless based software systems and applications. Norwood, MA: Artech House.

Lindmark, S. (2002). *Evolution of techno-economic systems: an investigation of the history of mobile communications*. Published Doctoral dissertation, Chalmers tekniska högskola, Göteborg.

Pau, L.-F. (1989). Failure diagnosis and performance monitoring (8th ed.). New York: Marcel Dekker.

Pesonen, M., Rossi, M., & Tuunainen, V.K. (2004, March). *Mobile Technology in Field Customer Service*. Presentation at Austin Mobility Roundtable, Austin, TX.

Wang, N., Zhang, N., & Wang, M. (2006). Wireless sensors in agriculture and food industry: recent developments and future perspective. *Computers and Electronics in Agriculture, 50*(1), 1–14. doi:10.1016/j.compag.2005.09.003

Chapter 13
The Egyptian National Post Organization Past, Present and Future:
The Transformational Process Using ICT

Sherif Kamel
The American University in Cairo, Egypt

ABSTRACT

Over the last 20 years, the international postal sector has changed drastically due to several forces, including globalization, changing technology, greater demands for efficient services and market liberalization. For Egypt, keeping up with the changing atmosphere in the global market meant investing in information and communication technology. The Ministry of Communication and Information Technology (MCIT), as part of its efforts to transforming government performance using ICT, chose the Egyptian National Post Organization (ENPO) as a model for ICT integrated government portal. The selection was due to ENPO's extensive network, the public's confidence and its trust in the organization. The case of ENPO, capitalizing on public-private partnership models, proved successful when reflecting ICT deployment for organizational transformation within the context of an emerging economy. In addition to its importance in providing eGovernment services to citizens, ENPO is evolving as a critical medium for effectively developing Egypt's eCommerce. This case study takes an in-depth look at how ICT has improved the quality and range of services offered by ENPO, while asserting the magnitude of its impact on the country's emergence as a competitor in today's global postal market.

INTRODUCTION

As Amr Badr Eldin, ENPO vice chairman for Information Technology and responsible for IT strategy, infrastructure deployment and utilization, approached his office at the headquarters of ENPO, which is ironically Egypt's oldest museum, located in Ataba square, one of the busiest squares in downtown Cairo; the first object that immediately grabbed his attention was the 1865 automatic stamp vending machine. In a country like Egypt,

DOI: 10.4018/978-1-61520-609-4.ch013

where automatic vending machines were scarcely found and were still considered innovative, one immediately realized the great role that this place once had in the establishment of modern Egypt. The postal sector, on an international level, had changed drastically in the last 20 years. Several key forces had driven this evolution in the postal market including but not limited to changes in the volume of supply and demand of postal services, globalization effects, market liberalization, changing technology, dynamic communications shift, and regulatory progress, amongst other factors.

One other primary reason was the ever-growing competition from the private sector threatening the comfortable monopoly enjoyed by public operators for centuries. The level of services offered by the private sector had grown dramatically forcing public operators to change to meet the demand of the globally and growingly integrated mail market. Postal organizations across the world had started to transform their business and use information and communication technology (ICT) in order to compete with the change in market trends. There were various successful models of services offered by various postal organizations. This included the United States Postal Service (USPS) offering email and eCommerce services; the South African Post Office offering hybrid mail services; and, Korea Post offering a synchronized information network (mobile, radio communication and RFID) whereby consumers have access to mail services and track the whereabouts of mail or packages anytime, anywhere. ENPO was determined to join that league in offering new services beyond the traditional mail services it used to over for decades.

Since 1999, Egypt had been implementing an aggressive ICT strategy as part of its national development plan; and ENPO was perceived as an integral part of such strategy. ENPO has been leveraging its capability to serve millions of consumers and trust to regain lost ground and compete, offering a plethora of services similar to those of other national postal organizations and pursuing further developments; thus turning it into a highly competitive organization. With images of newspaper titles racing in Badr Eldin's mind, celebrating ENPO's latest achievement in succeeding to become the only governmental institution to be part of Egypt's third mobile operator, he started to wonder where this organization once was, where it currently is and, most importantly, where it is going. The question immediately presented itself; was Egypt's National Post Organization rediscovering itself once again? The development of ENPO using ICT comes as an integral factor in the overall ICT development in Egypt. Table 1 demonstrates the evolution of ICT in Egypt.

BACKGROUND

ENPO volume and diversity of service had reached more than 18 million local customers and led to a long lasting trust between the national post and the local population from different segments and backgrounds. Such trust led to increasing the number of customers who took part in the financial services of ENPO to 2 million last year. A trust that has resulted in having *"Daftar Tawfeer"* (Arabic translation for a savings account) becoming the generic name for a saving account used by literally all segments of the society from all ages and from different social and economic segments and groups. ENPO was the main pillar for connecting Egypt with the outside world. This unmatched penetration within the Egyptian culture was made possible by the organization's large and extensive distribution network of more than 3,700 post offices located in every province and across the nation's 4,000 villages making ENPOs' distribution network the largest in Egypt coming in second to the network of national schools. One of the major characteristics of such a huge organization was the exceptionally large number of employees working in it; whereas, ENPO employed over 45,000 people of which 50% are located in remote offices in order to secure the quality and rate of

Table 1. Development of the information society in Egypt

Programs	Year
Open Door Policy	1974
Economic Reform Program	1985
Information Project Cabinet of Ministers (IPCOM)	1985
Information and Decision Support Program (IDSC)	1986
National Information and Administrative Reform Initiative	1989
Egypt Information Highway	1994
Ministry of Communications and Information Technology (MCIT)	1999
National Information and Communications Technology Master Plan	2000
Egypt Information Society Initiative (EISI)	2003
Egypt ICT Strategy 2007-2010	2007

the services provided to clients and enabling the same service on a nationwide scale.

SETTING THE STAGE

Organizational History

ENPO was established on January 2, 1865. Located in the heart of Cairo, it is thought to be one of the oldest and most prestigious governmental organizations in Egypt. Since its inception, Egypt Post was united with the ministry of occupation, under the British rule, which lasted from 1882 to 1954. Later on, its association was transferred to a number of ministries until 1965, when Egypt Post was under the umbrella of the ministry of finance, which issued a regulation specifying that the transfer of letters and issuance of stamps is exclusive to the government of Egypt. In March 1876, all employees working for the postal service were required to wear a uniform. In 1899, services that were offered since the Post inception were cancelled including salt and soda stamps, steamboat tickets and telegram and telephone services.

In 1919, the ministry of transportation was established and was given control of the Post authority. Later on that year, law number 9 was issued to set all postage fees; and the Post authority headquarters was moved from Alexandria to Cairo in Ataba Square. In 1934, the 10th conference of the International Post was held in Cairo, coinciding with the 70th anniversary of the established Egypt Post. After the 1952 revolution, Egypt Post was transformed into a cost center using revenue surplus to improve its services to the community. In 1957, Egypt Post was replaced by the Egyptian Post Authority (EPA), and in 1966, EPA was replaced by the General Post Authority (GPA). In order to regulate Egypt Post, law number 16 was passed in 1970. In 1982, the name was changed again to the National Post Authority (NPA) under law number 19. Finally, in 1999, the Ministry of Communications and Information Technology (MCIT) was established and became in charge of supervising the National Post Authority (NPA), Telecom Egypt (TE) and the National Telecommunication Institute (NTI). In 2008, Egypt Prime Minister, Ahmed Nazif, inaugurated the new headquarters of ENPO in the Smart Village. Table 2 demonstrates the development timeline of ENPO. This was perceived as a new phase in Egypt Post evolution where its role was being repositioned as a tool to avail eGovernment and as a platform for services provision that can reach all segments of the community.

Table 2. ENPO timeline

Time Line	Evolution Phase
1865	Egypt Post Established (associated with ministry of occupation under British rule)
1919	Egypt Post was associated with ministry of transportation
1952	Egypt Post was transformed into a cost center
1954-1964	Egypt Post was associated with a number of ministries
1957	Egypt Post was replaced by the Egyptian Post Authority (EPA)
1965	EPA associated with ministry of finance
1966	EPA was replaced by the General Post Authority (GPA)
1982	GPA name was changed to National Post Authority (NPA)
1999	NPA was transformed to ENPA and became associated with MCIT

ENPO TRADITIONAL SERVICES

Since its redesigned services and repositioning under MCIT in 1999, ENPO has been focusing its services on three main areas: postal services, financial services and social services. Exhibit 1 demonstrates details of the services offered.

Postal Services

The postal services are considered one of the oldest and cheapest methods of communication between individuals provided by ENPO and they include:

- Regular post, which has enjoyed price stability over the years that is not available by any other service offered.
- Fast/express mail, which is the fastest means of sending parcels and documents, within 24 hours in Egypt, and 48-75 hours outside Egypt to over 215 countries. The service is totally insured with door-to-door delivery and confirmation provided to ensure efficiency.
- Postal parcels, which is a service that allows the transfer of parcels, luggage, and gifts that weigh more than 2 kilograms with fees fixed by the authority according to the weight, distance, kind and value of contents.
- Public postal services, which includes private post boxes, postal cards and stamps

Financial Services

Despite some minor unreliability in its postal services, ENPO's financial services have always enjoyed a good reputation as a reliable financial institution, owing to the fact that it has never defaulted in payment of interest to the depositors. This has allowed it to earn the trust of all segments of the society for more than 100 years. ENPO started its financial services in 1905 by issuing saving booklets (Postal Saving PassBook), which is one of the oldest financial services offered in Egypt. In addition to the benefits of privacy, flexibility of depositing and monthly awards, the saving booklet's main advantage lies in the fact that the government guarantees its balances as well as the interests. Additionally, ENPO offers individuals, companies and organizations safe money transfer from one post office to another. With the number of depositors reaching 14.4 million and an enormous amount of deposits reaching 6.3 billion US dollars, it was becoming obvious that ENPO was a hidden treasure waiting to be discovered, or more accurately, waiting to be correctly invested. Other services include GiroNil, which has been introduced as result of cooperation between ENPO, Misr bank and Commercial International Bank, a private bank operating in Egypt. The GiroNil Company specializes in utility payments, allowing customers to pay their bills to large corporations and multinationals via ENPO offices. Customers also have the option of simply signing a document that allows ENPO to pay utility bills on their behalf, thus avoiding the risk of forgetting bills or queuing long hours to pay.

Social Services

ENPO's social service initiatives began in 1963 based on its conviction with its role in Egypt's socioeconomic development. For instance, the organization started paying pensions to around 3 million citizens with a total pension of 1.6 billion US dollars per year (www.egyptpost.org). The service has allowed pensioners to receive their pensions from 3,600 national post offices distributed all over the country, thus facilitating the process. This is part of the new services offered by ENPO to the government and a disbursing agent. It is important to note that this service has also removed the pensioner's burden of commuting from different areas in Egypt and queuing for long hours. Moreover, for customer convenience, some pensions are delivered directly to the customer's homes free-of-charge. More than 300 thousand pensioners (many over the age of 70, ill, and people with special needs) benefit from this service. ENPO also manages to deliver parcels to customer's homes through post offices nationwide. Another social service is the housing project where customers through the post office can benefit by buying application forms and making housing reservations provided by the state to young graduates. Finally, there is the fourth social service known as "lost property", which involves ENPO's cooperation with police forces to return lost property to the original owners.

ENPO ON THE NATIONAL ICT AGENDA

The main drive of the eGovernment initiative is to modernize the citizen's experience of public services and to improve the functionality of the government (Tarek Kamel, Minister of Communications and Information Technology).

ICT Sector Reform

In 1999, MCIT announced the formulation of the first national ICT plan. The plan was focused on upgrading the existing infrastructure in terms of information and technology and availing an ecosystem that can help diffuse ICT in Egypt in terms of laws and regulations. The success of this plan was intended to create more business opportunities through ICT-empowered products and services that can benefit all stakeholders. The convergence between information, media and telecommunications was one of the most important developments. Based on a number of amendments in 2000, 2004, and 2007, a new ICT strategy was developed focusing on three main pillars (a) ICT sector restructuring, (b) ICT for development; and, (c) innovation and ICT industry development. The strategy was formulated by MCIT in collaboration with leading expertise in the ICT sector. The aim is to continue the development of the ICT infrastructure to maximize its benefits, leverage public-private partnerships, create more community involvement and link Egypt globally.

ICT Sector Restructuring

MCIT stated that the ICT sector restructuring would be realized through the development of state-of-the-art telecommunications infrastructure and export of services, reform of the postal sector, and enhancing the framework governing the use of ICT networks and services.

ICT for Development

The use of ICT for economic development could be achieved through ensuring easy, affordable access to ICT for all citizens, diffusing education and lifelong learning through the Egyptian Education Initiative (EEI), integrating ICT in health services, supporting the production, use and distribution of Arabic digital content (eContent) and providing the necessary ICT support for the government. It is important to note that ICT for development is a collective effort by different government entities in collaboration with the private sector in an attempt to create industry-related opportunities using ICT.

Innovation and ICT Industry Development

Innovation and creativity using ICT is integral in the development process and MCIT formulated a plan to realize this objective through developing export-oriented IT-enabled services, developing the ICT capacity of Egypt, formulating strategic plans for research and innovation and promoting local and foreign direct investments in the ICT sector.

CASE DESCRIPTION

The Beginning

Since the inception of the eGovernment program, the government of Egypt was determined to deliver high quality government services to the public where they are and in the format, that suits them. The vision was guided by three main principles:

- **Citizen centric service delivery** where the program slogan is *"government now delivers"* reflecting government intention to develop a one-stop shop eSer*vices* approach focused at citizen's needs.
- **Community participation** where eGovernment is a project with nationwide impact, thus community participation is necessary. Citizens' demands are constantly being analyzed and reflected, and private/public sector companies are active participants in project's implementation and management.
- **Efficient allocation of government resources** where the emphasis was focusing on techniques to improve the level of efficiency, increase productivity, work on cost reduction, as well as the efficient allocation of resources.

When Egypt first launched its eGovernment initiative in 2001 in partnership with Microsoft Corporation to design and implement a web portal that would serve as a gateway to government services, the project was faced with much criticism for its low usage levels. According to the literature, the eGovernment initiative faced several challenges, illiteracy being the biggest challenge along with an Internet penetration rate of 20%, which endorsed the idea that eGovernment services are just for the rich educated segment of the society instead of being a nationwide targeting tool to improve the way governmental services are offered. There was a need for providing enabling technologies, products and services to underpin the development of Egypt as a knowledge economy in the global market. The initiative, which also aimed at crossing the boundaries between ministries by offering joint services, has been accused of being an unrealistic step towards improving governmental services, the reason being the low acceptance of the newly introduced eServices. Low acceptance is a problem, which, according to the minister of state for administrative development, Ahmed Darwish, lies in the word trust or more accurately, lack of it. According to Darwish, "*the process of gaining people's trust will take time, but we are working on building that trust by providing tangible results*". This is where the role of ENPO emerges.

Moreover, the emphasis of the role of ENPO was clearly highlighted in a speech by Tarek Kamel concerning Egypt's national ICT agenda, which included three objectives. First, developing and modernizing ENPO infrastructure; second, transforming ENPO as a delivery arm for financial and eGovernment services; and third, building on the trust with the citizens in maximizing the utilization of their postal savings. The positioning of ENPO was invaluable due to its constant interaction with different clusters of the community irrespective of the social or economic segment.

Reform of the Postal Sector

MCIT's investigation of ENPO showed that it was performing below potential. The services offered to both individuals and businesses were inefficient. The reform program was aimed at resolving these issues, achieving national development objectives and increasing national competitiveness. The following objectives were identified (a) to develop a worldclass postal service in terms of quality, innovation and accessibility, (b) to increase overall levels of private sector investment in the postal market through open and fair competition and progressive regulation; and, (c) to create a new export-oriented postal industry in Egypt. It is important to note that all these objectives were formulated with a platform that reflects the notion that ICT is an enabler for service improvement, government efficiency and economic development.

Development of a State-of-the-Art Postal Network

A fundamental component of the MCIT strategy for developing the postal sector is modernizing Egypt Post by availing state-of-the-art ICT. MCIT is working tediously with ENPO to facilitate the development of services and systems that support eGovernment and eCommerce. Moreover, MCIT encourages ENPO to form partnerships with businesses and the private sector through outsourcing business models, bringing innovative products and services to customers by making use of the ICT industry. By restructuring ENPO through ICT, citizens have greater accessibility to information and government services. This is especially true for citizens who reside in rural areas and underprivileged communities in Egypt allowing them to easily register with the government, apply for licenses and obtain tax documents by simply visiting their nearest postal retail office. MCIT works with ENPO to develop new and innovative products that combine digital and physical communications systems, such as "*hybrid mail*". This will eventually result in the development of systems for sorting, tracing, addressing and customer care. An additional area of interest for MCIT is eCommerce, which can benefit by utilizing postal networks and ICT to manage global supply chains and enhance delivery. It is important to note that since the inception of MCIT and the leadership at the helm of ENPO selected was long-time ICT professionals and experts with extensive experience in diffusing ICT, which clearly indicates the intention of transforming ENPO to be ICT-enabled in terms of services offered.

Regulating the Postal Market

Creating an open and competitive postal market has been met with some success, but has also been accompanied with complications. Given the fact that ENPO is the entity that is granted the sole authority for issuing licenses to postal operators, it could easily utilize its power to monopolize the postal market. However, this has not been the case. On the contrary, ENPO has a high level of competition with 12 operators providing various forms of postal services. ENPO actually promotes private sector participation in the market by forming partnerships with a number of private sector individuals and businesses to expand services and products in the market. This has been very promising and indicated vast potentials in a fast growing and competitive marketplace.

It is important to note that increasing private sector investments has been a more difficult process for a number of regulatory reasons. This is problematic because, as the Egyptian economy grows and mail-heavy industries such as financial services and utilities expand, there will always be an increasing need for an efficient postal network to handle advertising, bill delivery and payment, and goods and cash transfers. In order to meet this anticipated increase in demand, the level of partnerships and private sector investment must increase and keep pace with the developments taking place. However, in order to increase private sector participation in the market, reform of the postal sector and effective policies was necessary. MCIT is brainstorming incentives to encourage private sector participation in projects that could stimulate further market progress. The biggest barrier hindering private sector investment in the postal market has been the lack of an effective regulatory oversight. A study carried out by the Universal Postal Union (UPU) concluded that the "postal market in Egypt is performing below capacity and that there is room for expansion and additional private sector investment". This has been the ENPO focus over the last few years while assessing what emerging ICT tools and techniques can bring in to the postal services.

However, lack of transparency concerning ENPO's dual role as a regulator and an operator in the market, as well as legitimate regulations of the sector, have affected the willingness of the private sector to invest. Respectively, a number of measures were taken by MCIT, postal operators and other stakeholders between 2007 and 2008 to develop effective postal regulatory policies, laws and regulations; in addition to establishing a neutral regulatory mechanism responsible for monitoring ongoing growth and innovation in the Egyptian postal market and benchmarking the sector's progress against international standards. It is hoped that, with proper regulation and market definition, private sector participation will significantly increase in the postal market.

Creating an Export-Oriented Industry

Faced with opportunities for global expansion, ENPO has been adapting itself to new international postal regulations. International regulatory advances, such as the World Trade Organization's (WTO) General Agreement for Trade in Services, are rapidly reducing or eliminating trade barriers, creating new opportunities for Egypt Post's penetration into the global market. Egypt Post envisions itself as a "*hub*" in the region, managing supply chains on a regional level. Egypt's geographic position, in conjunction with its growing ICT infrastructure, puts the country in a favorable position for regional expansion. When postal networks work with customs counterparts, import and export channels are strengthened, supporting the growth of Small and Medium Enterprises (SMEs) and other businesses.

MCIT continues to build on its successful experience with multinational telecom and IT companies to generate international appeal for its candidacy as a regional "*hub*". Working on a national level to create the first postal free zone in the Middle East, MCIT is mediating with governmental agencies responsible for transport and trade including aviation, transport, finance and investment, the Customs Authority, Egypt Post, private operators and other stakeholders. The establishment of the postal free zone will require that the current regulatory frameworks being developed will parallel those in prominent regional and international free zones. New infrastructure and supply systems are being created to connect the zone to global and regional markets, while appropriate regulatory processes and inspection mechanisms being developed to facilitate transactions and promote business. To promote Egypt's free trade zone as a regional hub and logistics center on an international level, MCIT is cooperating with the General Authority for Investment and Free Zones (GAFI), as well as other international postal and supply-chain operators. However, despite the developments in its ICT infrastructure, MCIT recognizes the postal sector's need to develop the regulatory laws necessary for liberalize services and complying with international trade regimes. Regulatory reforms will need to be introduced to prepare the postal sector for future changes in international postal regulatory systems, allowing it to succeed as a global market player.

Modernizing ENPO

Starting his position as ENPO Chairman, Alaa Fahmy, currently minister of transportation, knew that his job was going to be anything but easy. As Chairman, he is mainly responsible for providing strategic directions, explore business opportunities and promote ENPO in different market, business and industry circles. ENPO's history of being the oldest, most widely used and trusted governmental organization in Egypt simply represented the missing link between the government and the people. Fahmy knew that ENPO was desperately in need of a restructuring process, whether organizational or ICT-related. Moreover, he knew the burden of being the executing arm of Egypt's new eGovernment initiative that was being placed as part of ENPO's agenda.

During Aly Moselhy tenure, the former ENPO chair, and current member of government and minister of social solidarity, he realized that there was a potential for ENPO to do more for its customers and most importantly, for Egypt. He had a vision that with a comprehensive ICT infrastructure coupled with availing easy-to-use services by the community, ENPO could be the gateway for an eService-oriented society in Egypt.

During this time, the seeds of a new business model began to be planted. The former team started the first phase of creating an image that would turn this ever regarded, highly trusted governmental institution into an institution with a corporate image, objective and most importantly organizational infrastructure that enables service provision in a smooth and easy way. ENPO's new business model had to capitalize on the new, healthy investment atmosphere in Egypt and the ICT infrastructure made available by MCIT, and integrating the new offerings with the former products in order to create a better value proposition for ENPO customers. Based on this plan, and under chair Fahmy, the image of ENPO started to manifest itself into a reality. In the early phase, some 920 post offices were modernized beginning with the main traffic offices located in Ramses, Alexandria, Tanta and Cairo international exchange centers at the Cairo airport.

The modernization process was done in three phases based on the location of the office. Many of the offices located in small alleys or villages, ironically, did not need to look too polished or else the regular customer would have started to have doubts and would have reconsidered dealing with ENPO, directly shattering the concept of trust, which is the edge that ENPO enjoys and intended to capitalize upon. However, in order to implement this business model and introduce a profit making culture to a governmental organization of more than 45,000 employees a lot had to be changed, and with an organization as old as ENPO, this was definitely a challenge.

ICT and Postal Services

A strong postal network reaches all residents, many of whom have no other means of communicating with the outside world. By providing this universal level of communication, posts can also provide the increased access to information that is essential to poverty reduction in the information age. But postal services do much more: they connect people and raise their level of social development and cohesion (Nemat Shakif, Vice-President, Private

Sector Development and Infrastructure, The World Bank Group).

Although the core of the postal business would remain paper-based for years to come, ICT had created a new realization to the world's postal services. By applying ICT-based infrastructure, postal services could improve the quality of their "*traditional*" services and introduce new, reliable, affordable products and rapid services to meet the growing needs of the community. New ICT offers enormous potential to post offices that are reinventing themselves to remain the primary means of communication and continue playing a significant role in the world's economy and information society. Merging ICT and the post was a big challenge waiting to be transformed into an opportunity that would effectively lead to a powerful entity, which provides accessibility, trustworthiness, security and privacy.

Accessibility

Combining around 700,000 worldwide postal outlets with ICT facilitate Internet access to people in remote areas. Posts are often seen as attractive partners in the provision of eGovernment services.

Trust and Security

The post office has always been trusted with people's mail. As people start to send messages through new communication networks they expect to deal with a trusted party who can securely and confidentially deliver information. Through ICT, post offices can provide innovative and secure services and products to continue honoring their role.

Privacy

Direct or advertising mail delivered physically to consumers could be converted electronically to customers who specifically request to receive such information.

ICT Back Then, ICT Now

The coming era will witness an expansion in the use of ICTs in developing Egypt's postal organization and providing the citizens with better services in a more efficient way

Badr Eldin, ENPO Vice Chairman for IT, sat in his office reviewing his IT infrastructure development plan as he realized the long way they had come, and the longer way they were yet to go. The deadlines were strict and tangible benefits were awaited. Prior to the new developments, the concept of using IT as a tool to improve the quality and efficiency of the services offered by ENPO was simply non-existent. All services, such as opening accounts, invoices and payments were offered using paper. This was unfavorable to customers due to the time wasted. The lack of communication between offices made the process inefficient and costly in terms of paper and time wasted. Moreover, there were no networks, PC's or application software present in any of the offices.

During the period 2001-2005, the mission was to connect all offices through the postal network. However, this was easier said than done. Starting from scratch, the former ENPO team managed to connect 640 offices because of the threat ICT posed to the employees especially in poor and underprivileged communities where these offices were located, and the lack of the infrastructure needed to support it. Sometimes offices had to be connected through satellite, which involved high costs. The 640 offices being connected on a network marked the introduction of a new culture to the staff and to operations of ENPO in general.

In February 2006, Fahmy urgently continued to build on what was previously accomplished. Respectively, the total number of branches reached 3,688 at the end of the second quarter in 2009. With more advanced ICT came an even greater responsibly. Now the ICT team has to make sure that the four pillars of ICT in ENPO were present and functioning in a reliable way. The pillars could be demonstrates as follows:

- With respect to networks, reliability needs to reach 5-nine (99.999%) availability as a target. Currently, availability stands at 75% with backups for emergencies.
- With respect to application rings, there need to avail databases and applications working in parallel and reliably.
- With respect to PC penetration and PC support, the target is to reach 24 hours a day, 7 days per week reliability.
- With respect to servers and database center, the objective is that if any of the servers fail, backups should be ready for replacement. However, the current problem is that some companies offer services using the ENPO network, which sometimes causes confusion when a company's server fails and people think it is ENPO's server.

Establishing the ICT Platform

In his plan, Badr Eldin decided that his department's main mission was to provide quality IT products and services to help ENPO reach its goals. Consequently, the organizational structure and strategic goals of the IT department should pave the way towards achieving the objectives of ENPO as a whole *"by transforming itself into a quality focused, highly productive, responsive organization supporting a market driven system"*. Realizing the importance of having a complete and up-to-date ICT infrastructure, major changes to the IT department's organizational structure had to be taken. Exhibit 2 demonstrates the former organization structure of the IT department. Depicted in the chart is a strategic business unit structure in which each department has IT as a complimentary tool integrated within the function itself. The functions, which are divided into finance, communication, human resources, security and IT sourcing all reported to the chief information officer (CIO) after passing through the steering committee. The vice chair for IT realized that several disadvantages were associated with such a structure, which would directly hinder the role of IT in the progress of ENPO. These disadvantages could be summarized as follows.

ICT Core Business

The structure is not placing IT as a core business behind and supporting the whole organization, which is the case in real life. Combining functions like communications, human resources and finance, which are not directly linked to ICT presents ICT as a tool rather than the main driver for the whole organization.

ICT Operational Methodology

Badr Eldin specifically objected to the idea of dividing the structure into small business units. In his own defense, he asserted that in a newly transforming giant, organizations like ENPO, a business unit structure (as demonstrated in Exhibit 2) is not suitable. It is only suitable for large multinationals where business units are set up like separate companies, with full profit and loss responsibility invested in the top management of the unit and the units are at a level to compete with each other. Such condition was neither present, nor currently requested within ENPO's ICT organization structure. Other disadvantages included the minimal strategy coordination that occurs across business units and the performance recognition, which is often very blurred. Making his point, the vice chair for IT suggested that the new organization structure would consolidate the role of ICT as the core business of ENPO. The structure mainly focused on creating six functions that are directly related to ICT including information centers, technical support, operations, services, infrastructure, and design and planning. Exhibit 3 demonstrates the new ICT structure. Each function has a vice president responsible for the subordinate sub-functions and eventually all departments report to the CIO, passing by the executive committee and reporting secondly to the financial services and human resources departments. The functions are further grouped into three phases, which represent the IT life cycle, beginning with design then implementation and lastly, technical support.

According to Robert Dailey, organizational behavior Professor at Drake University, there are numerous advantages to this structure. Firstly, the structure is based on specialization, which allows employees within each function to speak a common language. It also minimizes the extent of duplication and facilitates tight control. Badr Eldin was extremely convinced with the need to introduce a double reporting system in the ICT structure to ensure that the system will always function effectively. The department's manager will not only be reporting to financial services and human resources, but he/she will also be reporting to the main office, which ensures that each department is functioning as expected and that work is evaluated objectively. The new structure takes into consideration the maximum number of people that can report to the CIO which, in this case, are seven people reporting on behalf of their functions. Although the suggested changes were approved by senior management, there were a number of challenges that still faced the IT department including:

- ENPO is the largest organization (number of employees).
- Increased business dependency on ICT.
- Growth in business applications and storage requirements.
- National and international coordination.
- Increasingly remote workforce.
- Technology obsolescence cycles, which related to the employees' changing attitudes toward ICT requiring regular attempts to channel them from the inherited manual systems to the newly digital ones.

In an effort to instill this new concept in the minds of staff, and also to aid its transformation process, Fahmy had decided to take advantage of Egypt's new high-tech business district, the Smart Village, by relocating the Ataba office into a consolidated state-of-the-art premises at the Smart Village. Often referred to as Egypt's Silicon Valley, the Smart Village provides office space over 450 acres of land hosting over 55,000 IT professionals in more than the 120 Companies. With the growing number of new buildings and companies joining the Smart Village in Cairo, the village is expected to host more than 500 companies and more than 100,000 employees by 2014. The village is built using state-of-the art ICT infrastructure and high speed connectivity for integrated services, (whether data, audio or video) making it an ideal location for most of Egypt's ICT-based companies both national and international. In conjunction with the move to the smart village, ENPO planned to eliminate paper entirely within the organization in order to transform it into a digital workplace. Knowing the difficultly in implementing such a decision, the management decided to undertake dual operations during the phased implementation of the relocation. The management needed to develop an electronic file management system, including support and an operational system in order to process ENPO applications electronically allowing automated enterprise operations. In addition to this, a reliable IT infrastructure was needed to connect the main office in Ataba square with the Smart Village premises requiring the design, development, installation and testing of fiber connections, cable plants, data switches and telecommunications. The telecommunication services covered meeting room capabilities, electronic building directories, a facility help desk and a full-fledged security control system.

Human Capital: Investing in What Matters Most

With over 45,000 employees working at ENPO, the human factor was an essential part in the transformation process, if not the most important one. With such a huge number of employees, having a well-constructed organization structure was a necessity in order to manage and administer effectively, while also adding value to their development process. Unfortunately, that was not the case until the new transformations were introduced. In the former organizational structure, all functions reported to a single vice president who in turn reported directly to the CEO. According to El-Labban, director of international relations, *"how can anything be done effectively when you have 45,000 employees reporting to one person? It can never work"*. Based on this, the organizational structure itself was changed in order to achieve decentralization; with six vice presidents, each concerned with a certain function/area.

Moreover, the new management team has tried to introduce the idea of having a human resources (HR) department, which was not common in most governmental organizations and the customary approach was to deploy a personnel department that mainly handled employee files as a storage room. However, the new human resource department could manage the appraisals and elevate the HR function as a whole by managing employee performance and most importantly concerning itself with the training necessary in the coming transformational period. In an attempt to introduce a market oriented and customer-focused culture, the marketing campaign began functioning instantly, with new marketing material announcing ENPO's services, as well as its new image and entrance into the market. Marketing material included pens, mouse pads, newspaper announcements and advertisements highlighting the new services offered by ENPO in Egypt's most wide spread daily newspapers like Al-Ahram and Al-Akhbar.

Public and Private Partnerships

The Post is based on the concept of connecting two parties together which is the same concept on which communication is based, so if any organization should be part of the new operator it should be ENPO, the oldest supplier of the service (Amr Badr El Din, Vice Chairman, IT)

Etisalat

Outside the ICT Minister's office at the Smart Village, Fahmy and Badr Eldin were anxiously waiting to know the results for the long bidding process. The Minister came out and congratulated them, *"You took it"* he said, *"ENPO won the bid"*. They were thrilled. According to Badr Eldin, *"The feeling was indescribable, everyone was congratulating us and that is the moment when I felt that my job was most fulfilling"*. The next day the news was all over Egypt that ENPO would become the first governmental organization to be part of a huge corporation like Etisalat, UAE's number one operator and currently Egypt's third. The consortium included Etisalat, Egypt Post, National Bank of Egypt and Commercial International Bank; they won the bid for over 3 billion US dollars and a 6% of the annual revenues to be paid to the National Telecommunications Regulatory Authority (NTRA). Exhibit 4 demonstrates the public-private partnership (PPP) model and the shareholders in this consortium. This step was one of Fahmy's major strategies to transform ENPO into a profit-making organization by diversifying their offerings and introducing some major public private partnerships that could capitalize on ENPO's competitive advantage, manifested in its distribution network and credibility within the society. According to Badr Eldin, *"Public Private Partnerships are arrangements between the government and private sector entities for the purpose of improving public infrastructure and community facilities. Although these partnerships entail sharing investment, risk and responsibility, this long term partnership is very rewarding in many ways"*.

Referring to his partnership with Etisalat, Badr Eldin knew that ENPO's capabilities would be important for the company to penetrate successfully the local market. Some of the partnership benefits include: (a) using ENPO's network of offices that extends all over Egypt as a distributor of the company for selling their prepaid charging cards and accessories; and, (b) using ENPO office buildings to install the company's antenna instead of using resident buildings and paying hefty amounts of money in return. Another similar agreement was made with Vodafone Egypt, the nation's second mobile operator in terms of coverage, according to which ENPO would provide prepaid charging cards for Vodafone Egypt consumers in its offices.

Egypt Air

ENPO Chairman Fahmy and EgyptAir Chairman Galal signed a memorandum of understanding (MoU) to utilize the post office for booking airline tickets. According to the Minister of CIT, *"The MoU reflects the collaboration between different stakeholders and Egypt Post to facilitate the service delivered to the community"*. This MoU is part of ENPO's bigger strategy to turn their post offices into full-fledged service centers.

Jordan Post Company

After studying, the executive and legislative framework of Jordan Post Company in cooperation with EFG Hermes and Jordanian Riyada Ventures, Egypt decided to bid for a share in the Jordanian postal service. Egypt Post is competing with La Poste France, Aramex Jordan and British Consultative Post Services over acquiring a share in this company. This was another strategic move through alliances in order to develop the investment volume of Egypt Post and derive revenues with limited risks. These different steps indicated the intention of the government of Egypt to transform the role played by ENPO in providing a diversified portfolio of digital services while being one of the tools and platforms to diffuse eGovernment services.

International Agreements

ENPO extended their activities to regional levels. In February 2007, ENPO signed off 1.8 million US dollars *"Institutional Twining Program"* with the French National Post Organization, *"La Poste"*. It was aimed at developing the various departments at ENPO in order to match European and international standards applied. The program allows exchange of expertise, allowing ENPO to capitalize on La Poste expertise in marketing, service monitoring and quality assurance. Furthermore, to help boosting their bilateral economic and technical relations, they signed a cooperation agreement. The agreement that went into effect in 2008 would aid in developing postal and financial services and marketing tools for these services. This was another step in the integrated plan to utilize international expertise to upgrade Egypt Post. Moreover, another agreement between ENPO and the Italian Post (Poste Italiane) was concluded in order to develop the sector in both countries. The minister of CIT indicated that the postal sector is a key economic and service driver to Egypt's modernization system. It is important to note that the agreement involved training human resources and raising the value of its assets.

Moreover, Fahmy signed an agreement with 7 Arab countries to exchange the financial remittances through the congress postal conference which was held in July 2008 in Geneva, Switzerland. The seven countries included Egypt, UAE, Syria, Yemen, Morocco, Tunisia, and Qatar. Just recently, Fahmy signed a pact to run electronic remittances between both Egyptian and Jordanian postal services, which started functioning in January 2009. This new agreement will enable Egyptian citizens working in Jordan, around 1 million, to transmit their money through the post offices at extremely competitive prices.

Services offered by ENPO

A plethora of ICT-powered services was added to the "traditional" list of services after restructuring the ICT sector.

Government Services

With about 14 million Egyptians using the Internet, the comprehensive adoption of eGovernment services is still a far-fetched idea. However, as previously mentioned, being a delivery arm to the government concerning these services and having an integral role in servicing its community became one of Fahmy's priorities. Capitalizing on its intensive distribution network, ENPO currently offers 3 million citizens their pensions with a total sum that exceeds 1.6 billion US dollars through its 3600 offices, which are geographically dispersed across Egypt. Additionally, the service has also extended to reach citizens who are ill or having special circumstances and those aging 70 years and above, who cannot make it to any of the offices by offering them a delivery of their pension to their doorstep.

The *"Tamween"* card is another practical application of ICT that is utilized to help provide better services to the citizens. Previously, everything regarding food subsidies, which citizens were entitled to receive was documented on paper, which made the system more liable to fraud and human error. However, in this project and after installing the electronic platform, citizens are given smart cards; a person passes his card through the point of sale (POS) that is connected to the post network. This in turn, is connected to the network of the ministry of social solidarity, which holds the files of all citizens who are entitled to receive monthly nutrition subsidies. These records show the utilization of every item supplied based on the needs of citizens. This provides feedback that helps the ministry to decide, which items to increase depending on consumption. According to Badr Eldin, this was just a prototype that was tested in the Suez province, because of the limited geographical coverage making it more controlled and possible to monitor. Moreover, the limited population was a very important determining factor making Suez an ideal place to test the new electronic platform. If proven successful, the *"Tamween"* card will be implemented in all of Egypt's 29 provinces.

Financial Services

In 2007, as demonstrated in Exhibit 6 showing the postal financial services, there was the launch of a new postal savings account based on investment in the stock market. Minister Kamel stated that the new service aims to facilitate changeable high revenues through long-term investments. The new savings account requires a minimum of 18 US dollars with no ceiling while enabling citizens to invest in the Egyptian Exchange (EGX) and raises their investment awareness. The service has proven to be a success collecting 2 billion US dollars of small savings in less than a week. During the Cairo ICT 2009 Trade Fair and Forum several new financial services were launched which continue to meet the needs of the postal service wide spectrum of customers. The services included a variety of offerings including current accounts in US dollars and Euros and earning daily variable revenues for companies and individuals; *Hadiyati* (Arabic translation for *"My Gift"*), a prepaid electronic gift card that can be charged with up to 181 US dollars. It also include *Mahfazti* (Arabic translation for *"My Wallet"*), a prepaid rechargeable card that can be used for purchasing goods and services and to withdraw cash from ATM machines.

To complement the above, Egypt Post launched Universal Windows in 600 post offices in June 2009 in order to offer a diversified portfolio of financial services that meets the needs of the citizens including different types of money orders. More than 20,000 employees have been trained to aid in the dissemination of these financial services nationwide. In order to keep pace with the introduction of these financial services, ENPO signed a cooperation agreement with SAP, the enterprise resource planning company, who will provide business software applications to speed up the implementation of postal and financial services offered. In addition to this, there was also the introduction of the Misr Mail service.

Misr Mail

Now that our ENPO is entering the new era of ICT, certain applications cannot be ignored. E-mail is the new version of what we have always offered. Delivering mail is now done electronically, and we are doing it as well as anyone else (Amr Badr El Din, Vice Chairman, IT)

ENPO introduced *"Misr Mail"*, which offers a list of services that include 2GB capacity mail, a portal with exclusive news from Egypt and high security that ensures privacy. Moreover, Misr Mail offers career and training opportunities for all those registered on it. Misr Mail is complemented by another service called *"hybrid mail"*. The service makes use of the new electronic signature to track and confirm the delivery, acceptance and delivery date of emails sent, which is an essential step along the way to governing and encouraging eCommerce activities in Egypt. Furthermore, they extended to reach other services like EPEM service, which is crucial for eCommerce. The service allows electronic checking of all electronic documents involved in any electronic transaction by verifying and validating the signatures on the documents, as well as the date and time of signature and automatically stores this information for future reference. This helps eliminate fraud in business transactions taking place over the Internet, and thus encourages the citizen use of these services by establishing trust. With these security measures, eCommerce has a greater chance of flourishing in Egypt. Finally, Egypt Post has recently introduced the International Post Service (IPS) system allowing the sender to *"track and trace"* their message from the time it leaves their home until it is delivered to the addressee.

CURRENT CHALLENGES/ PROBLEMS FACING THE ORGANIZATION

Egypt has been gradually building its information society since the mid 1980s, adapting its strategy and approaches to the evolution of the global

Table 3. ENPO critical success factors

Critical Success Factors	Definition
Completing ENPO ICT infrastructure build-up	– Design, develop and implement a nationwide infrastructure connecting all ENPO offices across Egypt's 29 provinces
Availing value-added services	– Availing connectivity and develop value-added information networks between ENPO and the community of organizations and users
Linking Egypt globally-digitally	– Link Egypt to the growing information and postal networks across continents capitalizing on the outreach of ICT
Investing in human capital	– Invest in human capacities across different ENPO departments and units to be able to transform the organization and promote eServices
Building an online society	– Build an online electronically ready community that can appreciate and use ENPO services

ICT sector. The steps taken included supplying accurate and timely information, encouraging private investment, formulating effective economic reforms, improving productivity, providing programs for lifelong learning, making public services more efficient, improving health care, optimizing the use of natural resources and protecting competition. Despite the major progress in IT deployment, policy and regulatory frameworks and implementation levels, many milestones must still be achieved to reach the critical mass of ICT users and critical level of ICT utilization that can enable organizational such as ENPO that are transforming themselves to become ICT-enabled to be successful. There needs to be an overall strategy that promotes electronic readiness and help create a critical mass of ICT-literate users that can appreciate and use the type of services that ENPO is offering. Table 3 demonstrates the critical success factors for ENPO.

REFERENCES

Badr, E. A. (2007, April 1). *ENPO Vice President for IT Interview*.

Cairo, I. C. T. (2009). Retrieved March 20, 2009, from, www.cairoict.com.

Egypt Post. (2009). Retrieved March 25, 2009, http://egyptianpost.net/en/index.asp.

El-Labban, D. N. (2007, April 1). *ENPO Director for International Relations Interview*.

Gillingham, A. (n.d.). *Bulk Mail Takes Notes and Goes Hi-Tech*.

Roger, W. (1999) *Postal service getting wired.* Retrieved May 31, 2009 from www.interactive-week.com

Si-young, H. (2007, June 11). Korea post utilizes cutting-edge IT. *The Korea Herald.*

Universal Postal Union. (2003). The role of postal services. WSIS Summit, Bern.

World Summit on Information Society. (2009). Retrieved February 10, 2009, from www.wsis-egypt.gov.eg.

Zekri, N. (2006, April 16). Transformation of the post organization, *Al-Ahram Newspaper.*

ADDITIONAL READING

Editorial. (2006, April 16). Marketing and Electronic Signature as Latest Services Offered. *Etisalat Al Mostaqbal*.

Editorial. (2006). *Egypt's Third Mobile License to Benefit from the Egypt National Post Offices Distribution Network.* Retrieved July 7, 2009, from, www.dailystaregypt.com.

Editorial. (2007). *Egyptian-European Postal Institutional Twinning Program Under Way*. Retrieved May 3, 2009, from, www.mcit.gov.eg.

Information and Decision Support Center. (2008). http://www.idsc.gov.eg

Kamel, S. (1999). Information technology transfer to Egypt. In *Proceedings of the Portland International Conference on Management of Engineering and Technology (PICMET). Technology and Innovation Management: Setting the Pace for the Third Millennium,* Portland, Oregon, United States, 25-29 July (pp. 567-571).

Kamel, S. (2008). The Use of ICT for Social Development in Underprivileged Communities in Egypt. In *Proceedings of the International Conference on Information Resources Management (Conf-IRM) on Information Resources Management in the Digital Economy*, Niagara Falls, Ontario, Canada, May 18-20.

Ministry of Communications and Information Technology. (2005). Egypt Information Society Initiative (4th ed.).

Ministry of Communications and Information Technology. (2009). Retrieved May 5, 2009, from, www.mcit.gov.eg.

Ministry of Communications and Information Technology. (2009). *Egypt post launches new financial services at Cairo ICT 2009.* Retrieved April 22, 2009, from, www.mcit.gov.eg.

Ministry of State for Administrative Development. (2008). Retrieved September 3, 2009, from, http://www.ad.gov.eg.

APPENDIX

Figure 1. Detailed list of ENPO services offered

Postal Services	
Letters	Registered Mail
	Ordinary Mail
	Net Courier
	Banking letters
	Direct Mail
	Cassette Post
	FaxPost
	E-document exchange
	Publications
Express Mail (EMS)	
Parcel Services	Domestic
	International
Public Postal Services	Postal Cards
	Private Post Box
	Clearance Tools
Social Services	
Home Delivery	Delivery of Pensions
	Delivery of Parcels
	Delivery of Remittance
Housing Projects	Youth Housing Project
	Miscellaneous
Lost Property	

Financial Services	
Electronic Payment	ATM Cards
	Visa
	Mastercard
GiroNil	
Postal Savings Passbook	
Daily interest account	Golden Account >10000
	Silver Account <10000
Postal Investment Book	
Postal Remittances	Internal remittances
	Governmental Remittances
	Electronic remittances
	Cashed external remittances
Postal Proxy	

Figure 2. Former ICT structure

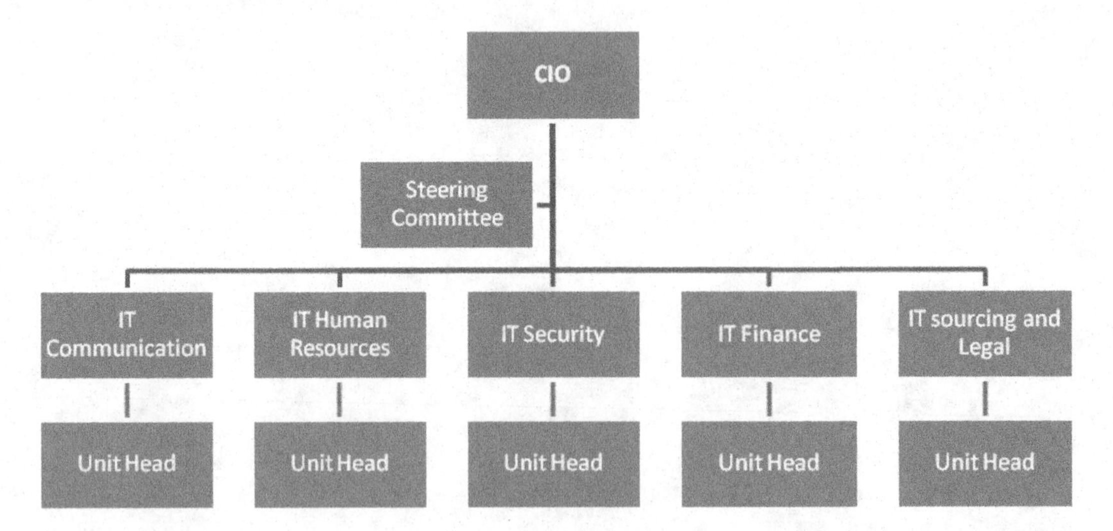

Figure 3. Current ICT structure

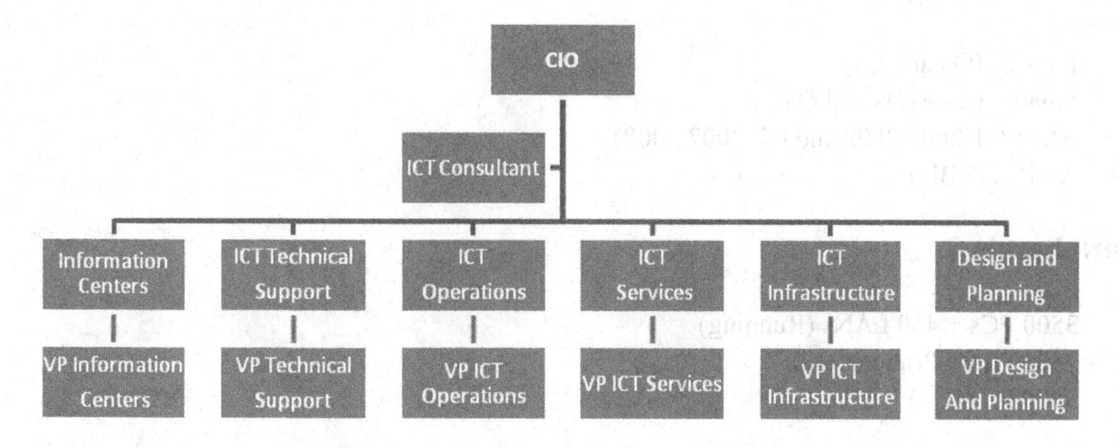

Figure 4. Public private partnerships

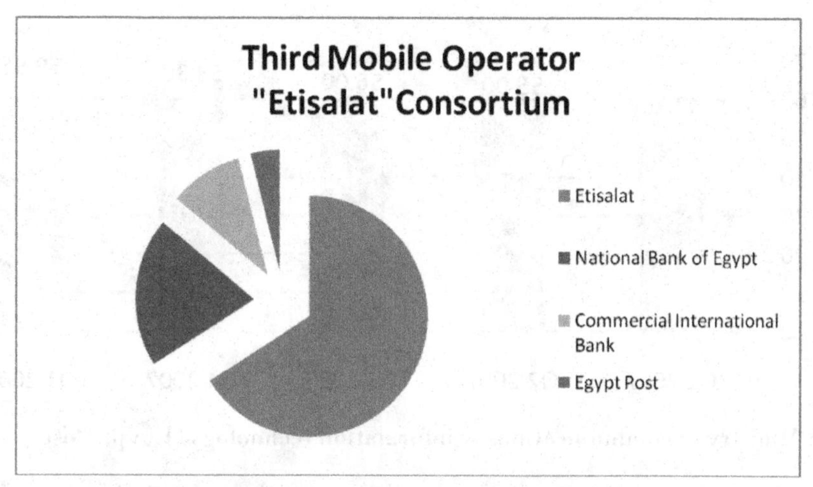

- Planning, designing, deploying, and managing ENPO IP network.
- Full responsibility of IP Network and voice services operation activities.
- Services will include network operations center, service desk, field operations, onsite support, customer care, third party support.

Action Timeline (2007)

Telecommunication

- Fiber cable, WiMax and Co-location (Q2)
- 640 Branch SDSL and ISDN (Running)
- Upgrade 512K (Q2)
- 450 Branch ADSL, VSAT and VPDN (Q3)
- Co-Location 40 TE POP (Q3 and Q4)
- 1000 Branch VPDN (Q4-2006/2007 and Q1-2007/2008)

Data Centers-Op and Mo

- Ramses (Q3 and Q4)
- Smart Village (Q3 and Q4)
- Alex (Q1-2007/2008 and Q2-2007/2008)
- Assiout (TBD)

Peripherals

- 3500 PCs + 450 LANs (Running)
- 1500 PCs + Printers (Q3)
- 3000 PCs (Q4)

Figure 5. Postal financial services

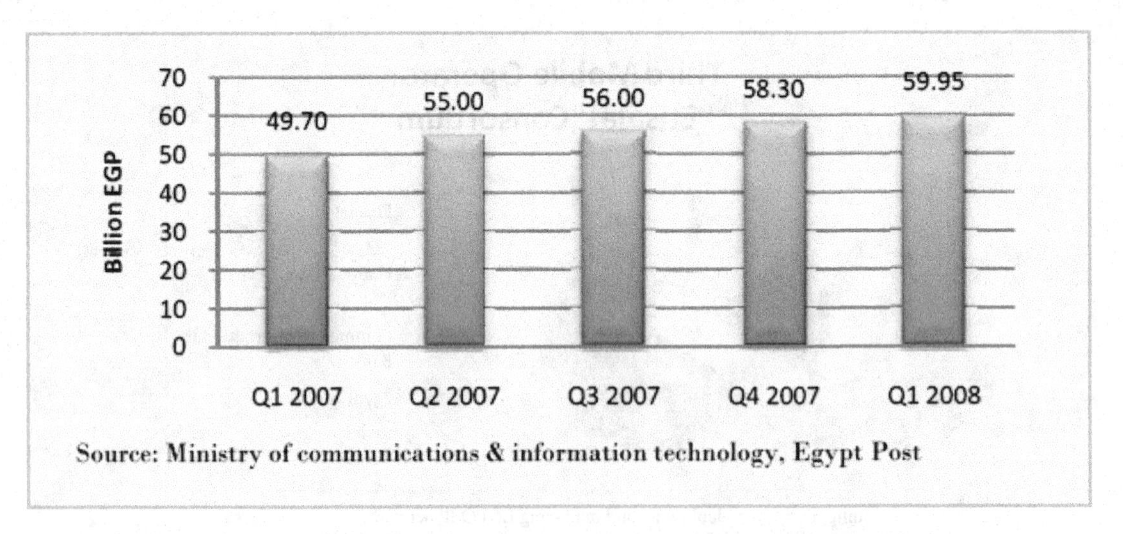

The total number of postal investment accounts invested in the Egyptian Exchange reached 33 thousand accounts since November 2007 with a total balance of 55.4 million US dollars until February 2008.

Chapter 14
Redefining Medical Tourism

Desai Narasimhalu
Singapore Management University, Singapore

ABTRACT

Dr. Wei Siang Yu, a medical doctor who is a compulsive serial entrepreneur and the founder of the FlyFreeForHealth, has created a comprehensive medical tourism service using a multimedia platform as the core engine. FlyFreeForHealth has started gaining traction in several countries across Asia including the Philippines and Australia. Dr. Wei has been very resourceful in exploiting business opportunities by leveraging emerging technologies for his company's products. The rapid growth of the company requires Dr. Wei to focus on developing an enterprise related information and technology architecture for FlyFreeForHealth.

BACKGROUND

FlyFreeForHealth (FFFH) is a company that has been built using multimedia technology for the medical tourism industry. Medical tourism is a term coined by tour agencies to facilitate travel across international borders to get either affordable or specialized healthcare. This specialized area of tourism is the focus of FFFH technology innovation, as presented in this case study.

DOI: 10.4018/978-1-61520-609-4.ch014

Medical Tourism

Medical tourism is fast becoming a national industry in a number of countries. This reflects international popularity of medical tourism; as also, the potential for large market growth over the next decade. Recently, countries have begun to establish a medical tourism component as part of their national tourism board or equivalent agencies.

Medical tourists come from every part of the world including first world countries such as the United States, United Kingdom, France, Germany, and other European nations. Some medical tourists travel to countries that offer healthcare services that are very specialized and fill a particular health or lifestyle niche. They may travel to other countries to undergo procedures that are banned in their own country. Other medical tourists choose to get elective procedures outside their countries because of the affordable prices in the destination countries. In addition, medical tourists may choose to travel to other countries to get required medical procedures where there would otherwise be a long delay in their home country due to service capacity limitations.

Challenges in the Medical Tourism Industry

The availability of medical tourism services generally spreads through word of mouth. As such, the industry is difficult to regulate in terms of quality control measures put in place to protect the consumer. Though unknown at this time, there is the possibility that some international service providers may offer less than desirable quality of service. Quality control, from an industry perspective, remains a challenge.

The actual cost of medical services of an international medical provider may differ significantly from anticipated costs. Unknown to the consumer, there may be hidden costs that are not communicated until after services are provided. While the cost of a procedure might be clearly articulated, the service provider might be silent on related costs.

Some of the elective procedures do not need the presence of accompanying friends and family to spend significant time with the patient. These companions would often want to spend time engaged in touristy activities such as shopping, sightseeing, entertainment, dining and other lifestyle services. However, even those hospitals offering good one stop health care and lifestyle services are not mandated to provide tourist-related services.

Some budget-conscious tourists would want to compare and contrast the costs and benefits of using service providers in various countries. Given that the industry is very fragmented, they do not have any help in determining the best service option in staying within their budget. Some tourists, with limited resources, might be traveling to a different country because of the non-availability of a procedure in their own country. For these tourists, the cost of air travel and international accommodations could be daunting when taking into account financial constraints.

These are some of the challenges faced in the medical tourism industry. FlyFreeForHealth entered the market to address as many of these challenges as possible.

FLY FREE FOR HEALTH

FFFH was the vision of its founder Dr. Wei who wanted to build the world's first comprehensive Business-to-Customer and Business-to-Business medical tourism hub providing a one-stop destination for health, travel, leisure and lifestyle services to the consumer, using multimedia technologies as the platform.

Dr. Wei is a medical inventor and entrepreneur. He is internationally known for his innovations in technology. He is known for his vision in expanding healthcare services in an international marketplace of medical tourism. Dr. Wei's background is presented in Exhibit 1, which is central to the current and future success of FlyFreeForHealth.

FFFH was built on the observation that today's technology savvy and mobile consumers are no longer focused on domestic health professionals as the only source for prevention, diagnosis, and treatment of conditions and diseases. Thanks to Internet-based information resources, such as WebMD.com, MayoClinic.org, NIH.gov, and others, consumers are becoming increasingly knowledgeable about preventive medicine, aesthetic enhancements, and lifestyle management.

They also have electronic access to healthcare information in a global marketplace to explore medical tourism as a viable option to manage personal healthcare needs. Many are recognizing that there is a choice between domestic or international health services taking into account personal factors such as income, medical costs, and leisure time. Increased mobility and availability of high-quality, affordable healthcare in an international marketplace, has enabled individuals to choose to be treated by doctors in different countries, depending on their personal needs and desires.

The company's vision is to become the 'global standard' for medical tourism, creating a lifestyle value chain with outstanding travel, leisure and lifestyle opportunities together with first class medical services.

The company's mission is to be the world's pre-eminent global health facilitator, using global brands and a multi-media platform to bring together renowned hospitals and medical specialists, thus creating a value chain of global medical tourism. It aims to facilitate world-class healthcare of the highest quality at very affordable prices. It wishes to be the "ONE-STOP HUB" for global-health consumers, enabling them to combine the best quality and best value healthcare with sensational lifestyle, leisure and travel options.

FFFH was started based on interest shown by hospitals and doctors from both Singapore and Thailand during the fall of 2007. There were already many medical tourists from Indonesia coming to Singapore and it made perfect sense for FFFH to organize a soft launch in Indonesia to take advantage of the existing market opportunity. The soft launch was in the December of 2007 on Indo TV, an Indonesian television station. Exhibits 2, 3 and 4 present the coverage of FFFH's soft launch in Indonesia.

The first half of 2008 was spent in identifying and engaging the different strategic stakeholders of the company. The first group of stakeholders of the company was the healthcare providers. This group included specialist centers such as the National Eye Center, private hospitals, and privately run clinics. FFFH also noticed that when patients traveled for healthcare services they did not travel alone. They came with friends who would spend their time in shopping and entertainment while the patient was undergoing treatment. The opportunity to service such needs naturally defined the second group of stakeholders to include service providers such as retailers, food and beverage outlets and large branded department stores and boutiques. The third category of stakeholders that FFFH was engaged with was tour operators.

In June 2008 FFFH introduced its services through the media in Singapore. The launch was organized at the Hotel Intercontinental in Singapore. It was a launch where Dr. Wei introduced his healthcare service provider partners from Thailand. FFFH was introduced as the abbreviated name of the company and the service provided was multimedia based on Web-interactive medical tourism.

The service started gathering speed. Initial service included a nurse who was trained as a Medical Butler. The nurse would receive medical tourists at the Singapore airport and would accompany them during their entire stay in Singapore. Since the Medical Butler was a trained nurse she would be the liaison between the medical tourist and the local health care service providers. In some instances the medical tourists could not speak the local language; and in such instances, the Medical Butler would also play the role of a translator. Once the patient was in the safe and trusted hands of the healthcare provider, the Medical Butler would switch the service mode to provide shopping and entertainment advice to members of the Medical Tourism entourage.

The Medical Butler service was followed by the launch of iMedical Butler in Singapore in the February of 2009. While a Medical Butler would cater to the needs of a medical tourist and his or her companions on arrival in a country, there were requests from some medical tourists for information on the different medical tourism options before they left their country. iMedical Butler service was in response to such market demand. Medical tourists would be able to contact an iMedical Butler through the Internet and find out the different options. They then could make a knowledgeable and informed choice of the service provider for their requirements. Most often, they made the selection in consultation with their local physician. In some cases, the tourist's physician contacted the chosen service provider through the iMedical Butler to discuss about the procedures. The physician was then be able to better advise their patients on medical options. iMedical Butler was one of the services offered by the FFFH.

Exhibits 5, 6 and 7 present materials released at a press reporting of the launch of iMedical Butler. The launch was so successful that Singapore's Workforce Development Authority was considering developing certification programmes for Medical Butler training.

BUILDING BLOCKS OF FLYFREEFORHEALTH

FFFH leveraged the endorsements of its celebrity patients as part of its marketing efforts. Movie and television celebrities who benefitted from FFFH services became the spokespersons appearing on television talk shows and other hosted programs. Such endorsements immediately caught the attention of consumers who were planning on benefitting from medical tourism. Comforting words and testimonials from the celebrities assured the medical tourists about the quality of care.

Dr Wei's previous work had been featured by international media around the world including Discovery Channel, CNN, BBC, Fox News, CNBC, ABC, Time, Wired, ZDF German TV, ARTE French TV, Japan TV, Yomiuri Shimbun, Korean SBS TV, Figaro, Asian Wall Street Journal, Washington Post, Guardian UK, LA Times, Channel News Asia, Age, Sunday Times UK, Newsweek, Tatler, Bazaar, Marie Claire New York, and Glamour Paris. Dr. Wei's prior contacts and relationships with the broadcast media program developers gave him easy access to celebrities who worked with him on previous programs.

A key selling point of FFFH's service was the airfare subsidies. FFFH was able to offer different levels of travel subsidies for all its members depending on their origin and destination. It also offered additional dining, entertainment and shopping discounts through its business partners for premium members. The discounts for non-health care related activities and services were certainly attractive to those accompanying the medical tourists.

Dr. Wei quickly realized the importance of customer relationship management in medical tourism. He sought out GMS (Global Marketing Strategies) Asia Pacific as a CRM partner for FFFH. Global Marketing Strategies (GMS) was the industry leader in the design, implementation and management of loyalty membership programs for five star hotels across Asia Pacific. Representing leading major hotel brands, they would continuously strive to be the best. This group was servicing the Starwood Alliance and hence had deep practical knowledge on dealing with customers to understand their needs and to record their experiences. Their services were important to get customer feedback to constantly improve the quality of services provided by FFFH.

The Medical Butler Concept

FFFH's key strategy was to provide a one stop service not just for the medical tourists but also for the accompanying members. The idea was to provide an arrival to departure service for the entire team and for the entire duration. The team would have the use of a limousine and a trained nurse cum Medical Butler during their stay. The availability of a person who had both medical and other local knowledge provided a complete package of services to both the medical tourist and his or her travel companions.

Medical Butlers provided services in the following areas:

1. General and Medical Tourist Administration
2. Health Guide
3. Tourism planning
4. Lifestyle Guide
5. Merchant Guide

iMedical butlers also invite potential FFFH customers to Webinars conducted by renowned doctors in FFFH's network of healthcare service providers. Such Webinars allow the viewers to enjoy unbiased information and help them work with the iMedical Butlers in the process of selecting healthcare service providers. Those medical tourists attending a webinar will be introduced to the alternative procedures for their health problems. These patients can then call up an iMedical Butler and ask them for details on the different procedures and the different healthcare providers including information on the cost of such procedures.

Medical Tourism Academy

The need for nurses to be trained in lifestyle and other tourism related information and services quickly led to the establishment of a sister organization called Medical Tourism Academy (MTA). MTA was set up in consultations with Singapore's Ministry of Manpower. This is a for profit organization which will generate yet another revenue stream for FFFH. MTA will admit candidates with nursing background and train them in a set of procedures and tasks that will prepare them to operate as Medical Butlers. This will include training in customer care, hospitality and entertainment services. Medical Butlers will be given personal digital assistants that they can use to interact with their customers. Each Medical Butler will be assigned the list of medical tourists they are supposed to service in advance. Such prior introductions promote interactions between the Medical Butlers and their respective patients long before the medical tourists arrive for their treatment. Such an engagement was expected to improve the rapport and trust between the medical tourists and the Medical Butlers.

iMedical Butler Concept

FFFH was keen to engage the medical tourists even before they arrived at their destination. This need to offer them information on the different options resulted in the creation of iMedical Butler service. An iMedical Butler service was available 24 x 7. This was possible by training a number of nursing professionals from different countries in the skills needed to become iMedical Butlers. They were typically individuals with nursing background who might have either retired or had time on their hands. Such individuals could come from any country in the world. They would have password controlled access to FFFH databases that they could use to respond to queries from potential customers. These nursing professionals were trained to brief a potential medical tourists on the different medical and lifestyle service options available for a specific medical or lifestyle request. The iMedical Butlers were instructed never to make any recommendations and to focus merely on sharing the factual information. This could be enforced given that the iMedical Butlers did not have access to comparison data across multiple service providers for the same healthcare procedures.

Community Building

FFFH also decided to link potential medical tourists with its past customers in order for them to understand their customers' experiences. This resulted in building a community of beneficiaries around FFFH who would readily share their observations, experiences and recommendations with potential medical tourists. Such a proactive community building exercise was important for the potential FFFH customers to make informed choices. The community was supported through an on-line forum where those interested in services could learn from the experiences of those who had undergone same or similar procedures across different service providers. The onus of deciding which healthcare service provider to choose in a particular instance was left to the medical tourist and / or his or her local physician. Those seeking the services of FFFH were encouraged to consult their local physicians to get their recommendations on selecting the right service provider and also to plan post-service recuperation plans once they returned to their own country after undergoing the procedure. This was important given that some procedures might require a prolonged recovery period.

Once a medical tourist made an informed choice of a service provider, FFFH enabled a dialogue and communication between the two sides in order to build and enhance the trust even before the medial tourist arrived in the foreign country. This remote consultation could be carried out either through phone or email. This interaction allowed either the medical tourists or their local physicians to discuss the proposed procedures, potential risks, post procedure recovery plans and in general establish a rapport amongst the patient, local advisors and the service provider. This communication was considered to be pivotal to the success of the services provided by FFFH's healthcare and lifestyle partners.

Healthcare Services Provided by FFFH

Eyecare was the initial service provided by FFFH. As mentioned previously, the impetus for starting FFFH arose due to a request from Singapore's National Eye Center (SNEC). The SNEC engaged FFFH to promote Lasik surgery amongst patients from the ASEAN region, particularly to medical tourists from Indonesia. This market generated demand was largely responsible for FFFH to start earning revenues from day one.

The set of services presently offered by FFFH is listed below. Details of the individual healthcare services are listed in FFFH's website (www.flyfreeforhealth.com).

- Aesthetic medicine including anti-aging medicine, obesity management and surgery, aesthetic dentistry, laser procedures, resurfacing, Botulinum Toxin (BOTOX) treatment, fillers, peeling techniques, plastic surgery, phlebology (varicose vein treatment), and sclerotherapy (spider vein treatment).
- Back care focuses on relieving back pain, which is considered to be the third most common healthcare problem after heart disease and cancer.
- Breast care treatments address early detection, screening and management.
- Cosmetic surgery includes procedures for fills, implants, lifts and tucks. World famous Brazilian plastic surgeon, Dr. Marco Faria-Correa, is one of the service providers for these procedures.
- Cell therapy uses a patient's own blood cells in roughly a one hour day surgery to rejuvenate, revitalize and smoothens facial wrinkles.
- Eye care programs including Refractive surgical procedures such as LASIK reduce and/or correct refractive errors such as myopia (nearsightedness), hyperopia (farsightedness), astigmatism (blurry vision) and presbyopia (focus difficulty).
- Hyperbaric medicine is used to increase oxygen levels in patients suffering from decompression sickness and air embolism.
- Neurosurgery and neurology programs include managing and treating severe chronic headaches.
- Reproductive wellness primarily addressing infertility related issues.
- Weight management addresses obesity related problems.

FFFH AND TECHNOLOGY

Dr. Wei is a firm believer in the use of technology in general and multimedia technology in particular to create a competitive advantage for the companies and products that he designs and develops.

FFFH has leveraged on multimedia platforms from day one of its operations. It has also benefited from media such as television for promoting its services. FFFH had identified the need for expanding its Medical Butler service for online, distance support using Internet and mobile technologies. It was clever in recruiting a proven partner, GMS for its CRM requirements.

FFFH had developed compelling application support on miniature Personal Digital Assistants (PDAs) for use by its Medical Butlers. This enabled the Medical Butlers to get the latest and most updated information from FFFH's servers. Exhibit 8 provides the current state of technology infrastructure at FFFH. There are several independently developed systems that loosely address the different functions of the company.

FFFH'S FINANCIALS

FFFH has been cash positive from day one. This was largely due to the demand from the healthcare service providers who engaged FFFH to reach out to customers beyond their own mailing list. A number of service providers would engage FFFH on a retainer basis to provide visibility to their services. The retainer fee paid by the providers allowed FFFH start their operations without having to raise money from any investors.

CHALLENGES IN FFFH'S GROWTH PHASE

The initial success of FFFH has attracted several healthcare service providers and other partners. Commonwealth Travel service Corporation (CTC), a leading regional travel company, had signed up as a business partner of FFFH to promote medical tourism through its subsidiary CTC Healthcare Travel Pte. Ltd. Several healthcare providers from Singapore and the region have also been holding talks with FFFH to establish business partnerships. This surge in the number of interested business partners has FFFH thinking about its growth strategy. It is currently inclined to establish a holding company and set up many of its central and country-specific services as subsidiaries. All these activities promise an exciting future for FFFH to achieve its vision of becoming the global, gold standard for medical tourism. The vision for FFFH's growth is presented in Exhibits 9 and 10.

Although the initial set of healthcare service providers came from Singapore and Thailand, there continues to be significant interest from service providers in other countries to engage FFFH as a business partner. In June 2009, Dr. Wei was invited by the government of Philippines to explore the establishment of a Medical Tourism Academy to train their nurses to become Medical Butlers. Several first world countries have also been following the growth of FFFH and are likely to start using its services in the coming months and years. The state of New South Wales in Australia approached Dr. Wei to arrange for a meeting in July 2009 to explore a business relationship.

Dr. Wei finds that it is much easier to globalize Medical Butler and iMedical Butler programs than globalizing other health care services. This is an opportunity that is waiting to be seized. Dr. Wei has to identify and manage the key drivers for the growth of FFFH. Such drivers will also include the appropriate technologies required for ensuring that FFFH becomes the global gold standard for medical tourism. He has to also identify how to structure and manage the growth of FFFH.

It is timely to review the technology infrastructure from a global perspective given the groundswell of interest for business relationships with FFFH from the different corners of the world.

What is needed is a scalable architecture that will form the backbone of global operations of FFFH. Dr. Wei would like to receive proposals for defining and creating a global scalable architecture that would support a holding company and several subsidiaries in individual countries.

REFERENCES

Altman, S., Shactman, D., & Eliat, E. (2006). Could U.S. hospitals go the way of U.S. airlines? *Health Affairs, 25*(1), 11–21. doi:10.1377/hlthaff.25.1.11

American Medical Association. (2008). New AMA Guidelines on Medical Tourism.

Bakic, O., & Hrabovski-Tomic, E. (2008). Informal education for aanagement in health tourism. *Tourism and Hospitability Management, 14*(1), 1–12.

Bies, W., & Zacharia, L. (2007). Medical tourism: Outsourcing surgery. *Mathematical and Computer Modelling, 46*, 1144–1159. doi:10.1016/j.mcm.2007.03.027

Blouin, C. (2007). Trade policy and health: from conflicting interests to policy coherence. *Bulletin of the World Health Organization, 85*, 169–173. doi:10.2471/BLT.06.037413

Bookman, M. Z., & Bookman, K. R. (2007). Medical Tourism in Developing Countries. New York: Palgrave.

Borman, E. (2004). Health Tourism: Where healthcare, ethics, and the state collide. *British Medical Journal, 328*, 60–61. doi:10.1136/bmj.328.7431.60

Brown, S. (2008). Medical tourism: nations vie for health dollars. *Hospitals & Health Networks, 82*(12), 49.

Bryant, R. (2002). Despite turbulent times, Dead Sea draws medical tourism. *Dermatology Times, 23*(2), 8.

Carrera, P. (2006). Medical tourism. *Health Affairs, 25*(5), 1453. doi:10.1377/hlthaff.25.5.1453

Castonguay, G., & Brown, A. (1993). Plastic surgery tourism proving a boon for Costa Rica's surgeons. *Canadian Medical Association Journal, 148*(1), 74–76.

Cheung, I., & Wilson, A. (2007). Arthroplasty tourism. *The Medical Journal of Australia, 187*(11/12), 666–667.

Connell, J. (2006). Medical tourism: Sea Sun Sand and… surgery. *Tourism Management, 27*(6), 1093–1100. doi:10.1016/j.tourman.2005.11.005

Dunn, P. (2007). An American in Bangkok. *Hospitals & Health Networks, 81*(11), 43.

Forgoione, D., & Smith, P. (2007). Medical tourism and Its impact on the US health care system. *Journal of Health Care Finance, 34*(1), 27–35.

Goodrich, J. (1993). Health-care tourism in the Caribbean. In D. Gayle & J. Goodrich (eds.) Tourism marketing and management in the Caribbean (pp. 122-128). New York: Routledge.

Gray, H., & Poland, S. (2008). Medical Tourism: Crossing borders to access health care. *Kennedy Institute of Ethics Journal, 18*(2), 193–201.

Hanson, F. (2008). A revolution in healthcare: Medicine meets the marketplace. *Public Affairs Report, 59*, 4.

Hay, B. (2001). New tourism market: health and well being holidays. *Countryside Tourism, 9*(2), 14–16.

Henderson, J. (2004). Healthcare tourism in Southeast Asia. *Tourism Review International, 7*(3-4), 111–121.

Herrick, D. (2007). Medical tourism: Global competition in health care. National Center for Policy Analysis

Horowitz, M. D., & Rosensweig, J. A. (2007). Medical tourism - health care in the global economy. *Physician Executive*, *33*(6), 24.

Judkins, G. (2007). Persistence of the US - Mexico Border: Expansion of medical-tourism amid trade liberalization. *Journal of Latin American Geography*, *6*(2), 11–32. doi:10.1353/lag.2007.0042

Kangas, B. (2007). Hope from abroad in the international medical travel of Yemeni patients. *Anthropology & Medicine*, *14*(3), 293–305. doi:10.1080/13648470701612646

McCallum, B. T., & Jacoby, P. F. (2007). Medical outsourcing: Reducing clients' health care risks. *Journal of Financial Planning*, *20*(10), 60.

Milstein, A., & Smith, M. (2006). America's new refugees—Seeking affordable surgery offshore. *The New England Journal of Medicine*, *355*(16), 1637–1640. doi:10.1056/NEJMp068190

Mirrer-Singer, P. (2007). Medical malpractice overseas: the legal uncertainty surrounding medical tourism. *Law and Contemporary Problems*, *70*(2), 211–233.

Mitman, G. (2003). Hay fever holiday: Health, leisure, and place in gilded-age America. *Bulletin of the History of Medicine*, *77*, 600–635. doi:10.1353/bhm.2003.0127

Pennings, G. (2002). Reproductive tourism as moral pluralism in motion. *Journal of Medical Ethics*, *28*, 337–341. doi:10.1136/jme.28.6.337

Rhea, S. (2008). Medical migration. *Modern Healthcare*, *38*(18), 6.

Ringer, G. (2007). Healthy spaces, healing places– sharing experiences of wellness tourism in Oregon, USA. *Selective Tourism: The Journal for Tourist Theory and Practice*, *1*(1), 29–39.

Rotenberk, L. (2008). Medical tourism. As the world flattens, US hospitals expand their global reach. *Hospitals & Health Networks*, *82*(6), 14.

Smerd, J. (2008). Large companies hopping aboard medical tourism. *Workforce Management*, *87*(11), 8.

Stephens, S. (2008). Healing the World: Reverse medical tourism. *Magazine of Physical Therapy*, *16*(9), 26–29.

Storrow, R. (2005). The handmaid's tale of fertility tourism: Passports and third parties in the religious regulation of assisted conception. *Texas Wesleyan Law Review*, *12*, 189.

Turner, L. (2007). Medical tourism: Family medicine and international health-related travel. *Canadian Family Physician Medecin de Famille Canadien*, *53*, 1639–1641.

Urology Times. (2008). Medical Tourists receive HIFU treatment abroad. *Urology Times*, *36*(9), 12.

Volz, D. (2008). Reverse medical tourism. *Hospitals & Health Networks*, *82*(6), 12–14.

APPENDIX 1

Exhibit 1

Background on Dr. Wei Siang Yu

Dr Wei graduated as one of the top students at Australia's Monash Medical School in 1995. He chose to go against the conventional career path of an honors student and became a medical inventor and entrepreneur in the industry space of digital bio-communication. He gained worldwide recognition in his work on social application of digital bio-communication and became the youngest nominee of CNN's People Choice Award in 2003.

Dr Wei's work has been featured by international media around the world including Discovery Channel, CNN, BBC, Fox News, CNBC, ABC, Time, Wired, ZDF German TV, ARTE French TV, Japan TV, Yomiuri Shimbun, Korean SBS TV, Figaro, Asian Wall Street Journal, Washington Post, Guardian UK, LA Times, Channel News Asia, Age, Sunday Times UK, Newsweek, Tatler, Bazaar, Marie Claire New York, and Glamour Paris.

Today, Dr Wei is a regular speaker at many medical conferences. Wei has his own TV and radio shows, magazines, and a host of bio-communication related wireless services. Dr Wei has worked with many health agencies in the world including the Dutch health promotion board, Singapore health promotion board and many renowned doctors/professors to bring about better reproductive wellness, sexual wellness, and population balance.

Dr Wei is the founder of the world's first multi-media medical tourism hub www.flyfreeforhealth.com where the best hospitals and doctors aggregate to provide the highest quality healthcare to the world.

Exhibit 2

A Convergence of Travel Value Chain and Medical Consumerism Globally

28th Nov 2007, Jakarta - The global trend of mobility has not only ignited the explosive growth of air carriers, it has also changed the mindset of medical consumerism as well. Today's highly mobile individuals are no longer engaging the health professionals for disease diagnosis and treatment; they are pro-active in preventive medicine, aesthetic enhancement and lifestyle management. In addition, medical consumers are getting used to the idea of combining the use of local healthcare and overseas healthcare to optimize their choices. It is becoming inevitable for today's highly mobile individual to be managed by different doctors in different countries as well as a local doctor to manage many international visitors.

Www.flyfreeforhealth.com is the world's first multi-media platform that allows internet signup for membership where adopters of premium health programs can enjoy air ticket rebate and travel/lifestyle incentives via SMS vouchers. Www.flyfreeforhealth.com has already enlisted regional partners from the health to travel/lifestyle industries including **Jetstar / Valuair**, renowned hospital and doctors like **Singapore National Eye Centre, Phyathai Hospital, Melinda Hospital** in the region, travel/lifestyle partners like **Vayatour, Watson, Hourglass, Healthy Choice, Nirwana Garden Resort, Spa Boutique and Lawrys Restaurant etc.**

During this launch period, the public can sign up to be members of www.flyfreeforhealth.com for free. Members of www.flyfreeforhealth.com can then choose a range of health programs from renowned hospitals and doctors and arrange appointments with the listed hospitals or clinics directly. Upon completion of the health program, members will receive an air ticket rebate which is pegged at least at S$100. In addition, members can receive SMS vouchers with exclusive offers from FlyFreeForHealth's travel and lifestyle partners.

Www.flyfreeforhealth.com aims to provide today's highly mobile health consumers a global option of premium health programs and reward them with lifestyle/travel incentives that are borderless.

In addition, www.flyfreeforhealth.com will also hope to focus in supporting doctors and other health professionals in co-management of health consumers which will be inevitable in today's geographically independent healthcare landscape globally.

Www.flyfreeforhealth.com will first be launched in Asian countries like Indonesia, Malaysia, Singapore and Vietnam. The second phase launch will be in United Kingdom, Germany and the United States of America. The membership signups in Indonesia will be free during the launch period. Membership fees will be implemented at a later stage.

Please email **admin@meggpower.com** or call Alia at +628561024777 if you have any queries. Photographs and soft copies of press release are available in enclosed CD. Video Footage and English Press Release are available on request.

Exhibit 3

Young, Affluent Middle Class Reshapes Regional Travel Tourism Market

By Channel News Asia's Indonesia bureau chief Sujadi Siswo | Posted: 29 November 2007 2123 hrs

JAKARTA: Rising affluence of Asia's younger middle class is reshaping the region's growing market for medical travel and tourism.

Each year, about 200,000 Indonesian patients head to regional medical hubs such as Thailand, Malaysia and Singapore for medical treatments, spending a whopping US$600 million in the process.

The bulk of such patients have traditionally been medical travelers who travel for serious medical reasons, often for better or more complex treatments that are not widely available in their home country.

But healthcare entrepreneurs said the trend is changing "there is a growing number of affluent, younger middle class patients who can afford to have elective procedures done overseas.

These yuppie medical tourists, as they are called, like to combine healthcare visits with holidays.

"We know that they also understand preventive healthcare, from aesthetic, weight management to Lasik, so on and so forth. So this whole group at this time hasn't been targeted at all," said Dr Wei Siang Yu, developer of FlyFreeForHealth.com.

He wanted to tap into this growing market, and his online platform provides medical tour packages to those seeking preventive and aesthetic treatments in Bangkok, Jakarta and Singapore through tie-ups with airline companies, retailers and healthcare providers.

Laser surgery to correct short-sightedness is one of the lifestyle treatments that have grown in popularity.

"We are certainly trying to grow this segment as we enhance our lifestyle options. As we are more affluent, I think people are looking beyond simply treatment of diseases and into lifestyle options, wellness options," said Singapore National Eye Centre's chief projects officer Lee Kai Yin.

In fact, patients are being wooed not only with shopping vouchers, dining and other entertainment options, but also airfare rebates. - CNA/ac

EXHIBIT 4 (Figure 1)

Figure 1. FFFH's launch of service in Indonesia

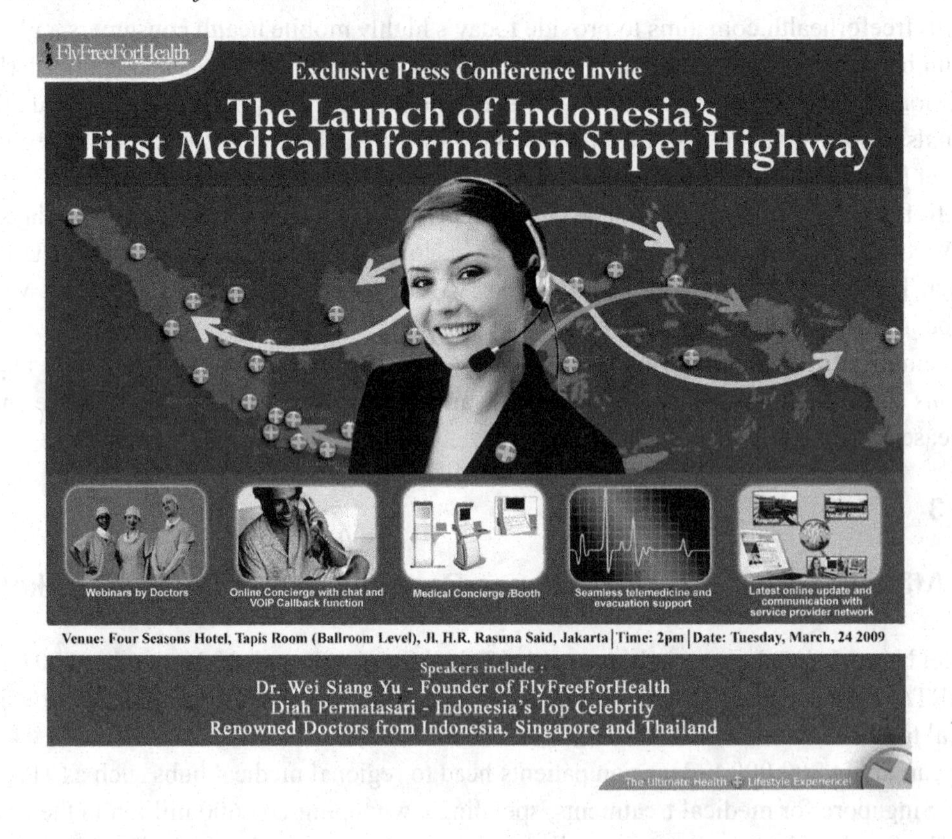

Exhibit 5

Singapore and Thailand Launch World's First Medical Tourism Concierge Network - The New Platform is a Promising Solution to the Healthcare Challenge in Today's World of Economic Downturn

Singapore – February 24, 2009 – FlyFreeForHealth, the world's first comprehensive Business-to-Customer and Business-to-Business medical tourism hub with the support of Tourism Authority of Thailand, today launched the world's first medical concierge network which signifies a new trend in consolidating borderless healthcare globally. Inter-country medical tourism concierge network is starting to be established between major medical hubs in Asia to offer medical travelers around the world immediate access to personalized medical tourism information in various languages at a 'click-away'.

Speaking at the launch, Dr Wei Siang Yu, founder of FlyFreeForHealth said, "the new business to consumer shift (B2C) is seeing the behavioral pattern of medical travelers changing – they look for more options; they also want more interaction before selecting the doctor." Merely looking for information on the internet is no longer sufficient.

"The current model of hospital's agents/facilitators just assisting medical travelers in hospital booking has been completely revolutionized by FlyFreeForHealth's digital Medical Butler – iMedical Butler. Medical travelers look for unbiased information platform that can assist them in the process of doctor selection before having direct contact with the hospitals or doctors. Our nurses are trained in hospitality, IT and tourism to facilitate this selection process. iMedical Butlers act as an online researcher enabling medical tourists to obtain personalized information for health and lifestyle services, invite them to webinars by renowned doctors for a content rich pre-selection interaction", Dr Wei further added.

In addition to the B2C trend, businesses are also looking for more cost efficient and integrated solutions for all the partners. In this Business to Business (B2B) shift, all the players such as the hospitals, government bodies, airlines and travel agencies are looking at providing a more cost efficient and seamless experience to their clients, and hence the formation of synergistic medical concierge network for end to end service. With the support from the Tourism Authority of Thailand, partnership with medical service providers and travel agent in Singapore – CTC Holidays, FlyFreeForHealth is setting up an unprecedented medical tourism concierge network.

The network will consolidate existing medical tourism destinations in Asia as one region with seamless integration in health and lifestyle services. iMedical Butlers will assist the medical travelers from the beginning till the post-procedure management after they return to their country seamlessly. Such end-to-end solution which includes post-op cross border patient management will further improve the patient's quality of recuperation. Medical travelers for the first time will be engaged in an impartial communication during their doctor selection process.

Speaking at the occasion, Mrs. Porntip Makornpan, Director, Tourism Authority of Thailand said, "Thailand is the largest health and wellness destination globally with more than 1.5 million medical tourists last year. The recent launch of Seven Amazing Wonders by Tourism Authority of Thailand, will not only integrate existing health and wellness tourism market but also with other aspects like – meditation, eco-tourism, shopping etc. Tourists can expect amazing value from a broad range of wellness products from the Land of Smiles."

The attraction of affording high quality healthcare is attracting patients annually to Asia for healthcare needs in Thailand, Singapore, India, and Malaysia. According to Deloitte Centre for health solutions, the number of Americans travelling abroad for care is expected to reach 6 million in 2010 and 15.75 million by 2017. $16 billion in cross border revenues is expected to grow by 325% to $ 68 Billion in 2 years.

"FlyFreeForHealth has solved my entire problem and given me assurance, now, I can chat with the iMedical Butler and have the butler help me research on my desired aesthetic services I want before I select the doctor. Not to mention, I am bringing some friends along who can't wait to lay their hands on my shopping vouchers", said Lucy Tan who is looking at tapping great value provided by health service provider in Thailand. Singapore to Thailand traffic for medical tourism can look extremely favorable now, with the Singapore government setting the green light for Medisave claims, which was recently featured in the news.

iMedical Butler has not only transformed the consumer interface for medical tourism, hundreds of nurses and parahealth professionals are now signing up to be part of this powerful online concierge network.

About FlyFreeForHealth

FlyFreeForHealth (www.flyfreeforhealth.com) is the world's first comprehensive Business-to-Customer and Business-to-Business medical tourism hub, a one-stop destination where health, travel, leisure and lifestyle converge, giving freedom and wonderful experiences to the consumer.

The global medical tourism space will be revolutionized through the creation of a truly global brand for health, travel, leisure and lifestyle. FlyFreeForHealth has its own TV shows, health columns at major newspaper and portals. The focus of FlyFreeForHealth in multi-media platform, customer relationship management technology and human resource has placed FlyFreeForHealth at the forefront of medical tourism by creating a borderless healthcare eco-system adopted by global brand and business value chain in health and lifestyle services.

FlyFreeForHealth with its eco-system working in close partnership with government agencies as well as health and lifestyle service providers is set to revolutionize the global landscape of medical tourism.

About Tourism Authority of Thailand

The Tourism Authority of Thailand (TAT) is a state enterprise with the key responsibility of promoting and encouraging tourism in Thailand. With 21 branch offices around the world, the Singapore office is also responsible for the Indonesia and Philippines market. The "Amazing Thailand" promotional campaign rolled out by TAT has been extremely well received and is reflected in the undisputed popularity of Thailand as a travel destination for the Singapore market. In 2008, TAT recorded a figure of 800,000 travelers from Singapore (By Country of Residence).

About Commonwealth Travel Service Corporation

Incorporated on March 2, 1990 as a Singapore-based tour operator, Commonwealth Travel Service Corporation popularly known as CTC Holidays has evolved into a leading travel and tour operator today. CTC has expanded beyond Singapore with a paid up capital of US$7.55 million and have dealings with tour operators all over the world including China, Europe, America, Australia, New Zealand, Middle East, Africa and many more.

CTC Healthcare Travel Pte. Ltd was incorporated in August 2005 and is wholly owned by CTC Holidays. Through CTC Healthcare Travel, company aims to be in the forefront of the medical tourism industry in Singapore and in the region.

CTC Healthcare Travel offers a full spectrum of hassle-free healthcare tour packages that include medical services, recuperative treatment and leisure activities. These packages include a comprehensive and exclusive range of healthcare services that give clients access to the most advanced medical technology and top level medical expertise in Singapore as well as to Korea, Thailand and India.

CTC is now embarking on expansion plans to promote Thailand as a medical tourism hub.

For further information, please email us at info@flyfreeforhealth.com

EXHIBIT 6 (Figure 2)

Figure 2. FFFH's Medical Concierge

EXHIBIT 7 (Figure 3)

Figure 3. FFFH's iMedical Butler

EXHIBIT 8 (Figure 4)

Figure 4. Technology Architecture of FlyFreeForHealth

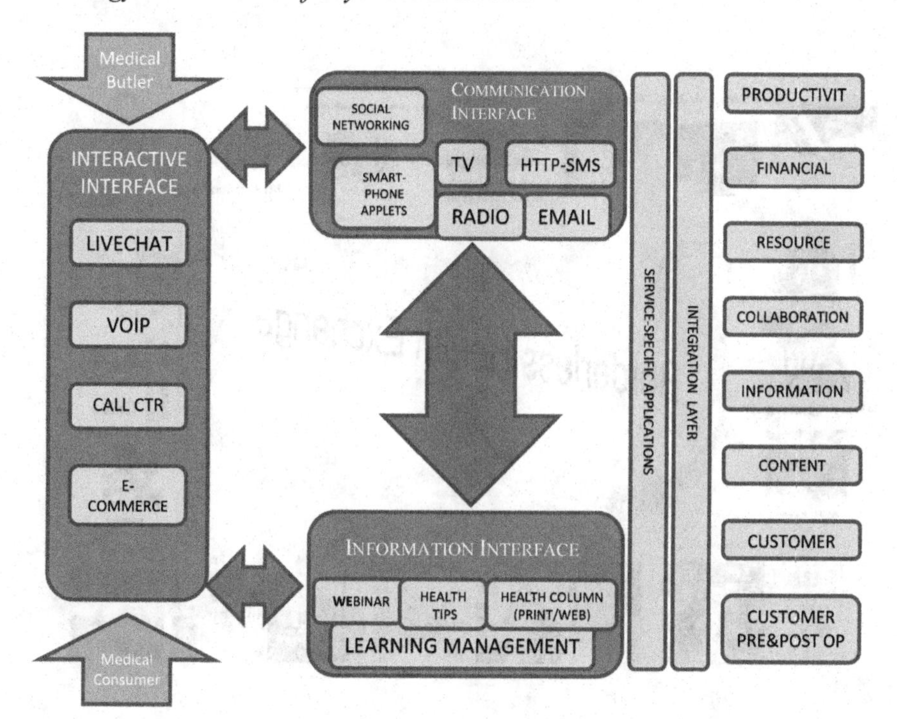

EXHIBIT 9a and 9b: FFFH's borderless Health Exchange concept (Figure 5, Figure 6)

Figure 5.

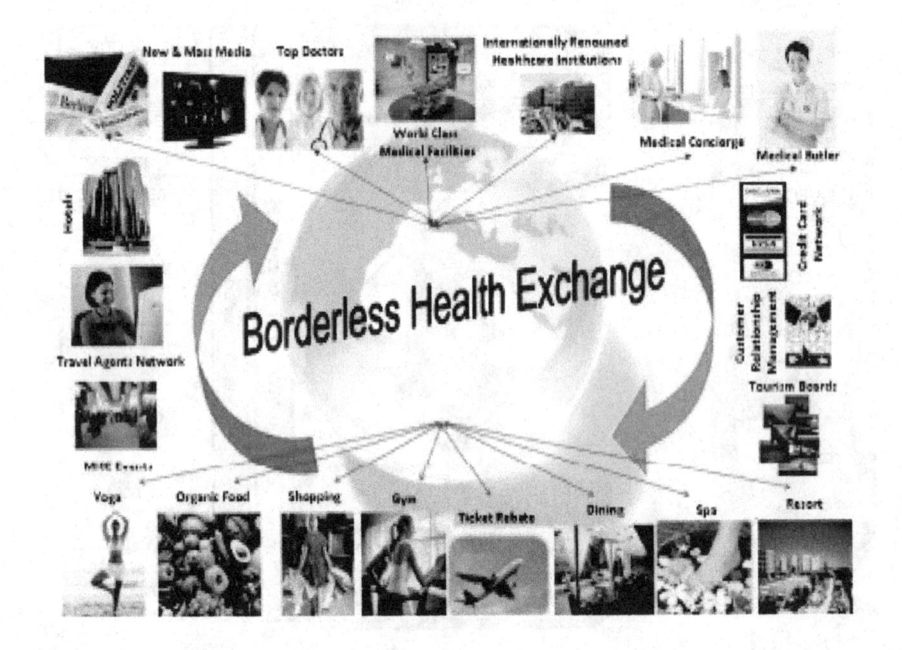

Figure 6.

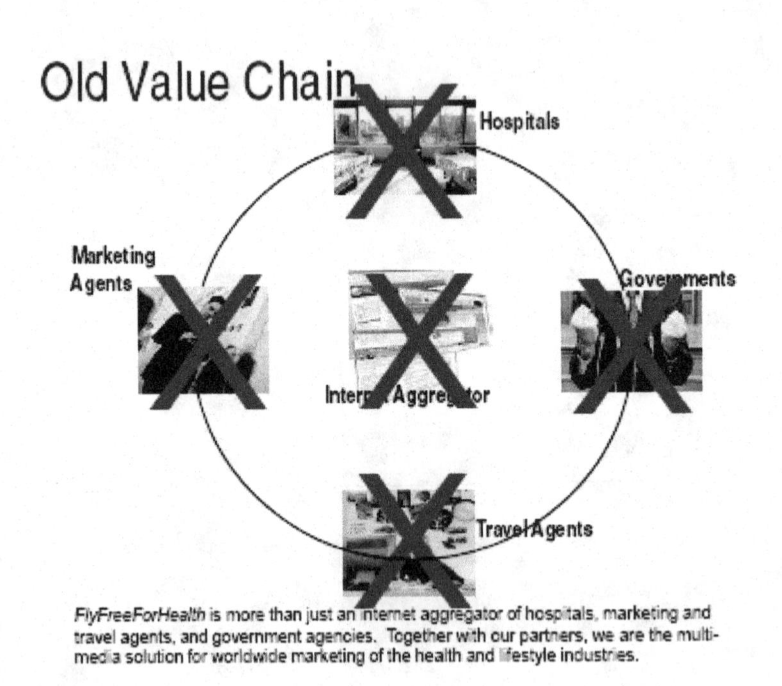

Exhibit 10 (Figure 7)

Figure 7. FFFH's Consumer-centric Emotive Brand

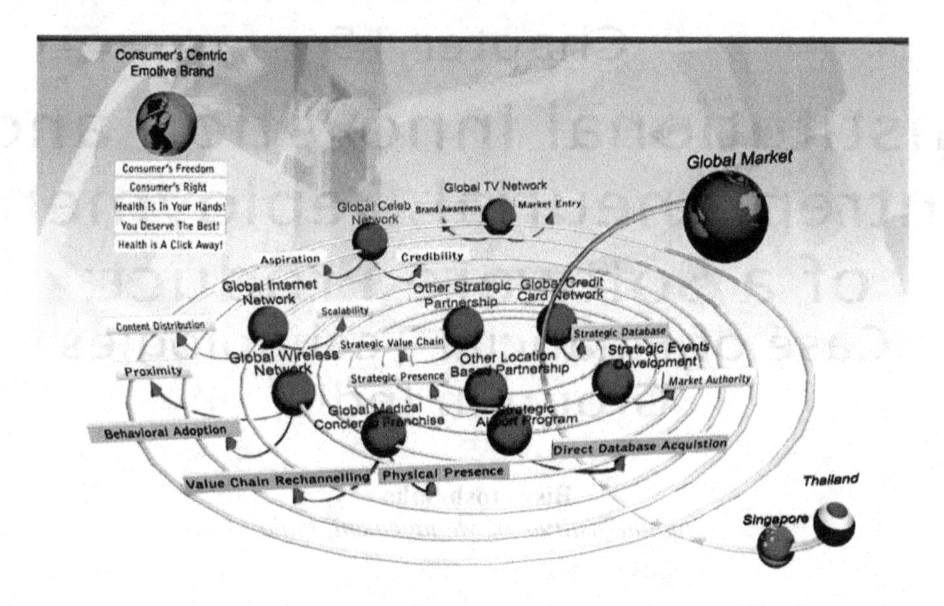

Chapter 15

Institutional Innovation and Entrepreneurial Deployment of a Software Product:
Case of Financial Technologies Group in India

Biswatosh Saha
Indian Institute of Management, India

ABSTRACT

This chapter represents entrepreneurship as a temporal evolution of the creation and control over assets. The value of the asset lies in its transactional relations with other assets in the ecosystem or in other words being part of the architecture of related assets. It is argued that the deployment of financial trading software, as a product in brokerage houses in the emerging securities trading ecosystem in India by the software firm called Financial Technologies (FT), hastened institutionalization of new rules governing transactions embedded in the software design. As a result, FT implicitly collaborated with the regulator and other ecosystem participants who coordinated the innovation in design of the ecosystem. The software firm went on to expand the market for its own products (trading software) by incubating exchange ventures. This was achieved through a strategy of spawning of linked subsidiaries that led to both a growth of the trading ecosystem and further entrenchment of the innovated ecosystem.

INTRODUCTION

Is it ours, this dream in us,
I make my way alone and multiplied
Am I myself, am I another
Are we but imagined beings.
(Geo Libbrecht, quoted in Bachelard, 1971: 105)

All of human intelligence and endeavor has a 'value'. It is the intrinsic power vested in a thought, pulsating with life. It lives within an idea, waiting patiently for those with the imagination to unlock it. And when liberated, it has the power to change the world view (Jignesh Shah, in Financial Technologies Annual Report FY07-08).

DOI: 10.4018/978-1-61520-609-4.ch015

Narratives of entrepreneurship have mostly emphasized 'individual creativity' or have stressed the elements of competition. A typical account would identify a snatching of opportunity that takes an entrepreneur ahead of the pack. This account of growth of the *software product* company Financial Technologies (acronym FT from hereon) provides, instead, a narration of a temporal unfolding of an entrepreneur's idea, that materializes into creation of assets as part of a shared ecosystem. While rivalry is important to the story, cooperation is equally important and our account focuses the narrative gaze on the cooperative aspect without denying the role of rivalry. Rivalry and cooperation possibly inheres as part of the same process. Entrepreneurship, it is argued, may be viewed more pragmatically as creation and control over novel assets; novelty being understood with respect to the *particular ecosystem* where an entrepreneurial play occurs.

Assets generated by entrepreneurs derives its value as part of wider ecosystems, which in turn can be viewed as an architecture of inter-related and inter-connected assets. Value of an asset is thus relational, flowing out of specific relations and transactions with other connected assets. The transactions between assets (or asset-holders) occur under specific rules that define the relation between the assets. The specific rules also define the governance within which asset-holders transact, identifying roles and partitioning risks. Such rules must have a temporal stability, at least over a horizon for it to work. Following North (1990), we can visualize such temporally stable rules as 'institutions'. The software *products* of FT embody such rules that connect different categories of asset holding organizations in the contemporary financial securities market in India, where securities transaction occur over an information highway with intense information exchange between multiple actors under specified constraints (rules). The software is thus like an *infrastructure* both in a sense of structuring the exchange of digital information and providing the infrastructure of underlying rules. Adoption of a software *product* embodying the *standard* rules hastens the cooperative subscription to a set of rules within the ecosystem. The *product* stance in contrast to the *service* one becomes valuable for a software writer within this context.

Ecosystems, constituted thus, would be in *contest* with each other. This *contest* is distinct from familiar notions of competition; it is more like a contest between several proposed architecture of governance rules. A winner ecosystem has a larger following – its specific governance architecture has dominance over a larger terrain or business space. Assets within a winner ecosystem are generally more valuable than assets in ecosystems that have lost out in the *contest.* (Banerjee, 2008). This contest between ecosystems defines an element of rivalry that an asset owner participates in. This rivalry is distinct from the familiar sense of competition in a defined product market which has been extensively dealt with in economics, particularly within the structure-conduct-performance paradigm. Regulator's vote in favor of a particular ecosystem often becomes crucial. The market making of the software products of FT piggybacked on a regulation driven inter-ecosytem contest.

The temporal dimension too is crucial. We must view an entrepreneurial narrative as indelibly etched in time. The entrepreneur must take note of the *particularity* of descriptions of important actors in an ecosystem at every point of time with respect to the assets held by each, the inter-relations between the assets and a possible temporal profile of evolution of such assets driven by the entrepreneurial motives of asset holders. Unlike managerial coordination, entrepreneurial agency and coordination lies in this space of dynamic evolution of an envelope of inter-related assets. An entrepreneur, as an asset generator, looks towards increasing the value of owned asset, either through change in inter-relations with other assets or through expansion of the ecosystem and establishing links with other ecosystems that so far had remained non-linked to the asset-holders' own eco-system.

An asset is thus, inter-temporally transformed through entrepreneurial acts. Entrepreneurial agency too enjoys a possible transformation inter-temporally as the extent or depth of coordination sought by the entrepreneur within the ecosystem changes, armed as she is with a newly transformed asset piece. In other words, there is a change in the cognition of the desired ecosystem over which an entrepreneur would seek a play of dynamic coordination. This possibility of transforming own assets as well as other assets that are in transactional relations with her own-assets represents a manifestation of an entrepreneurial power or a 'feat'. The process is possibly close to a dramatic unfolding. Understanding of the entrepreneurial phenomenon thus depends on locating the spatial and temporal unfolding of the process of transformation on which an entrepreneurial asset buildup rests.

Since a software *product* assumes wide adoption of a standard logic of operation across a large business space, the *product* business must either itself enjoy the power or be close enough to a powerful cohort (in this case the regulator) to drive the adoption of the standard through. In this case, the 'fiat' of the regulator and the software product of FT (and other software product providers) together aided rule adoption. While the regulator's 'fiat' imposed boundary conditions setting up the expectation of a possible ecosystem design, the software materialized rule following at the level of routines of transactions. Most accounts differentiating between software *products* and *services* have missed this dimension of coordination required in a software *product* stance. The two stances, as implicit in this narrative, differ more along dimension of entrepreneurial 'feat' rather than a technological one. Innovations were institutional rather than technological. Role of technology (the software) was akin to a device that aided in formation and stabilization of the 'innovated institution'. This narration, by adopting a frame of capturing the dynamics of temporal evolution, attempts providing such a description.

FINANCIAL TECHNOLOGIES GROUP: SNAPSHOT 2008

...our vision of creating one of the largest global exchange networks across the fast-growing economies of Africa, Central Asia, Middle East, India, China and other Asian countries. (Jignesh Shah, CEO, Financial Technologies, in Annual Report, Financial Technologies, FY07-08).

Credit markets haven't developed beyond our cities. Enterprise and entrepreneurship in India is limitless. There are some 14 million SMEs in India. They require growth capital. Can't we arrange a fantastic market infrastructure for them? (Jignesh Shah, CEO, Financial Technologies, in excerpts of interview published in Mint[1], April 16, 2009)

Financial Technologies (FT), the flagship enterprise of the FT Group is a US$473 million (as on December 31, 2008) enterprise, listed in Bombay Stock Exchange (BSE) and National Stock Exchange (NSE) in India. FT is an IP based software technology enterprise that sells, on a licensing mode, software products that run the central trading (or price-time order matching) engines of electronic exchanges as well as the broker/investor software that allows brokers to connect to exchanges. The software also enables Straight-Through Processing (henceforth STP) allowing automation of post-trade processing of clearing and settlement operations, which requires exchange of information and matching/reconciliation acts, across several organizations. By 2008 it was at the centre of a cluster of inter-linked subsidiaries and joint ventures, which included 10 exchange ventures, 5 of which were outside India – in Singapore, Mauritius, Dubai, Bahrain and Botswana, and six ecosystem ventures that provided products and services that complemented the market ecosystem around the exchanges.

FT group was trying to create the trading highways for transaction in commodities and securities linking Mumbai (and India) with growing economies in Asia, Middle East and Africa. Its standalone revenues and profits in FY 2007-08 stood at INR 13,375 million and INR 9,612 million respectively, while consolidated revenues and profits stood at INR 14,195 million and INR 8,608 million.[2] (See Annexure 1 and 2) FT group achieved this remarkable growth in a short span of just a decade – it was in 1998 that FT had logged its first commercial sale, having started as a garage start-up three years earlier by first generation entrepreneur[3], Jignesh Shah. *Institutional Investor,* in its 2008 rankings featured the FT group as one of the leading 30 e-finance technology innovators alongside NYSE-Euronext, NASDAQ OMX Group, Instinet, CME Group and other leading global exchange markets and electronic trading networks. In 2008, it won the NASSCOM Innovation Award[4] and the Delloite Technology Fast 500 Asia Pacific nomination. From November 2005 it got included in the BSE-IT Index as well as BSE Madcap India and BSE SENSEX, as an Index stock.

THE EARLY YEARS: AN ENTREPRENEUR IN THE MAKING[5]

In early 1991, a young Gujarati electronics and telecommunications engineer spent his evenings in the Seth Chunilal Library of the Bombay Stock Exchange (BSE), pouring over old manuscripts and accounts of the stock and commodity exchanges of India. He was an employee of the Bombay Stock Exchange, picked up along with a dozen other young engineers in 1990 to work on the BOLT (BSE On-Line Trading) platform – an INR 1 billion project to introduce on-line automated electronic trading into the country's then largest and the oldest stock exchange. He joined the BSE project giving up the option of moving to the US to complete his masters' education – a more popular track among his peers. Early 1990s was heady days at BSE.

The young man's interest, however, lay in the history of the exchange. He came to know about the rich history of commodity trading in India. There were around 300 commodity exchanges thriving in India during the Second World War – till government action in banning gold futures trading in 1962 and futures trading in all commodities thereafter in 1969, put an end to all that. Surviving transactions moved to the underground illegal markets. Biographies of individual traders – of which there were several documented in the Chunilal Library – were of special interest to the young man; especially their trading linkages and the heuristics of their trading logic.

The account of Premchand Roychand was particularly interesting. Roychand was a famous figure in the Kalba Devi Cotton Exchange in Mumbai, the hot spot of cotton trading then. Trading would continue in Kalba Devi until midnight, as prices from the US Cotton Exchanges would flow in before traders could square up their positions. Cotton trading generated enormous amount of wealth in Mumbai – in some sense, the city was built on cotton money. Roychand was the center of that activity in the second half of the 19th century (he lived until 1906). He was connected to the global price discovery points in the international cotton circuit – Tokyo and New York – and through trusted messengers got information on prices there ahead of other traders in Mumbai. Roychand provided leadership to the Mumbai cotton trading community and the trust reposed on him by the broker/trader community was crucial in building the exchange.

'An exchange, after all, was an institution of trust.'

Roychand also was an architect of several modern urban institutions of Mumbai during the second half of 19th century when the city was transformed from a set of islands to the commercial hub that it is today – his endowments started several departments in Universities of Mumbai and Calcutta. Yet, the young engineer was surprised that very few people knew about Roychand.[6] The post-independence era economic policy in India, under a mandate of nationalistic modernization, had given a short shrift to trading interests; policy sought conversion of traders into industrialists with investment commitment to fixed assets, such as manufacturing factories.

Trading was not limited to the metropolis of Mumbai – there were almost 300 exchanges around the country and several other dense trading locations. Surendranagar, in Gujarat, was one such cotton trading and production hub. An enterprising man there set up a private radio station to disseminate cotton price information in a radius of 50 km and became rich enough to buy real estate in Mumbai and building under the name Tarwala Mansions. (*tarwala* translates into 'telegraph man') Information, commodities and finance flowed across a dense network of trading relations. There were numerous such accounts.

In 1993, as part of his assignment, the young engineer was sent by BSE on a world tour of exchanges (Hong Kong and Tokyo stock exchanges as well as NASDAQ) to study its trading system, its rules, and technologies of operation. On their return, however, shocking news awaited them. BSE had transferred the project mid-way to CMC Limited, an Indian software service firm. The young man - Jignesh Shah - quit his job at BSE along with Dewang Neralla and four others. They co-promoted and registered a start-up private limited company. On the New Year's Day in 1995 work started from a 250 square feet mezzanine floor office with five working terminals in the Fort area of Mumbai (the financial hub) with an initial capital of around INR 0.5 million, bulk of which went to pay off the bond with BSE that the team had signed before they went abroad on the study trip. They had an internal benchmark of 1000 days of incubation of their product idea, within which they wanted to achieve commercial success. For three long years, the small team led by Neralla as the software architect and Jignesh Shah as the developer of the business algorithm, dug in to design the core of their first trading software. The software engine was christened ODIN and it powered the front end of brokerage houses linking them to the automated trading engine of electronically operated stock exchanges. It was in 1998 that the product was ready for being taken to the market.

A small Gujarat based brokerage company – Growth Avenues, became their first client. ODIN was sold as a branded service – with a pay-as-you-use model, where license fees were linked to the turnover of transactions over the software. FT shared the risk of software product investment of the client. This was a departure from the then dominant model of high-priced product sales by IBM, TIBCO and TCS that dominated the Indian brokerage software market.

The price of TIBCO offering, the market leader, was around US$ 1 million, which was too costly for smaller brokerage houses. FT offered their product at less than 1/10th the going price as upfront payment - the rest being a transaction-based fee.

ODIN helped Growth Avenues quickly roll out a retail chain and register a quick volume expansion of their client list as the investment in software became an expense stream extending over time, instead of a one-time up-front investment.

ODIN represented the core (*the seed*) of the idea, in terms of both technology design as well as design of the revenue model. This would remain as a guiding principle of the group over the next decade. They made then a few basic choices:

- Jignesh Shah and Neralla decided that they would build Financial Technologies as an IP based software enterprise, departing radically from the service mode of growth adopted by the Indian IT peer group.

- Technologically, the software would be designed to work optimally under widely different (or multiple) telecommunication linkages between the broker/investor and exchange ends, especially under constrained bandwidth conditions, thus taking an early bet on telecommunication channel convergence and wide differences in connectivity conditions in the country.
- The solutions must offer multi-exchange, multi-asset class trading access, thus taking an early bet on convergence of asset classes and exchanges that would eventually define the evolving architecture of financial markets.
- Generation of cash flows on the IP would be linked to growth in trading turnover, both at broker/client and exchange levels. Therefore FT, as an IP owner, had a strategic interest and a thrust on innovatively driving growth of the trading eco-system.

The eventual story of growth of Financial Technologies is an unfolding of the core idea contained in these basic decisive stances taken at the formative moment of the enterprise – *a process of multiplying and morphing* as it extended itself into new transaction ecosystems within the wider interconnected trading networks. ODIN, in 1998, was still a very simple product – it just linked the trading clients of brokerage houses to the electronic exchange. It was, in fact, quite early days for the electronic financial trading systems in India. It was in 1994 that National Stock Exchange (NSE) first adopted electronic order-matching trading system. The ecosystem designs regarding working of related organizational entities such as custodians, depositories etc, including the rules that would partition risks amongst these organizations, were put in place through regulations only around 1996-97. ODIN, as a software product, would be continuously developed through addition of features and layers that would aid connectivity of the software deployed at the trader workstation (end customer point) to different organizational entities in the evolving trading ecosystem working under different classes of rules/regulations. The specificity of the rules/regulations generated a specific logic of trading and had to be incorporated in the software. Rules also required compliance systems and risk management systems to be developed in line with the risk partitioning that regulation was striving to achieve. ODIN morphed into a network product and provided FT the opportunity to become familiar with the wider ecosystem. Even in 2008, ODIN was the major revenue line for FT. Rooted in the trader workstation at the periphery of a complex interconnected financial trading network, FT slowly moved into the center – so to say. It moved into the exchange business designing exchange software as well as running exchange business; and in another sense moved close to the regulator – the point that provided ecosystem-wide coordination for novel rule formation in the evolving financial space.

INDIAN FINANCIAL MARKET ECOSYSTEM IN 1990S

Indian securities market in the early 1990s was dominated by the Bombay Stock Exchange (BSE), which was a broker controlled and managed exchange established as a mutual benefit organization. BSE controlled almost 75% of the total stock trading turnover in India and was the most liquid secondary market. There were other stock exchanges in the country as well. The market was divided based on notions of exclusive territorial jurisdiction of respective stock exchanges in each region – a model of geographic local monopolies. However, stock exchanges also had a sub-broking system. So a BSE broker could have sub-brokers spread across the country. Clients in non-BSE areas could link up to BSE through local sub-brokers and thus competition between exchanges across territories got organized through sub-brokers. Trading through sub-brokers, however, posed a few problems for investors who did not have direct access to the price-setting process in the exchange.

Most exchanges were member controlled as well. Although, in form, a few were incorporated as non-profit-making companies, brokers effectively controlled management of exchanges. Membership of the exchange was limited – exchanges globally worked on such models of limited club-based memberships that were tradable. The only way for an outsider to become member of an exchange was to purchase a 'trading seat' from a current member in the secondary market. It was operationally a result of limitations of the physical trading floor space (the trading pit) where brokers bid to set prices (called bid-driven price discovery). The need to organize clearing and settlement processes that preserved the sanctity of the trade in the pit also led to reliance on a small cohesive group of stock trading community, where norms of governance could be enforced. Brokers and the community of brokers (through the exchange) *bundled several risk-provisioning services* on which outside investors could depend. Among major exchanges, NASDAQ was the only one possibly, which did not historically ever have limitation on dealer membership (Ingebretsen, 2002).

Over the 1990s, around the world, pit based trading systems were under stress, partly due to advent of electronic trading networks that provided a technological possibility of breaching the asymmetry of real-time information access that was the pillar of pit-based and bid-driven exchanges. These networks operated on order-driven system, where investor orders rather than broker bids, discovered prices. (See Weber (1999) for details). Electronic trading networks in the US were corporate entities decoupled from traders, unlike the model of stock exchanges such as NYSE, where the exchange was broker controlled. The growth of these networks posed a threat to the exchanges; NYSE held out for the longest period trying to work on the old pit-based model, but ultimately embraced technology based trading and decoupling of ownership and control of exchange from direct trader/broker interest in 2006 (Lucas, et al., 2009). 'Technology enabled trading' also allowed handling of very large trade volumes at reduced costs of trade processing and tolerable risks of error (or value-at-risk). A new trading ecosystem emerged around the world through the 1990s – one key feature of which was voluminous trading. One of the key tenets of this design was unbundling of different risks. Each risk category became the specialist domain of a particular kind of organization - that had corresponding regulatory approvals and implemented compliance systems as part of its risk management and fraud prevention procedures. Unbundling of risks around a single trade into multiple organizations also generated the requirement for *traceability (audit trail)* of every trade through the lifecycle from order generation to settlement.

Stock exchange ownership was separated from direct trader control, clearance of trades was concentrated into a clearing corporation different from the exchange, deeds of ownership of stocks were maintained in dematerialized (and immobilized) form in central depositories, custodian services were separated from investment or fund management activities.

The advent of electronic trading technologies also influenced stock trading in India. The path of adoption of technology and the modes of coordination required, however, differed substantially. To appreciate that, we need a brief background on the context in which Indian market adopted the new securities trading ecosystem design. Prior to 1990s, stocks/securities had very little penetration as an investment/savings vehicle in India. The role of the regulator, the Ministry of Finance, was quite weak and several attempts to introduce changes in the secondary capital markets were earlier thwarted by the 'broker/trader controlled' exchanges that continued to remain the bundled risk provisioning point.[7] The Economic Reforms Program initiated by Government of

India in 1991 mandated a new focus on financial market reforms – one major target of which was to create a wider and deeper financial market so that dependence of the corporate sector on bank based debt financing could be reduced. Foreign Institutional Investment was also sought into the secondary market – both to provide a supply of foreign exchange as well as to generate liquidity in the domestic securities market. Very early on the reform path, however, the stock market witnessed a major scam.[8] It brought to light the inadequacies of the existing institutional structures in handling a rising volume of trade transactions. Foreign institutional investors, who were used to a different institutional arrangement back home, also raised demands for alignment of trade processes across borders for investments to flow.

The Ministry of Finance devolved powers to the Securities Exchange Board of India (SEBI) and enacted the SEBI Act in 1992 to strengthen the independent exchange market regulator. It was during this period that a host of financial institutions in India (who had suffered heavily during the scam) took the initiative to promote an alternate stock exchange ecosystem under radically different rules of operation, where the basic logic of unbundling of risk was adopted. The subsequent evolution of the stock market in India is partly a story of regulatory innovation led by SEBI, with National Stock Exchange (NSE) emerging as a working model of such changes, whose success finally led to introduction of changes within even the broker controlled BSE. Between 1994 when NSE went live and 2000 when ICICI Direct was launched providing electronic trading facilities to retail investors, Indian securities market was transformed. R.H.Patil, the first Chairman of NSE believes this to be unprecedented in the history of global securities markets, especially in terms of the pace of change (Patil, 2006). This required a hastened coordination and a political drive to create a new architecture of the securities trading ecosystem that worked out through the regulator.

The NSE Model

NSE was set up as a demutualized stock exchange where the promoters (banks and financial institutions) did not have any trading interest. Trading interests were decoupled from strategic control as well as operations of exchange. It adopted a Central Limit Order Book as the price matching system, deviating from the broker-quote driven (or bid-based) system of BSE. It adopted an electronic order routing and trading/matching system, with trading terminals spread across the country. An experienced multinational technology vendor provided the central trading engine with Tata Consultancy Services doing the customization of the system. NSE did away with broker membership restrictions as well. Within a few years, NSE terminals were available in close to 500 cities in the country and several traders, who were not members of the old exchanges, took up membership along with the old brokers. The electronic order routing system meant that everyone across the country would have access to the same information and access to quote (or bid) without asymmetries under anonymous conditions, thus doing away with the special bidding role of the broker. It thus attempted to move towards the creation of a central liquidity pool. It restricted listings to only a limited set of blue chip securities that were well known across the country, since that was essential for its model to function. It set up a depository and a clearing corporation to provide central third party clearing and settlement of trades, thus taking up the counter-party risk of trade failures from the trading parties. Such moves were in close coordination with the regulatory moves from SEBI in introducing new regulations defining the risk management protocols and compliance norms to be adopted by these new types of entities. Traditionally, as discussed before, all these were the functions of stock exchanges provided in a bundled form (Patil, 2006). Unbundling then was the rival proposal of architecture of governance by the NSE milieu to contest the 'old BSE mode' of bundling.

NSE went live with equity trading in November 1994. The NSE model, thus, was a radical departure from the established institutions of stock trading. It introduced changes in different stages of the life cycle of a trade. The change in post trade stage was concentrated on watertight settlement processes to reduce risks of trade failures; while the change in the 'trade' stage of the life cycle aimed at creating a central liquidity pool and an automated matching system. The first change led to the generation of the demand for what is known as Straight-Through-Processing (STP) that would enable a very short settlement time (T+1 day). When NSE began, settlement cycles were 15 days; which was gradually reduced to T+1 rolling settlement. FT's journey began by meeting this demand. Next section discusses this in greater details.

The second change led to the creation of 'thick markets' with extremely high transaction densities creating new requirements of the order-routing system that managed the 'pre-trade' and 'trade' stages of the 'trading lifecycle' of an 'order'. High density real-time markets meant that trading software began to be benchmarked at levels of fraction of a 'milli-second' taken by the order management system to execute orders – delays were extremely costly to the trader. High density of transactions is important to exchanges working on central liquidity pools, since it reduces costs of transaction measured by bid-ask spreads. The post trade surveillance and settlement guarantee (along with very short settlement times), on the other hand, reduces the post trade failure risk and thus brings down costs of transacting for the traders/investors. Such transaction cost reductions are crucial for exchanges in competing with each other as provider of a central 'market-place'. FT's entry into exchange software business was sometime later – in early 2003.

STP AND DEMAND GENERATION OF AUTOMATION SERVICES

A contemporary securities market, based on unbundled risk partitioning, is an inter-linked network of several types of organizations/entities with differing mandates set by securities regulation in a country. Each security transaction or trade, thus, passes through all such organizations with accounting (or book-keeping) transaction at each entity in what is called a 'trade life-cycle'. A trade is finally settled or consummated only when all such accounting reconciliation succeeds. Trading, or front-office activity, at broker, exchange and investor levels, requires significant back-office accounting and bookkeeping activities across several organizational entities for every trade. The 'trading life-cycle' has three main stages – trade confirmation, trade clearing and trade settlement.

Broadly, the different entities (or entity types) are as follows:

a) Investors, either retail or institutional, are the originators of a trade. The communication from the investor to the broker is called a pre-trade communication. Once a broker has a confirmation of an order from the exchange, it is passed on to the investor (or order generator), usually through a 'contract note'. The investor or order generator has to confirm the 'contract note' and send it back to the broker for the order execution process to be completed and be legally binding on both the parties. Institutional trades, such as that from an asset manager (AM), has a more complex trade order structure since the AM manages multiple client accounts with different mandates. Order for trade execution of a particular security to the broker is thus a pooling of transaction on several client books of the AM, which after successful trade execution are allocated to different client accounts. The interface between AM and broker, thus require information exchange at the level of the 'pooled trade order' and then at the level of disaggregated, client-wise allocation after the pooled order execution information is sent to the AM. So the trade confirmation stage requires additional layers of information exchange and confirmation in case of an AM intermediated institutional trade.

b) Investors usually access the securities exchange through a trading member (TM) of the security exchange. TMs are also called broker/dealers.

c) Clearing institutions are entities where netting of trade occurs – so that settlement can be organized for the net rather than the gross trade volumes. Clearing institutions also stand between the but-sell sides taking upon itself the counter-party risks. Traditionally, clearing functions were part of the stock exchange activities. Regulatory changes in India over 1990's separated clearing function into a separate entity.

d) Custodians are entities that hold securities on behalf of investors. Institutional investor who wants to separate safekeeping of securities from the investment manager or broker/dealer, use custodian services heavily. Separation of roles is aimed, as indicated earlier, at minimization of the risk of fraud.

e) Depositories are entities that maintain a double entry account book of 'dematerialized' securities of client account holders. A trade in a security finally leads to an updation of the accounting book at a depository – signifying the transfer in ownership of the security from the seller to the buyer.

f) The payment system and the banking system that handles the payment that is required to be carried out as the other leg of the change in ownership of a security.

A security trade passes through all these organizational entities in a temporal sequence. The 'life-cycle' consists of three stages – Trade Confirmation (between the Asset Manager/investor and Trading member/broker), Trade Clearing (involving communication between the broker, asset manager and clearing house), and Trade Settlement that involves the Depositories, Custodians and the Banking/payment systems. Figure 1 provides a graphic description of the information flows, including request and approval flows across multiple entities that are required for the completion of a single trade in Indian securities market. In several cases, such approval flows would be around 40 for a single trade.

Regulation sets up a limit of the maximum time for completion of a trade through its complete 'life-cycle'. A T+3 rolling settlement, for instance, means that all trades are settled within 3 days of the trade day. Contemporary financial security markets aim at compression of the settlement time to a T+1 regime in a bid to push up the intensity of trading, which can increase the liquidity of a security and hence parameters of efficiency of the security exchange as defined in the current mode of organization of the securities market ecosystem. Within the context of such temporal compression of the trade 'life-cycle', manual intervention in the post-trade process becomes a major bottleneck as efficiency of back-office processing of information becomes important. The demand for STP arose within this context.

The Securities Industry Association of the US has defined STP as:

The processing of a trade, whose data is compliant with internal and external requirements, through systems from post-execution through settlement without manual intervention (quoted in SEBI, 2002).

Securities markets across the world have invested over long periods in slowly automating different stages of back-office processing. Given the systemic nature of the post-trade processing involving regular information exchange and reconciliation between multiple entities, automation

Figure 1. Typical trade transaction cycle in India in 2001 under proposed T+3 settlement (Source: SEBI, 2002)

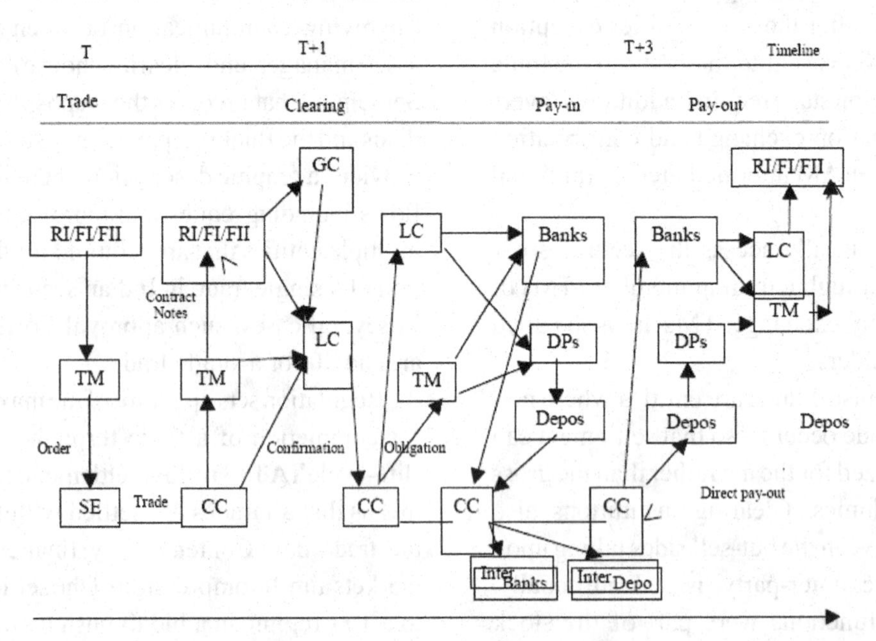

RI: Retail investor; FI: Indian Financial Institution; FII: Foreign Financial Institution; TM: Trading member/broker; LC: Local Custodian; GC: Global Custodian (FIIs availing of global custodian services have to handle the Indian settlement processes through a local custodian with whom the global custodian has agreements);CC: Clearing Corporation; Depos: Depository Institution; DP: Depository participant, who is a retail agent of depository institution through whom investors access depository services.

of the process requires coordinated automation adoption across entities and standardized protocols of information exchange under secure conditions. The software assets relating to securities transaction automation, in this sense, are systemic assets. It has to adhere to several shared protocols to enable inter-working. Different software pieces also constitute together an architecture of assets inter-linked to each other which together makes up the infrastructure on which the front-office activity of high throughput trading can occur at acceptably low levels of risk of post-trade failures and low transaction cost of back-office processing.

In India also such automation initiatives started from the mid-1990s when the basic institutional architecture was built up around NSE with the regulatory support of SEBI. Trade confirmation part of the 'trade life-cycle', however, remained a fully manual process even when brokers and clients were highly automated till even 2004. The process involved writing of the contract note (in physical form) and its transmission/retransmission through fax and the re-entry of the data on the same trade into computer systems at multiple sites in different entities. Transmission through fax immediately means that processing can be batch based at best and not real time, while manual re-entry of essentially the same data point would lead to increased error possibilities. Possibility of errors leading to non-settlement of trades was thus quite high. A high probability of trade failure constitutes an important risk in trading, since a failed trade has several direct as well as opportunity costs for the investor, especially when security prices fluctuate.[9] Manual intervention and re-feeding of data is quite common in these trade cycles. Errors and thus lack of quick settlement occurs because of several reasons – such as a) lack of information with counter-party, b) counter-party short of shares, c) counter-party has not delivered shares, d) clearing broker or principal broker information or broker commission incorrect and so on.

Several observers considered Indian securities market, even in 2001, as risk prone, particularly in post-trade stage of the trade cycle. A survey from a global custodian quoted in the Report of the SEBI Committee on STP (SEBI 2002, 2004) suggested that on-time settlement of trade (within the settlement period) was as low as 63.5%, compared to an average of 87% in emerging markets and close to 90% in major markets, even when the market was operating at longer settlement windows (SEBI, 2002). Adoption of STP does away with multiple manual feeding of the same data point and allows automated information flow, information matching/reconciliation (as between orders provided and executed, or information from different entities such as from broker and clearing institution that needs to be matched by the custodian), reducing the role of manual intervention and hence the chances of errors as well.

Manual intervention is still required in post-trade settlement processing, but only in a few cases of errors, where the automated settlement system flags off a mismatch of any particular field of information. Such manual processing is restricted to only a small share of the total transaction volumes and the costs of such 'exception processing' does not bear down heavily on total back-office processing costs. The ability of the inter-connected settlement system to handle explosions in volumes of trade crucially depends on making the manpower requirement in post trade cycle independent of trade volume.

Thus, as the NSE ecosystem attempted creating a new kind of market infrastructure and institution under the regulatory support of SEBI, demand for a host of new services in the IT space in financial services got generated. The NSE initiative succeeded remarkably. The success of NSE in increasing trading volumes and drawing liquidity to its exchanges enabled SEBI to move ahead with several regulatory changes (such as compulsory demutualization of exchange, reduction in settlement cycle time, introduction of STP) as a working NSE provided the regulator with enough clout to take on the 'old broker lobby' around BSE. It was within this milieu that FT operated – building technologies that would enable transactions in an alternate financial services space that was unfolding over the 1990s.

PRODUCT DEVELOPMENT AT FT: UNFOLDING OF AN ENVELOPE

ODIN – the front-end product suite for brokerage houses was the first product launched by Financial Technologies in 1998, when it went live on NSE connecting the broker trading workstation to the NSE trading engine. In a few years BSE, Australian Stock Exchange, and Singapore Stock Exchange certified ODIN as well. Closely following the ODIN launch was an enhancement to the broker suite through a back-office software product – Match – that performed back office settlement, accounting, and risk management functions of brokerage house. In 2000, NSE opened its systems for on-line direct internet based trading following SEBI Guidelines. ICICI Direct and Investmart were then in a race to be the first to begin internet trading. Financial Technologies won the order to provide technology to ICICI Direct ahead of IBM who was the other powerful contender for the contract. ICICI Direct went live first, with technology from FT – providing FT with its first breakthrough with a large institutional client. Soon, several other internet-based securities trading enterprises, such as Sharekhan, became FT clients.

A focus on improving ODIN, a front-end broker solution, led the FT technology group to inter-work with several developments in related areas to seek inter-operability and inter-connection following the evolving regulatory/technology standards. The value of ODIN to the broker lies in the connectivity that it can provide to multiple market sites across asset classes. FT thus got involved with STP technology and standard development processes both in India and globally. From early 2000, it participated closely with the SEBI STP Committee,[10] which was developing the roadmap and standards for moving the securities trading ecosystem towards adoption of STP technologies. It developed STP solutions (STP-Gate) in tandem and was amongst the first to provide a solution to the emerging space. In 2003, it became the first STP provider to obtain SEBI approval as an empanelled STP technology vendor. It created an STP Market Advisory Group – to bring together industry participants on STP technology and with SEBI organized several workshops on Shorter Settlement Cycles (T+1).

Automated exchange of financial transaction information and its reconciliation, required standardization of the messaging formats. Standard definition of fields carrying information regarding a trade transaction was required. SEBI chose ISO 15022 as the messaging standard for the Indian securities market (Refer to SEBI (2004)). As the SEBI STP Committee noted, there was during that time a lack of globally convergent standard of financial information exchange protocols. Automation of financial information exchange had evolved globally through coordination in microcosms rather than the whole ecosystem. Coordination efforts were, moreover, national – with significant inter-country differences. Therefore, interface definitions were different between different stages of the trade cycle or between trades in different asset classes such as debt and equity. Standards depended on what kinds of organization coordinated the standard setting process.

The Financial Information eXchange ("FIX") Protocol, for instance, was initiated in 1992, as a bilateral communications framework for equity trading between Fidelity Investments and Salomon Brothers. It was a series of messaging specifications for the electronic communication of trade-related messages. It has since evolved to become the de-facto messaging standard for pre-trade and trade communication globally within the Equity markets, and is now experiencing rapid expansion into the post-trade space, supporting Straight-Through-Processing (STP) from Indication-of-Interest (IOI) to Allocations and Confirmations. From this foundation, the protocol is also seeking expansion into Foreign Exchange, Fixed Income and Derivative markets. FIX did not extend into post-trade stage, however. (www.fixprotocol.org)

In contrast, Omgeo was formed in 2001, as a joint venture between DTCC (Depository Trust and Clearing Corporation) and Thomson Reuters to provide certainty in post-trade operations through the automation and timely confirmation of the economic details of trades executed between investment managers and broker dealers. Currently, Omgeo enables an efficient community of more than 6,000 financial services clients in 45 countries to manage matching and exception handling of trade allocations, confirmations, and settlement instructions. Omgeo has also extended its trade lifecycle coverage to include counterparty risk management, which supports end-to-end collateralization and reconciliation across multiple asset classes.

FT invested simultaneously in working with different global standardization initiatives – as an Asia (regional) partner of OMGEO and a member of the FIX protocol group. Three products from FT group had obtained FIX certification (network product FT-Net, applications ODIN Institutional and FIX Connect, which is a bridge technology[11]) by 2009. FT had also become a certified member of Oasys Global (an OMGEO platform) offering global connectivity to post-trade surveillance/monitoring systems of OMGEO. These interconnections enabled FT to provide a solution that would inter-connect seamlessly between Indian and foreign securities transactions, where buy and sell sides are located across borders.

Table 1. ODIN installations (Source: Various annual reports of financial technologies)

	FY04-05	FY05-06	FY06-07	FY07-08
Brokerage clients	400	650	791	862
Number of installations	33,000	86,815	164,000	320,000

It was also an early participant in the Keystone Alliance (with Microsoft, Intel and HP) in the financial services segment, which dealt with technology standards related to the hardware used in systems deployed in the financial markets, especially given the challenges of high-density trade-flows under real-time environments in an STP regime and under the alliance became one of the two independent software vendors to successfully develop a roadmap (blueprint_ towards STP conversion.[12] By 2008, it was the undisputed leader in STP service space – with 175 institutional members, more than 375 trading members, all custodians, and a few Foreign Institutional Investors subscribing to its fully hosted STP-Gate service in India. It has virtually created the de-facto STP standard in India, compatible with ISO standards and fully inter-workable with two of the other leading global standards.

ODIN installations increased steadily. In 2003, commodity futures exchanges were launched in India and trading in the new asset class generated demand for trading software from commodity trading houses. ODIN installations expanded from 7500 in 2002-03 to 15,000 licenses in 2003-2004, after FT launched ODIN Commodity that could work with both Multi-Commodity Exchange (MCX) and National Commodity Derivatives Exchange of India (NCDEX) – the two commodity derivatives exchanges in the country. In 2004 itself, ODIN had a penetration in more than 100 cities across the country; its geographical reach has increased further over the last few years with an explosive growth in installations.

Kotak Securities, one of the larger brokerage houses and an early customer of ODIN would have offices spread across more than 300 cities, with more than 80 trading stations and over 300,000 customers. (Information from website at www.kotaksecurities.com)

Growth in ODIN installations was driven primarily by growth of brokerage houses. From 2002, however, FT also concentrated on retail customer sales, apart from institutional selling through brokerage houses and achieved considerable growth on the retail account as well. Table 1 below provides figures on the growth of ODIN installed base. ODIN has continuously been enhanced with addition of new functionalities and modules. In FY07-08 for instance, FT released version 9.0 of the software with functionalities of Margin Funding and Collateral Management. Since different categories of investor, such as domestic or foreign institutional investor, internet-based retail investor operated under different regulatory norms and compliance rules, ODIN was packaged as differentiated suite of functionalities for different client segments. ODIN continues to be the backbone of the FT product portfolio, contributing substantially to its product license revenue, making up close to two-thirds of the total operating revenues of the enterprise in FY 07-08.

Figure 2 provides a snapshot of the temporal evolution of the main products and the enterprises operated by the FT group. From its base in broker front-office software, Financial Technologies expanded into other spaces, particularly the exchange software space and eventually went on to promote several exchange subsidiaries and ecosystem ventures. Between 1995 and 2002, FT was a pure play product company in the brokerage solutions space. From 2002 to 2004, it entered the software

Figure 2. Temporal evolution of the FT group and products of Financial Technologies (Source: Investor communique, Financial Technologies, December 2008)

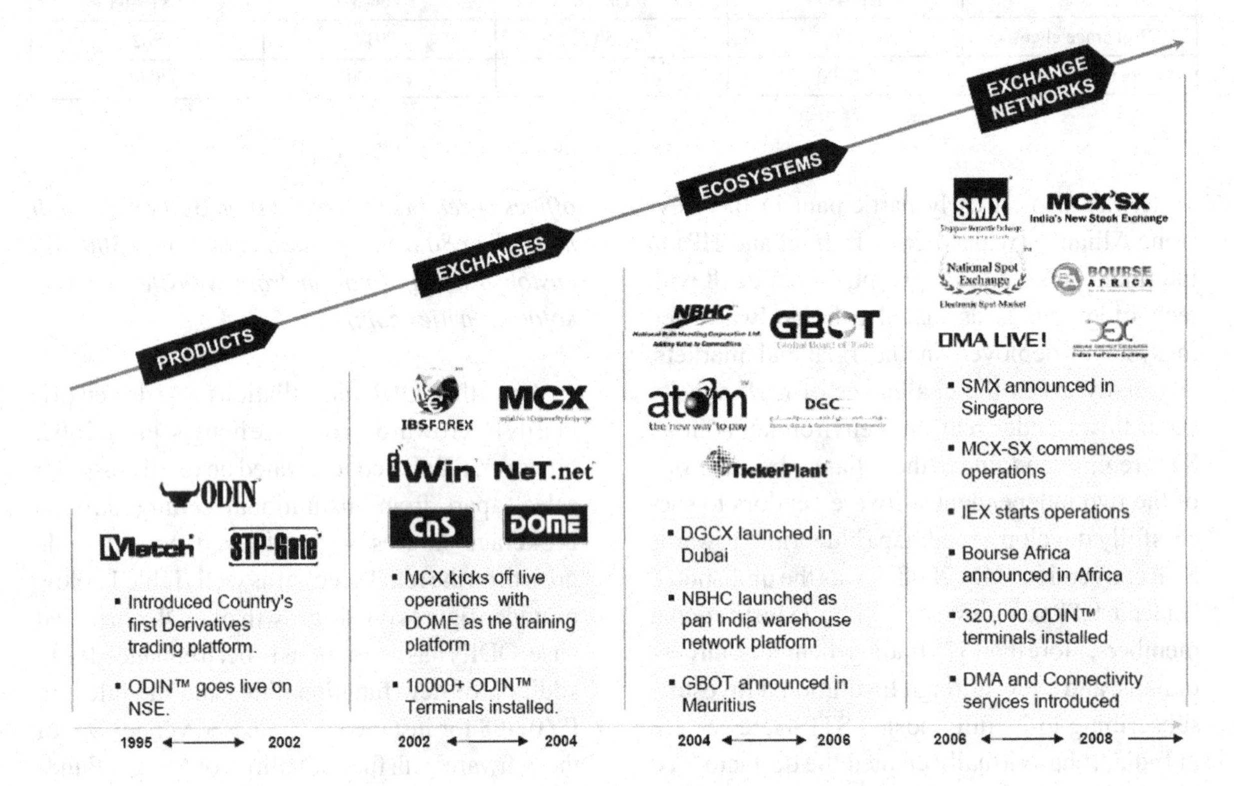

product space in exchange technology along with diversification into exchange business – a step that is akin to entrepreneurial market making for its products. From 2004 to 2006 it promoted several businesses around the exchange venture, which it called ecosystem ventures. As an instance, Tickerplant Infovending was set up to disseminate the price and other information generated in the exchange trading activities. Tickerplant is thus a vehicle to monetize information assets generated in the exchange as well as a vehicle that strengthens the exchange based trading activity by providing information broadcast services to investor clients. In that sense, with innovative information dissemination Tickerplant can improve operations of the exchange in a synergistic way. Between 2006 and 2008, FT group entered several trading markets and promoted a series of exchanges across Asia and Africa, along with moves into exchange business in trading of asset classes other than equity based securities (such as electricity). Each exchange business creates a position for FT in a specific geographical or asset class market. For instance SMX (Singapore Mercantile Exchange) creates a position in Singapore (a geographic market), while MCX-SX creates a position in a different asset class market (currency derivative market). To the extent that convergence across some of these different markets evolves through regulatory guidance, the portfolio of market positions can create a network. The moves made by FT group between 2002 and 2008 are a clear departure from accounts of growth of its global peer group in the e-finance space. *FT group, in terms of the portfolio of assets that is controls, does not have any global comparator.*

Entry into Exchange Software

In 2002, FT launched their first exchange software through a subsidiary IBS-Forex, jointly promoted with foreign exchange broking houses. It provided the first indigenously developed forex-trading exchange software (FX-Direct) for the inter-bank market, contesting the monopoly of Reuters[13] who was the only technology as well as information (forex trade information) provider in the inter-bank forex trading space then. FT attacked Reuters on pricing structure – substantially reducing the initial expense from the high levels of Reuters offering and shifting to use-based payment models. FT offering made the trading software accessible to a wider range of smaller banks and forex trading broker houses that found the Reuters offering too costly to finance as upfront investment. The rationale for two of the top forex broker houses, Kanji Pitambar & Co and FR Ratnakar & Co, to join and co-promote the venture with FT was to introduce software that would expand the forex trading market. FX-Direct had facilities for anonymous trading (and price-time based automated order matching), separate channels for Negotiated Dealing as well as special window for trading in smaller lots (US$0.25 and 0.5 million, compared to the standard lot size of US 1 million).

The big thrust in exchange software development came, however, when FT received the license to promote MCX as one of the first commodity derivative trading platforms in the country in early 2003 after government decided to reintroduce commodity futures trading under the regulatory supervision of Forward Markets Commission (FMC). DOME (or Distributed Order Management Engine) was developed by FT and was first deployed in MCX. The software went live in a record period of just nine months. Exchange software was eventually developed for Dubai Gold and Commodity Exchange (DGCX) at Dubai on a project contract and licensed out with an annual license fee arrangement. Several other exchanges promoted by FT group – Safal National Exchange (SNX), a national electronic spot market for horticultural products, National Spot Exchange Limited – a national electronic spot market for metals and other agricultural and industrial commodities work through licensed versions of DOME software suite. Figure 3 provides a mapping of the products of Financial Technologies with the trade cycle and the ecosystem linked to different stages of the trade cycle, while Table 2 provides short descriptions of different products currently offered by FT.

FT's product portfolio currently covers a wide terrain of the IT infrastructure required in a contemporary financial market; i.e. it has products for several microcosms within the trading ecosystem. Each of the software products (such as ODIN, DOME STP-Gate, FT-Net) may be seen as different asset pieces that inter-work with each other. Yet at the same time, they all can connect up to third-party systems (software) as well, as they were built based on open interface protocols. The software can be used as independent pieces as well as part of an integrated suite.

Figure 3 classifies FT products into Exchange and Brokerage solutions – depending on whether it is deployed at the broker end or the exchange that is at the center of the trading network. Each of these segments are further divided into the front and back office; front office activities generally relate to the trade transaction, while back office refers to settlement, payment and compliance systems. ODIN itself has several versions of client interface – ODIN Diet, ODIN Institutional, iWin (mobile based last mile access), ODIN TWS (Trader Work-Station) – for last mile connectivity to the investor client. Match handles broker related settlements while CnS handles settlements at the exchange level. Post trade settlement works through the STP engines that the reader would find located (in Figure 3) at several stages where communication with the broader settlement and depository system is required. Both sets of products are designed to work under Internet-based, leased line or VSAT based connectivity. Each

Figure 3. Product portfolio of FT mapped to the trade cycle of securities trading (Source: FT Annual Report, FY 2007-08, p. 41)

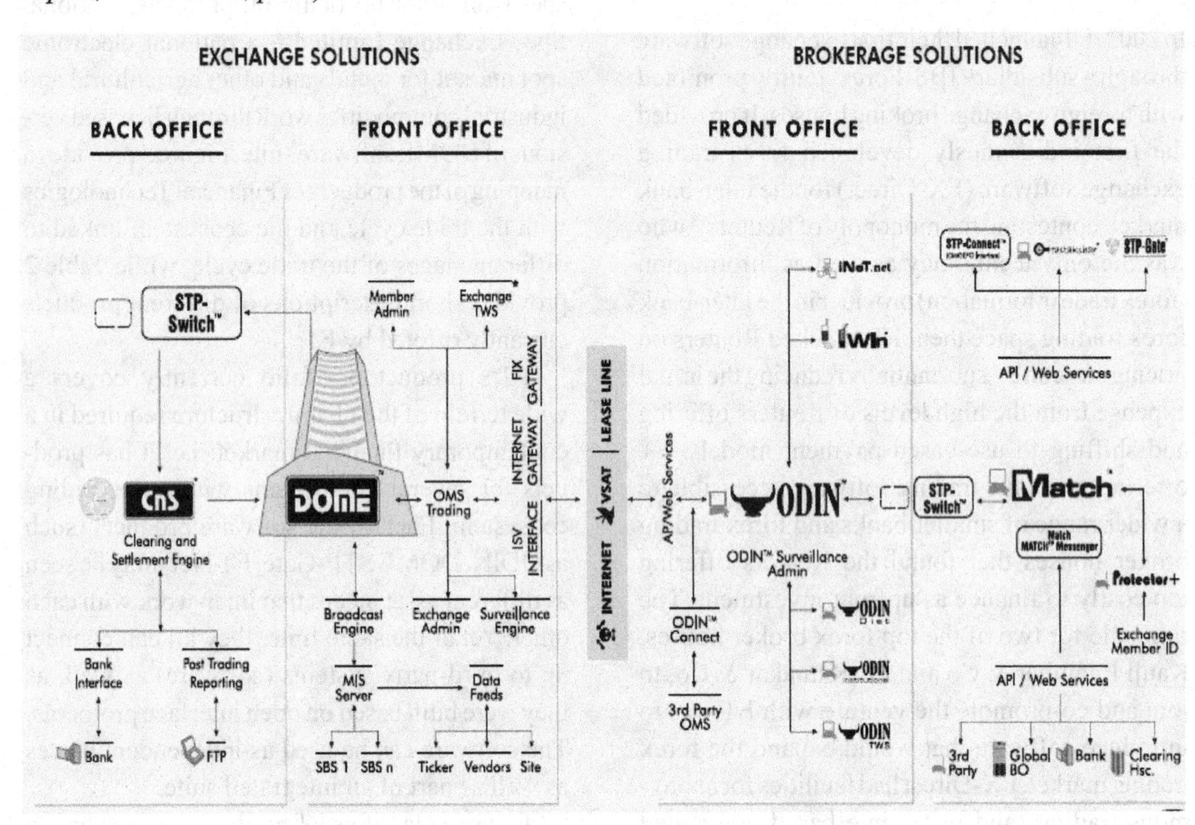

of the products can also inter-work with third-party software. For instance, a brokerage house can have order routing from third-party Order Management System and integrate it with order flows from ODIN installations by implementing an intermediate layer called ODIN Connect.

The product portfolio of FT evolved, in a different sense, along with the regulation led evolution of market niches. As regulation created the framework for novel linkages and transactions, FT always came out prepared with a software solution to address the newly arising transaction demand. Proximity to regulation allowed FT early access to the unfolding envelope of regulatory innovation. If we take, for instance, the case of mobile payment systems, FT Group Company Atom Technologies (about which a brief description is provided in later section) had a line-up of products certified for security standards of VISA, with initial customer group tie-ups almost immediately after the release of the mobile banking rules by the central bank of the country. Such rapid positioning in an evolving demand space has been a feature of its evolution all along. STP products such as STP Gate, Direct Market Access software (DMA Live) were released just as regulatory norms were announced for the new market transactions to begin or new compliance norms to be met. As the FT Group CEO argues, the new generation markets are regulation driven – regulatory innovation is coupled with product design innovation in these market places.

Table 2. Short description of a few key products of FT (Source: FT Annual Report FY -7-08, p. 41)

Exchange Solutions: A complete suite of trading and clearing solutions supporting a wide range of asset classes	
Front Office	
DOME:	Distributed Order Matching Engine. The heart of the exchange that matches buy-sell orders on a price-time priority based algorithm launched in 2003 with MCX as first client
Back Office	
CnS:	Clearing and settlement solution
FT-Prime:	Portfolio based risk management engine
FX-Direct:	Inter-bank foreign exchange dealing system
Brokerage solutions: End-to-end Straight Through Processing (STP) suite supporting high density of transactions, covering all stages of the trade life cycle	
Front Office	
ODIN	Integrated multi-exchange, multi-segment order management and trading solution available as ODIN Retail Trading Work Station and ODIN Institutional Trading Work Station
ODIN Diet	Application based internet trading front-end
Inet.net	Browser based online trading solution
iWin	Internet based real-time trading solution for mobile or handheld device launched in 2005
Protector	A post-trade, pre-acceptance online risk management solution that enables broker members to mitigate risks in equity and derivatives market
Back Office	
Match	An integrated multi-user, multi-exchange, multi-currency back office accounting and settlement system
STP-Gate	High capacity messaging solution that can manage multi-client, multi-segment transaction processing in post-trade stage of trade lifecycle
FT-Net	Fully managed and hosted network that provides FIX protocol enabled connectivity requirements between buy-side and sell-side applications. It provides Direct Market Access (DMA) to domestic and international institutional clients for cross-border trades

STRATEGY OF LINKED SUBSIDIARY PROMOTION AS MARKET MAKING

One of the key tenets of the revenue model adopted by Financial Technologies was linking the valorization of the software IP to growth of transactions in the ecosystem driven by technology/IP adoption. From its base in software technology design, it diversified into promoting exchange enterprises and other ecosystem ventures that would provide supporting services to exchanges. The core technology company would benefit financially through several distinct kinds of intra-group financial flows.

a) Success of exchanges and rise of transactions would expand the trading milieu, generating demand for brokerage solutions of FT. This would be an indirect synergy mediated through growth of the customer base of FT products.
b) Sale of exchange solutions/licenses to exchange ventures would generate a direct revenue stream. Service contracts with exchange ventures in different asset classes would also allow FT to test, design, and implement robust multi-exchange, multi-asset class trading functionality in its product suites – both in exchange solution as well as brokerage solution space, thus strengthening its core design philosophy.
c) Financial success of the exchange would lead to back-flow of financial streams as dividend payments to the core technology

Figure 4. FT asset portfolio – technology products & linked subsidiaries/associates (Source: http://www.ftindia.com/aboutus/BusinessOverview.htm)

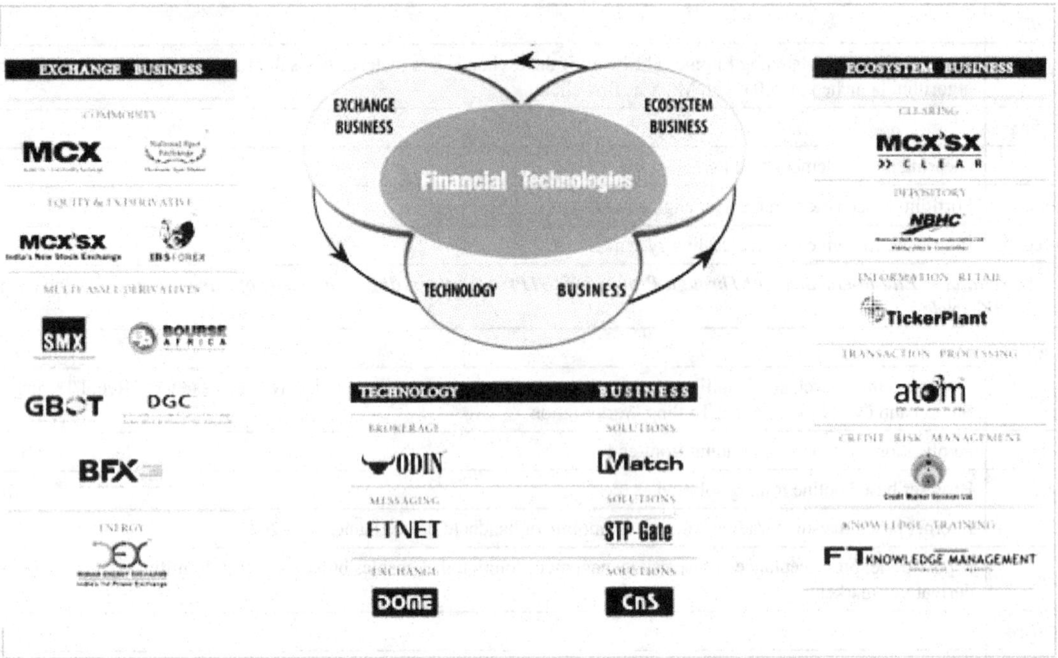

enterprise.

d) Success and growth of the exchange and ecosystem ventures can lead to attracting strategic equity investors at a later stage, to enable the core technology enterprise to make strategic partial exits, leading to one-time large pay-offs in the nature of project divestment incomes. Such equity dilutions were also required because of demutualization norms/regulations of exchanges that required a level of fragmentation of equity holding of an exchange. This financial stream would be in the nature of sudden spikes in novel profit realization, on the mode of venture financial payoffs. The core technology group can then be also seen as a venturing entity, providing incubation to numerous linked spin-offs.

This strategy was especially important for the expansion of the market for the trading/exchange software, since FT's growth was linked to the gradual emergence of the ecosystem – its gradual institutionalization as a set of inter-related and inter-connected assets, institutions, and practices amongst the ecosystem participants. This systemic nature of the technology demand required adopting a system view as a market expansion philosophy.

Figure 4 depicts the asset portfolio of FT. At the center is the parent company – the software product firm Financial Technologies. The technology business of FT has synergistic benefits with subsidiaries/associates in 'exchange' as well as 'ecosystem' businesses. Technology, exchange and ecosystem businesses feed on to each other in direct and indirect ways. Financial flows to technology business either through product sales/licenses to the other two segments represent a direct benefit, while a synergistic growth of exchange and ecosystem business that leads to growth in volume of trading and the number of trader/client software installations leads to an indirect revenue effect on technology business. Control over the

Table 3. Financials of few key subsidiaries/associates, as on March 2008 (INR million) (Source: Various Annual Reports of FT and Red Herring Prospectus filing of MCX for postponed public issue)

	Paid up Capital	Reserves and surplus	Total Assets	Total Income	Profit after tax
IBS Forex	40	(10.78)	29.4	4.15	0.55
NSEL	110	(16.71)	39.4	0.13	(52.9)
GBOT	1248.9	(152.69)	1099.1	44.87	(134.1)
SMX	852.9	(72.1)	956.7	18.7	(67.99)
Riskraft	50	(47.27)	11.47	5.12	(21.73)
Atom	75.4	(62.93)	23.64	5.64	(29.86)
NBHC	770	12.79	1141.9	590.75	29.62
Tickerplant	100	(99.81)	40.41	4.79	(74.85)
FT KM	0.5	(10.92)	10.43	8.16	(10.92)
Indian Bullion Market Ass. Ltd.	0.5	(0.87)	0.1	nil	(0.87)
Global Payment Networks	10	(11.94)	6.82	nil	(11.74)
FT ME	184.9	(95.55)	146.8	96.6	(43.7)
SMX Clearing	289.4	3.69	293.7	4.1	3.48
MCX *	390.8	2523.3	6689.4	2063.1	932.4
IEX *	0.5	(0.19)	0.5		(0.19)

* MCX figures are for 2006-07. Since MCX ceased to be a subsidiary (less than 50% stake of FTIL), MCX figures are not disclosed in Annual report of FTIL in 2007-08

portfolio of assets has created a potential for FT to benefit from convergence of markets – either in a geographical and spatial sense or in the sense of differences in asset classes. With its control over trading software at the client end (end-user point) and a portfolio of assets in different markets that have a possibility of converging, FT possibly is well positioned to benefit from convergence of markets.

INCORPORATION OF SUBSIDIARIES: A SHORT TEMPORAL PROFILE

The descriptions below provide an account of incorporation of subsidiaries and the investments made by FT group in subsidiary companies.

- Financial Technologies' expansion through linked-subsidiaries and joint ventures started in 2003, when MCX was promoted by FT as a 100% subsidiary with a subscription of INR 150 million to its equity capital in lieu of technology support provided to MCX.
- In 2004-05, two more subsidiaries were set up – Tickerplant Infovending in the financial information retailing space and IBS Forex in the exchange space for foreign exchange trading transactions.
- In 2005-06, the joint-venture derivatives exchange DGCX (with the Dubai Government) and five new subsidiaries were incorporated. Total equity investments during the year in all subsidiaries were a little above INR 300 million.
    - Atom Technologies (for developing mobile payment systems),

  - Riskraft Consulting,
  - National Spot Exchange Limited (NSEL) for developing commodity spot markets that could link up to the derivative markets,
  - National Bulk Handling Corporation (NBHC) for providing post-harvest warehousing and collateral risk management services,
  - FT Middle East DMCC (FTME) for growing the Middle East market for FT products.
  - It also invested another INR 100 million in MCX, enhancing its equity base.
- It was followed by an even more aggressive expansion in 2006-07, when nine more subsidiaries were set up. Three of them were exchange ventures:
  - Indian Electricity Exchange (IEX)
  - Singapore Mercantile Exchange (SMX)
- SMX Clearing Corporation (subsidiary of SMX) as an ecosystem venture of SMX
  - Global Board of Trade (GBOT), a commodity derivatives exchange in Mauritius
- In 2007-08, four more subsidiaries were created, important among which was:
  - FT Knowledge Management Company, set up for handling the ecosystem training requirements,
  - Indian Bullion Market Association – set up as subsidiary of NSEL to grow the spot trading of gold.
  - In 2008-09, four more subsidiaries were incorporated –
- MCX-SX (subsidiary of MCX) as an exchange for trading in currency derivatives and MCX-SX Clearing Corporation, for undertaking the clearing function
- Bourse Africa in Botswana (Spot and Derivative multi-asset Exchange dealing in commodity, currency, bonds and diamonds for the Pan-African region)
- BFX (Bahrain Financial Exchange) for trading of financial derivatives.

Few of the initiatives, particularly in Africa, materialized through support of the local governments, who invited FT group after the success of MCX venture in successfully restarting commodity derivative trading in India.

Financial Technologies supported the capitalization process of the subsidiaries through subscription to its equity. Such investments stood at approximately INR 790 million, INR 2380 million, INR 300 million, INR 28 million and INR 150 million in FY07-08, FY06-07, FY05-06, FY04-05, FY03-04 respectively. Cumulative equity investments in subsidiaries and associate companies stood at around INR 3700 million in FY07-08.

As Table 4 suggests, in each of the ventures, FT group brought in outside involvement, either through equity participation as strategic investor and/or through hiring of key management people from the ecosystem either as CEOs or as member of Board of Directors and Advisory Boards. Senior people from banking were brought in, for instance, to lead the initiatives at MCX-SX and NHBC, which would deal with currency derivatives and post harvest collateral and warehousing risk management for agricultural commodities respectively. Both ventures would require partnerships and linkages with the banking system as the key client/alliance group. A brief description of the business of NBHC would clarify issues a bit.

NBHC started off with the mandate to connect established agricultural warehouses in a virtual information network, so that it could provide risk management services to banks and financial institutions to enable them to undertake post-harvest financing – where bank involvement has been very low in India. Coupled with quality standard implementation on warehouse stocks,

Table 4. External stakeholders and relations with FT of a few key subsidiaries (Source: Compilation from respective websites)

Name	FT Stake	Started in	Strategic investor	Key leaders
IBS Forex	76%	2002-03	Forex brokerage house	CEO, Ganesh Rao, worked with SBI as a forex trader
NSEL	99.99%	2005-06	-	CEO, Anjani Sinha, former MD of BSE and Ahmedabad Stock Exchange
GBOT	100%	2006-07	-	Deputy MD, Joseph Bosco, former MD of OTCEI Stock Exchange
SMX	100%	2006-07	-	Ang See Twan as Chairman, former President of SGX, Thomas J.McMohan as CEO,
Atom	87%	2005-06	-	Acting CEO, Dewang Neralla
NBHC	86.5%	2005-06	Started in 2005, strategic investors took stake in 2007	CEO, Anil Chaudhury, formerly with SBI, was deputed from the bank to MCX after SBI took stake
MCX	37%	2003-04	NYSE-Euronext and several banks	Joseph Massey, MD & CEO (former MD of Interconnected Stock Exchange of India, associated with Vadodara Stock Exchange) Lamon Rutten, Deputy MD (Former Chief of Finance, Risk Management, and Information in the Commodities Branch of UNCTAD, Geneva)
DGCX	39%	2005-06	DMCC (Dubai Govt. entity)	CEO Malcolm Wall Morris, former head of business development at LIFFE, derivative arm of NYSE-Euronext
India Energy Exchange	44%	2006-07	Power Trading Corporation and several others	CEO, Jayant Deo, former member of Maharashtra State Electricity Regulatory Commission
MCX-SX	51%	2008-09	MCX	CEO, U.Venkataraman, former head of Treasury at IDBI Bank

certification systems of warehouse practices, NBHC's information network could possibly provide banks the ability to handle collateral risk if they lent on the basis of warehouse receipts. Regulatory restrictions had prevented banks in India from developing capability in managing post harvest credit risk; yet post harvest credit flow can happen only if banks can have access to a risk management system. Flow of post harvest credit can potentially unbundle the crop output market from post harvest credit market, releasing the growth of spot and futures trading market in agri-commodities. FT ecosystem ventures can benefit from such growth. Since NBHC had to work very closely with banks to create the working model of risk and information sharing, a former public sector banker was appointed as the CEO to take the venture forward. (website of NBHC at www.nbhcindia.com)

FT group initiatives also actively sought collaboration of the hinterland trade association – of traders in agricultural and metal commodities in the metropolis as well as in the hinterlands (up-country traders) by entering into Memorandum of Understanding with such bodies to expand the transaction space. Product design – both the software solutions, as well as contracts traded in the exchanges also reflected concerns of the trader/ user groups that were sought through interactions with the trade bodies. The 'old trader groups', as repositories of product knowledge on commodities, were thus sought to be connected to the 'new generation market' spaces, which adopted

*Figure 5. Growth in Income from Sales, Financial Technologies, FY04-FY09 * (Source: Annual Reports of Financial Technologies)*

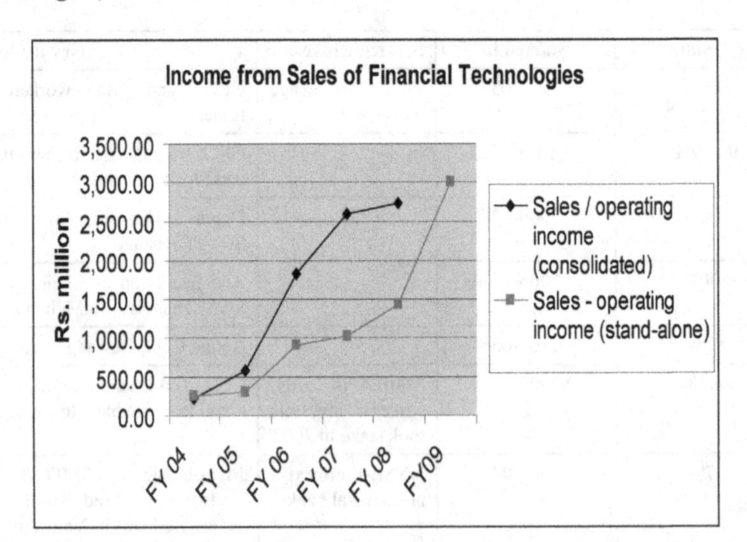

* FY09 annual figure of stand-alone sales is based on unaudited nine-month result, extrapolated pro-rata; consolidated results are not declared in quarterly investor releases

new technology and operated on different risk sharing principles.

FINANCIALS OF THE ENTERPRISE

Appendices 1 and 2 provides the financial details of Financial Technology, both as a stand-alone entity as well as after consolidation of the financial results of its subsidiaries and joint ventures. The financial results report two streams of revenue – an operating income that arises from the licensing revenues generated through royalties, transaction based payments from the licensees, and software service revenues, while a second category of earnings arise from one-time project divestment income. While the first stream is a continuous stream that would have a smooth growth depending on growth of the licensing market and number of software installations, the second stream is venture capital like one-time gains that would generate periodic, often unpredictable spikes after the success of risky venturing and incubation process.

Table 5. MCX Financial Results (in INR million) (Source: Various issues of FT Annual Report

	2004-05	2005-06	2006-07
Paid up Capital	215.1	389.6	390.7
Reserves	52.5	2567.8	2523.3
Total Assets	1257.9	6498.6	6689.4
Turnover	336.54	1043.9	2063.1
Profit after tax	111.2	458.9	932.4
Proposed dividend	21.5	77.9	-

Note: MCX financial figures for 2007-08 onwards are not available, since it is no longer a subsidiary of FTIL. Trading volume in MCX exchange, however, increased by 37% in 2007-08 and by over 50% in 2008-09 over previous financial years. Revenues of the exchange, which is a proportion of the trading volumes, therefore, also would have increased at least proportionately.

Unlocking value through equity stake sale and partial exits from promoted ventures has been realized only in the case of MCX and DGCX (the exchange venture promoted in Dubai). Table 5 below provides the financials of MCX, while the details of the transactions related to divestments are noted below.

a) One percent stake sale in DGCX in 2006-07 for US$ 12.5 million, reducing Financial Technologies' stake to 49%.
b) Stake sale in MCX and DGCX in 2007-08 yielded a profit (after meeting directly attributable expenses for the sale) of INR 11163.89 million, of which INR 10445.58 came from the MCX stake sale reducing FT's stake from 64% to 37% in MCX. On the P&L Account, this has been recorded as Project Divestment Income. NYSE Euronext, Citigroup, Merrill Lynch, ICICI Ventures, IL&FS, Kotak group brought in as strategic investors in MCX. The investors were partners and were often clients within the trading ecosystem. Financial Technologies' equity investment in MCX was around INR 250 million – INR 150 million in 2003-04 (in exchange for technology support provided by FT) and another INR 100 million in 2005-06. The venture pay-off was substantial.

The exits led to a substantial spike in FY07-08 revenues. Given the dictum within the group of commercializing ventures in rough cycles of 1000 days, several other ventures might be close to possibilities of yielding venture profits within a short time frame from now.

Apart from the project divestment income, regular operating income was also driven by direct revenue flows back to the core technology enterprise, either through software service contracts with the subsidiary, or through annual maintenance and other product license revenues (of exchange software). Dividend income from the successful subsidiaries also constituted a substantial income after the subsidiary achieves commercial success. Dividend income from subsidiary, as shown in Table 6 below, was particularly high from MCX. Project based service revenues are mostly driven by subsidiary contracts, where specific technology is developed on a project mode. Such revenues constituted a large share of total revenues when exchanges such as MCX were being set up. The license based product revenue, however, is not directly driven through inter-company sales. Product revenues depend on expansion in installation base of ODIN, STP-Gate usage (which is a hosted service based on a product).

Table 6. Standalone sales break-up of Financial Technologies (in million INR) (Source: Annual Report 2007-08, p.120)

Sales	FY07-08	FY06-07	FY05-06
Products (IPR based-License)	939.52	711.5	542.3
Services (Project based)	433.01	282.44	356.9
Sale of Goods (trading of software and hardware for connectivity)	55.99	28.95	-
Sub-Total	1431.52	1022.89	899.24
Less: Excise	55.59	34.55	-
Total	1375.56	988.34	-

COMPETITION IN THE EXCHANGE SPACE AND LOOKING AHEAD

Financial Technology as a technology vendor was providing one asset within the financial trading ecosystem, enabling the transactions and interconnecting actors in the transaction network. Its' entry into the exchange business space has, however, put it in a new league. As a software product company, it was a departure from more familiar paths trodden by its global peer group. Several observers doubted its ability when it applied for a license to set up a commodity derivative exchange when expression of interest was invited by the government for setting up futures trading after a long hiatus in 2003. It had virtually no experience running exchanges. The other applicants were entrenched players in the exchange business, including NSE. When it got the license eventually, it was the first to go live in a record time, and over the years has captured bulk (85% by 2008 figures) of the commodity

Table 7. Segment reporting Financial Technologies (consolidated) FY07-08, INR million (Source: Annual Report, FT, 2007-08, p. 105)

	STP Technology Solutions	Exchange based Revenue	Storage & allied services	Others	Elimination	Total
External Revenue	1154.46 (770.38)	962.99* (1718.08)	413.2 (97.67)	194.6 (15.4)		2725.25 (2601.53)
Less: Excise duty	55.95 (34.55)	-	-	-		55.95 (34.55)
Net External Revenue	1098.5 (735.84)	962.99 (1718.08)	413.2 (97.67)	194.6 (15.4)		2669.3 (2566.99)
Inter-segment revenue	249.13 (235.75)	1.49 (-)	22 (18)	39.58 (27.75)	312.2 (281.49)	-
Net sales/income from operations	1347.64 (971.58)	964.48 (1718.08)	435.2 (115.67)	234.19 (43.15)	312.2 (281.49)	2669.3 (2566.99)
Segment result	484.99 (374.99)	221.96 (1130.33)	-23.17 (12.62)	4.35 (35.58)		688.14 (1457.11)

Note: Figures in braces are last year's figures

* Exchange based revenue shows a decline in consolidated accounts since MCX ceased to be a subsidiary of FT since October 2007

Table 8. Inter-company transactions within the group (in INR million) (Source: FT Annual Reports, various years)

	FY07-08	FY06-07	FY05-06
Sales – services (project based)	373.1	204.3	316.4
Sales – Products (IPR based license)	61.5	46.3	34.5
Sale of traded goods	50.9	29	
Dividend received from MCX	125.25	601.2	15.1
Of which a few main transactions were:			
Sales – Products (IPR based license) - MCX - FTME	21.42 35	25.49 20.78	42.67 20.79
Sales – services (project based) - MCX - IEX - DGCX - NBHC - SNX	217.4 12.14 - 50 45.3	180.7 - 11.6 - -	136 - 179.35* - -
Sale of traded goods - MCX - Tickerplant - NSEAP - SNX	16.03 0.4 7.68 26.16	22.2 3.5 - 2.9	- - - -
Other Income - IEX	7.2	-	

* MCX received INR 45.8 million as 'content development fee' from DGCX during FY05-06 for providing technical support in exchange technology

derivative trading volume leaving NSE promoted NCDEX (National; Commodity and Derivative Exchange of India) a distant second. Exchange businesses provide central pools of liquidity and hence it experiences strong increasing returns to the first successful entrant – it is a winner takes all game – at least in a particular asset class. IEX promoted for electricity trading by FT group captured within a short span, after operations began in October 2008, almost 90% of the traded volume, while currency futures trading volumes appeared equally split between NSE and MCX-SX (as in March 2009). FT group, moreover, has grown financially as well – its 2008 turnover was higher than the turnover of NSE Group. It had already set up a slew of inter-connected global exchanges; in contrast NSE group operates only in equity and debt securities and in commodities (through NCDEX) in India.

NSE retaliated in 2008, revoking the empanelment of FT software – ODIN – as approved front-end software for NSE brokers, citing quality and technical problems. FT responded through litigation in the courts appealing against the move and the court has ordered independent third party audit of the software. NCDEX, the subsidiary of NSE in commodity trading has earlier attempted to attack MCX, by reducing transaction fees during evening hours for trade in gold contracts. Since gold futures are heavily traded only during evening hours and MCX volume was driven largely by trading on gold futures, the move was predatory. The commodities market regulator, FMC, disallowed the move on appeal by MCX.

FT meanwhile applied for listings rights, as a stock exchange, for its exchange subsidiary MCX-SX, which currently trades in currency futures. SEBI, the stock exchange regulator, was considering the application. In an interview in Mint (quoted at the beginning of the case), Jignesh Shah had argued that financial markets and banking institutions in India have remained elitist, with large sections of enterprising population being cut off from access to growth capital and risk finance. The lack of financial inclusion and under-financing of SMEs have been key themes running across several Government of India reports in the last two years. What would be the nature of the market for risk financing for SMEs? Would an electronic market designed for high transaction density in centralized liquidity pools work for trading of stocks of SMEs? The NSE model has to remain restricted to large well-known stocks – it is essential to maintaining liquidity. Small stocks will not have large trading volumes. Globally too, the limits of efficiency arguments of high-density liquidity driven anonymous markets are being understood. NASDAQ, the market of innovative small start-ups today does not attract small cap stocks anymore. As the exchange moved away from dealer-quoted systems to order-driven automated price matching, dealers lost incentives to specialize in small stocks and migrated to trading of large-cap stocks. (Ingebretsen, 2002) Alternative market sites have emerged, such as the Alternative Investment Market (AIM) in London, to attempt to address the gap for small firms.

As G.N. Bajpai, former Chairman of SEBI and currently on the Advisory Board of FT argues:

..providing risk finance to SMEs would require creation of a whole market infrastructure – market makers, risk bearers, analysts, trading exchanges with the incentive to specialize in these small cap stocks, which 'the few investment bankers' today are not interested in – all such asset holders can stand together as an ecosystem. An exchange never succeeds on its own (FT Annual Report, FY07-08).

Would that ecosystem be very different in its design, its technology, its regulatory underpinnings and modes of coordination required to generate the market *infrastructure*? FT, with a strategic link between technology based revenues and transaction explosion in markets using its technology, probably has a strong drive to search for the new design. If this market gets going, it would possibly be an explosion, an even more radical innovation than what FT has experienced so far.

Over the years, I have learned that the 'Value' is in the journey. And I believe the journey has just begun (Jignesh Shah, Financial Technologies Annual Report 2007-2008).

ACKNOWLEDGMENT

The author wishes to acknowledge the support of Dr Parthasarathi Banerjee of NISTADS, Delhi in hosting me as a Visiting Scholar during the period when I worked on the manuscript. Discussions with Dr Banerjee and with Dr Nilanjan Ghosh of FT Group over long periods provided theoretical insights that guided the case narrative. Dr R.K.Kakani of XLRI also provided valuable insights from his deep familiarity with the securities trading ecosystem in India. Comments from anonymous reviewers also helped in bringing clarity of thought and its articulation.

REFERENCES

Bachelard, G. (1971). On Poetic Imagination and Reverie (C.Gaudin, Trans.). Dallas, TX: Spring.

Banerjee, P. (2008). Biomedical Innovation in India: With a comparison to China and others. New Delhi, India: Har Anand.

Ingebretsen, M. (2002). NASDAQ: A History of the Market that changed the world. USA: Forum, Random House.

Janakiraman. R. (1992). *Interim report of the Janakiraman Committee to probe into the securities scam*. Government of India.

JPC. (1992). Report of the Joint Parliamentary Committee to probe into the securities scam of 1992. Government of India.

Lucas, H. C. Jr, Wonseok, O., Bruce, W., & Weber, B. W. (2009). The defensive use of IT in a newly vulnerable market: The New York Stock Exchange, 1980–2007. *The Journal of Strategic Information Systems*, *18*, 3–15. doi:10.1016/j.jsis.2009.01.003

North, D. C. (1990). Institutions, Institutional Change and Economic Performance. Cambridge University Press.

Patil, R. H. (2006). Current state of the Indian capital market. *Economic and Political Weekly*, *41*(11), 1001–1010.

SEBI. (2002, May 31). *Approach paper on implementation of STP in Indian markets*. SEBI. Retrieved from www.sebi.gov.in.

SEBI. (2004). *Revised scheme for implementing Straight-Through-Processing in India.* SEBI. Retrieved from www.sebi.gov.in.

Weber, B. W. (1999). Next-generation trading in futures markets: A comparison of open outcry and order matching systems. *Journal of Management Information Systems*, *16*(2), 29–45.

ENDNOTES

[1] Mint is a new Financial Daily published from Mumbai. Mint has an alliance with The Wall Street Journal.

[2] INR refers to the Indian currency 'Indian Rupees'. On December 30, 2008, 1 US$ traded for approx. INR 49.7 in inter-bank market.

[3] Jignesh came from a traditional Gujarati trading family. His father was an iron and steel merchant/trader who moved from Gujarat to Mumbai in search of business, gradually establishing a small-scale trading outfit in Mumbai. Gujarati traders also dominate the trading of securities and commodities in India, especially the trading in hubs in the Maharashtra-Gujarat corridor in the Western part of the country.

[4] NASSCOM is the association of software enterprises in India

[5] Most of the early history is reconstructed from two interviews of Jignesh Shah – 'The amazing story of Jignesh Shah and MCX' in Business Standard, October 13, 2005 and 'The serial 2.0 entrepreneur' published in SiliconIndia, November 2, 2006 accessed

from archives of the two publications

6 Archive based biographical account of Premchand Roychand can be found in *'Premchand Roychand – His Life and Times'*, written by Sharada Dwivedi in 2006. Rudimentary descriptions are also available at www.premchand-group.com, the website of the brokerage-industrial group founded by Roychand in mid-nineteenth century. An authentic historical account of the Kalba Devi Cotton Exchange can be found in *'Saga of The Cotton Exchange'*, Madhoo Pavaskar, Bombay Popular Prakashan, 1985

7 A high powered committee led by GS Patel on capital market reforms made several suggestions for changes in 1985, while another committee led by Pherwani recommended opening of new exchanges under different rules of operation and modes of governance. Both attempts were severely opposed by the broker groups. (quoted in Patil, 2006)

8 According to reports of committees enquiring into the stock market scam of 1991-92, popularly referred to as Harshad Mehta scam, poor settlement and surveillance processes at the exchanges enabled brokers to use funds diverted from the banking system, to speculate in stocks. Use of a trading institution called *'badla'*, which enabled rollover of stock trades without settlement, was blamed as well. *Badla* was banned after the scam and efforts were made from then on to bring in clearer separation of spot and futures transactions, with stronger settlement processes for spot trades. The problem was framed as emanating from the *bundling* at the broker point and a stance of *unbundling*, along several dimensions that we discuss later, was thus called for. (See Janakiraman, 1992 and JPC, 1992 for details)

9 One reason for the retention of manual process of handling contract notes was the lack of legal recognition for digital signatures on electronic contract notes in India. Electronic notes were legalized in 2000 through SEBI directive, under provisions of the Information Technology Act, 2000. Adoption of electronic contract notes in trade practice, however, took much longer with SEBI clarifications and notices coming out till end of 2004 to remove various legal/regulatory confusions. It also required changes in stock exchange bye-laws recognizing electronic contract notes. (See SEBI Circulars dated December 15, 2000, April 29, 2003, February 3 and February 25, 2004 all available at www.sebi.gov.in)

10 STP Committee of SEBI submitted its draft report in 2002 and from mid-2004 STP was mandatory for all institutional trades – one important and large segment of the securities market.

11 See http//www.fixprotocol.org/vendors/5942

12 See http://www.datamicron.com/Company/about.aspx for a brief description of the alliance

13 Reuters was the first provider of technology to the nascent inter-bank forex trading market in India in 1999.

APPENDIX A (TABLE 9)

Table 9. Consolidated Income Statement of Financial Technologies India Limited (In INR Million; Financial Year April-March) (Source: http://www.ftindia.com/investors/FinancialSnapshot/ProfitandLoss.htm, accessed on 20.3.09)

	FY 08	FY 07	FY 06	FY 05	FY 04
INCOME					
Sales / operating income	2,725.25	2,601.53	1,820.47	574.03	215.55
Less: Excise duty	55.95	34.55	0.62	-	-
	2,669.30	2,566.99	1,819.86	574.03	215.55
Other income	11,525.82	605.86	192.45	53.41	15.19
	14,195.51	**3,172.85**	**2,012.31**	**627.44**	**230.74**
EXPENDITURE					
Operating and other expenses	2,396.29	1,538.62	773.64	290.22	155.28
Interest	123.28	8.46	0.96	0.10	1.96
Depreciation / amortization	97.47	88.88	55.84	32.24	23.05
	2,646.55	**1,636.43**	**830.68**	**322.56**	**180.29**
Profit before tax	**11,548.96**	**1,536.42**	**1,181.63**	**304.88**	**50.45**
Profit after tax	**8,608.73**	**1,035.50**	**838.81**	**225.80**	**46.45**
Net profit for the year	**8,694.60**	**714.28**	**649.90**	**196.45**	**46.45**
Earning per share					
Basic (before exceptional / non recurring items)	193.70	15.85	15.14	4.77	1.23
Diluted (before exceptional / non recurring items)	187.66	13.77	15.01	4.77	1.23
Face Value per share	2/-	2/-	2/-	2/-	2/-

Note:
* FY08 Total Income includes Project Divestment Income, which was substantial
* Consolidated Income consists of revenue from Technology, Exchange and Ecosystem business, project divestment income (Inter company revenue netted while considering the consolidated revenue)
* Set up cost of the exchanges is charged as expenses to the P&L as and when incurred

Ratio Analysis	**FY08**	**FY07**	**FY06**	**FY05**	**FY04**
EBITDA Margin	83%	51%	62%	54%	33%
Aggregate Employee Costs / Total Income	7%	18%	14%	15%	27%
Depreciation / Total Income	1%	3%	3%	5%	10%
PBT Margin	81%	48%	59%	49%	22%
Net Profit Margin	61%	32%	42%	36%	20%
Growth Ratios					
Total Income	347%	58%	221%	172%	-
EBITDA	620%	32%	267%	347%	-
Net Profit (before exceptional items)	745%	19%	279%	386%	-

APPENDIX B (TABLE 10)

Table 10. Standalone Income Statement of Financial Technologies India Limited (In INR Million; Financial Year April-March) (Source: http://www.ftindia.com/investors/FinancialSnapshot/ProfitandLoss.htm, accessed on 20.3.09)

	FY 08	FY 07	FY 06	FY 05	FY 04
INCOME					
Sales	1,431.52	1,022.89	899.24	301.55	253.36
Less: Excise duty	55.95	34.55	0.62	-	-
	1,375.56	988.34	898.62	301.55	253.36
Other Income	12,099.21	752.98	63.95	33.54	23.19
	13,475.17	**1,741.32**	**962.57**	**335.09**	**276.55**
EXPENDITURE					
Operating and other expenses	881.09	531.24	319.86	191.34	135.14
Interest	109.31	2.78	0.03	0.00	1.42
Depreciation / amortization	23.53	15.38	11.35	9.82	8.57
	1,059.74	**572.85**	**331.24**	**201.16**	**145.13**
Profit before tax	**12,415.43**	**1,168.47**	**631.33**	**133.93**	**131.41**
Profit after tax	**9,612.52**	**989.71**	**482.35**	**99.13**	**127.41**
Earnings Per Share					
Basic (before exceptional / non recurring items)	214.15	22.48	10.02	2.26	3.38
Diluted (before exceptional / non recurring items)	208.10	20.54	10.01	2.26	3.37
Face value per share	2/-	2/-	2/-	2/-	2/-

Note:
* FY08 Total Income includes Project Divestment Income
* Standalone income mainly consists of Technology business revenue, project divestment income and other income

Ratio Analysis	**FY08**	**FY07**	**FY06**	**FY05**	**FY04**
EBITDA Margin	93%	68%	67%	43%	51%
Aggregate Employee Costs / Total Income	4%	18%	18%	23%	21%
Depreciation / Total Income	0%	1%	1%	3%	3%
PBT Margin	92%	67%	66%	40%	48%
Net Profit Margin	71%	57%	50%	30%	46%
Growth Ratios					
Total Income	674%	81%	187%	21%	89%
EBITDA	957%	85%	347%	2%	943%
PAT before extraordinary items	871%	116%	370%	(22%)	386%

APPENDIX C (TABLE 11, TABLE 12)

Table 11. Manpower and R&D expenses by Financial Technologies group (Source: Various Annual Reports of Financial Technologies)

	2000-01	2004-05	2005-06	2006-07	2007-08	2008-09 *
Group manpower	200	-	900	1451	1760	3135
R&D expense in INR million (declared on revenue account)	9.2	21.4	21.3	32.8	70.97	-

Table 12. Sources of EXTERNAL funds: Equity/Loan funds for Financial Technologies

Year	External fund infusion; amount, type
Feb-1995	IPO of 65.3 million
Mar-2004	Preferential issue to Reliance Capital; amount raised was INR 226 million, at a premium of Rs. 61 per share of face value Rs.2
Mar-2005	Preferential issue of shares to FIIs (T.Rowe Price and Goldman Sachs) at a premium of Rs. 260 with a lock-in period of 1 year; amount raised INR 752.9 million
Dec-2006	Zero coupon FCCB issued for US$ 100 million; listed in Singapore, amount on balance sheet as unsecured loan of INR 4344.2 million
Oct-2007	Global Depository Receipt issue of US$ 115 million, listed at Luxembourg; 7 GDR for each share of Rs.2, issued at US$9.88 per GDR; total amount raised is INR 4530.9 million

Chapter 16

Proving the Science: Opportunity Identification to Research Contract

Carolyn J. Fausnaugh
Florida Institute of Technology, USA

Mary Helen McCay
Florida Institute of Technology, USA

ABSTRACT

This chapter is about the process by which an inventor (a physician) secures the expertise he needs to determine if his observations and resulting patent have commercial value. It is also about the process by which the university accepts the engagement. A physician with an unproven patent contracting with a university for market and scientific research that would establish the commercial viability of his invention. It explores patterns of social networking, searching, communications, and negotiations theory to describe an inventor's quest for evidence that his invention worked. The chapter outlines the process by which the physician searched his network to find resources outside of his field of expertise that could guide his next steps in evaluating the commercial potential of his invention. In addition, it describes the information gathering and negotiation process leading to a university contract. The case illustrates that the issuance of a patent does not represent either technical proficiency or market potential for an invention.

INTRODUCTION

Much has been written about universities and technology transfer. Less has been written about universities and the processes for integration of knowledge between university specialties and the transfer of that integrated knowledge to parties outside the university. And, even less has been written about the issues in communication and contracting present when an inventor, from the community, seeks out a university in their quest to determine the commercial viability of their invention.

The term technology transfer gained prominence after the U.S. Congress passed of the Bayh-Dole Act of 1980. This Act and its subsequent amendments grant to the university the right to secure patents on discoveries made under federally funded research. The purpose of the law is to encourage the com-

DOI: 10.4018/978-1-61520-609-4.ch016

Table 1. Sample of news articles on American inventors

Headline	Date	Publication
Inventor's refrigeration system for planet shows promise, but scientists are skeptical	December 21, 2008	McClatchy – Tribune News Service, Washington
Inventor gets the wood out: pencil made from scrap tires wins space on Staples' shelves	March 9, 2009	Indianapolis Business Journal
So you want to be an inventor?: It'll take money, patience to get patents, raise capital and market ideas.	April 12, 2009	The Atlanta Journal – Constitution
Energy-Efficient Engine Technology Wins Ford Team National Inventor of the Year Award	May 28, 2009	PR Newswire. New York

mercial application of research findings in pursuit of economic growth (Remington, 2005). Thus, the term technology transfer relates to discovery and invention from within the university. The term university entrepreneurship has also been linked to these concepts (Rotheraermel, Shanti et al, 2007).

Golish, Besterfield-Sacre and Shuman (2008) studied academic and corporate technology development processes and found there was little commonality among the elements in the concept maps of these two groups. The research also found that the elements missing from the concept maps of the academic inventors were related to market identification, cost evaluation, and changing customer needs/market requirements.

The passage of Bayh-Dole ushered into the university an awareness of opportunity to not only patent, license, and create new firms based on discoveries within the university, but also to collaborate with companies as a means for decimation of expertise in the university. However, around the globe, cultural differences between the university and corporations have been identified as providing obstacles to be overcome if this desire to transfer knowledge is to be fully realized (Valentín, 2000).

In both the above scenarios, invention within the university, and the transfer of knowledge through cross-collaboration between university faculty and corporations, the entities are sufficiently large to garner the attention of academic researchers. The same is not true in the case of the individual inventor. Where the popular press publishes numerous articles reporting the insights and activities of both corporate and independent inventors (Table 1), we were not successful at finding any academic research study linking an independent inventor to a research contract with a university – the subject of this case study.

Interestingly, the same themes that are linked to the areas of technology commercialization and knowledge integration across disciplines appear in this case of a highly educated inventor seeking assistance from the university and negotiating the necessary contract for services through the university Office of Research. That is, we see the topics of communication, negotiation and commercial viability of discovery manifested in the case and its .

BACKGROUND

MDH, a highly skilled and well-known neurosurgeon has a reputation for keen observation, quick opportunity identification and an ability to easily visualize solutions. In 2003, MDH observed that his home sprinkler system washed down one air conditioner unit and not the other. The unit being washed down was in surprising good condition while a second air conditioner sitting right next to it appeared to be in poor condition. On December 23, 2003, MDH was successful in having a patent issued for an "air conditioning system including liquid washdown dispenser and related methods." The purpose of the described technology (the washdown dispenser) was to reduce the corrosion and thereby extend the life-time of air conditioning systems.

MDH's work as a medical device consultant provided him with knowledge of the patenting process, an appreciation for the necessity to seek patent protection for novel ideas before attempting to commercialize the idea, and a network of contacts that included a patent attorney. This contact enabled him to secure the patent with minimal effort. It should be noted however, that although MDH had experience with the patenting process, it was as a medical professional inventing within his area of expertise. MDH had little or no knowledge of the science of corrosion.

TU is a relatively new university, having been founded in 1958. It is an accredited, coeducational, independently controlled and supported university which combines education and research in the sciences, engineering, technology, management and related disciplines. There are currently over 4000 students enrolled with more than 3300 at the main site campus. TU houses a number of research institutes and centers that, in collaboration with academic departments, aid in the student's training and research.

Its graduate courses offer engineers the opportunity to obtain graduate degrees and keep up with the advancements in science and technology that were taking place daily.

The Research Office (RO) is the organization within TU that responds to research needs of the university. The Vice Provost of Research directs this organization and reports directly to the Provost who reports directly to the President. Housed within RO are the university research institutes and centers as well as financial and budgetary personnel who assist in preparing and assisting with grants and funded programs. Their experience with developing Intellectual Property was almost totally related to that generated within the TU community. MHD represented an opportunity for the university to extend its research expertise to members of the outside community.

Table 2.

Key Player	Title or Organization's Name
BRF	Business Researcher
CRM	Corrosion Researcher
MDH	Medical Doctor Inventor
MOU	Memorandum of Understanding
RO	Research Office
ROK	Research Office representative K
ROM	Research Office representative M
TU	Technological University
WDD	Washdown dispenser

Table 2 sets forth the key players in the project – both university organizations and people.

SETTING THE STAGE

The importance of networks of personal contacts, observation and communication skills as components of an ability to innovate can be seen throughout this case. These three facets of innovation are seldom linked in the academic literature on the success or lack of success in innovative projects.

In MDH's situation, his years of work as a surgeon and consultant have already established a network he can access. However, apparent from the case, even a person as skilled and established as MDH did not immediately know who, if anyone, within his social and professional network, had an answer to his dilemma of "what are the next steps if I am to commercialize my patented idea?" Who has the expertise I need?"

As we proceed through the case we will see numerous theoretical underpinnings to the events as they unfold. Descriptive theory from the field of communication informs us about the interplay of verbal and non-verbal transactions as the doctor and the university seek to reach agreement on the nature of their relationship, where that relationship will be hosted within the university structure, and the research outputs required. Communication moves from informal to more formal and ultimately a research contract is signed. Negotiation is taking place as the named parties in the contract are identified and a team is put together to perform the work of the contract.

Cohen (2002) observes "the reality is that whenever people interact to reach a decision or agreement, some form of negotiation is taking place. Similarly Craver (2009) states "business transactions are almost always structured through bargaining discussions." Negotiation is a skill highly dependent on communication skills and understanding of negotiating techniques. Knowledge of common negotiating techniques, how and when they are employed and possible counter techniques allows one to consciously choose which tactic to employ in a given situation and to effectively counter techniques used against them.

Craver identifies thirteen such techniques of which five are technical aspects of the ultimate agreement and eight are behavioral and counter approaches. Among the technical aspects of launching negotiations are consideration of the starting position or beginning offer, the level of demand indicated by that offer and the degree of flexibility communicated. *In a meta-analysis of studies of simulated negotiations, Guthrie and Orr (2006) found the initial opening demand or offer had significant impact on the contractual outcomes.* Indications of the authority of the negotiator to bind the party being represented are important as an indicator of the time required for offer review between verbal acceptance and signing of documents, if any. These technical and behavioral aspects of negotiation are associated with six distinct stages of a formal negotiation process: (1) preparation; (2) establishment of negotiator identities and the tone for the interaction: (3) information exchange; (4) exchange of items to be divided; (5) closing the deal; and (6) maximizing the joint returns.

In every negotiation both sides have needs and interests. Preparation for bargaining includes identification of each participant's factual, economic, political and legal issues and the goals associated with each. Goals, or desired outcomes, can be categorized as essential, important, and desirable.

The authority of negotiators at the table to legally commit to the terms becomes important because it embeds in the negotiating process the timeframes needed for approval of terms once agreed to by the bargaining parties. If the bargaining parties have legal authority to make the deal and commit to an agreement, time may be shortened. If they do not have such authority, a particularly generous offer may require enough time that the offeror reconsiders and withdraws. Thus, selecting the authority level of negotiators at the table is strategic to the negotiations.

Information gathering both before negotiations begin and during negotiations is an essential component of arriving at an agreement that is wise and considered fair by all parties. There are multiple purposes to information gathering. Among them are knowing something about the style of each person sitting at the negotiation table, being familiar with indicators of each persons beliefs about what makes a "good" negotiation. What does each party seek? Often a variety of solutions can be found that will meet the request. In negotiating business transactions there is 'pull' of wanting the work represented by the transaction, the terms to be met during the transaction, and criteria for completion of the transaction. There is a need to allocate risk among the parties and minimize risk from factors not identified and specifically negotiated.

As the information gathering phase draws to a conclusion the parties approach the "exchange of items to be divided" phase. The pattern of bargaining that follows is grounded in both the value system of the bargainers and their skill to respond to the bargaining conditions. Under distributive bargaining approaches a more competitive stance is taken with one side getting more than it would have settled for and the other getting less. In the book *Getting To Yes*, Fisher, Ury, and Patton (1991) advocate integrative bargaining where taking a cooperative stance to negotiation and seeking ways to meet the parties interests is advocated. Distributive bargaining has been described as claiming value while integrative bargaining is described as creating value.

At some point there will be agreement on a course of action. Negotiation now enters the phase of closing the deal. Here the parties focus on summarizing details, tying up loose ends and laying the groundwork for the relationship during the execution phase of the agreement.

A final stage in negotiations can occur after the parties have achieved tentative accord. Known as the cooperative stage or maximizing stage, willing parties may now emphasize cooperation and revisit earlier agreements to determine if their agreements were merely acceptable or if there are now opportunities for each party to maximize their satisfaction in some way – making the "pie larger" – achieving higher total satisfaction for all parties to the agreement. Brainstorming techniques may reveal options not previously discussed. As with all the stages of negotiation, care must be exercised not to contradict earlier posturing or misrepresentations as this may lead to a decrease in credibility and re-open negotiations on the entire accord. At the conclusion of this final stage, all specifically agreed terms should be reviewed to insure there are no misunderstandings to appear after the contract is drafted.

In well structured negotiations draft contracts will be reviewed for wording clarification, but substantial modifications of intent represent a re-opening of negotiations and may signal a backing away from the agreement.

In the stage model described by Craver (2009) distributive bargaining takes place earlier in the process and is concluded with integrative bargaining. In the model advocated by Fisher, Ury, and Patton (1991) the stance of the entire process is addressed. It is interesting to note that Craver is writing for attorneys and Fisher, Ury and Patton are writing for a more general audience based on the work of the Harvard Negotiation Project (http://www.pon.harvard.edu/)

CASE DESCRIPTION

Having successfully gained patent protection for his idea, MDH began telling other acquaintances of his innovation and the resulting patent. Among his friends was the Provost at TU.

The Provost first suggested that MDH contact a faculty member in Ocean Engineering who performed research on corrosion and antifouling coatings. Learning that attempts at contact were unsuccessful, the Provost asked CRM, a faculty member with experience in corrosion prevention, to make the contact. MDH was pleased to finally receive an e-mail providing CRM's qualifications and expressing interest in meeting to discuss how the university might assist MDH with product development and new product commercialization.

In late February 2007, MDH and CRM met in CRM's office at TU. Four items in particular were discussed. (1) CRM described some of the corrosion mechanisms that could affect air conditioners; (2) MDH expressed concern that the patent expiration date was approaching; (3) CRM recommended the participation of a business faculty; and (4) CRM suggested that since the project could be a valuable learning experience for students, it should include support for students as well as faculty.

Since time was running on the patent, CRM also suggested that an experimental apparatus (fog chamber) to accelerate corrosion be either purchased or built as means to test his idea in a reasonable time frame. Fog chambers are a method for accelerating corrosion, since corrosion often can take years to do its damage. Accelerated corrosion testing would shorten the time necessary for testing the science behind MDH's patent. In addition, a technique would have to be developed that would satisfactorily determine the effectiveness of the washdown device in such a manner that the information could be used to promote the device to potential customers.

In early April another meeting was held and a preliminary project description was discussed. MDH agreed to pursue the project. In general, the project involved both engineering and a business component. The engineering component would develop an accelerated corrosion testing chamber, design and build prototype washdown dispensers, test the washdown dispensers to determine the most appropriate design and obtain its characteristics as to reduction of corrosion and improvement of air conditioner performance. The business component would investigate competing products, determine the perceived need for a device such as the washdown dispenser (related to air conditioner life time and cost of the device), and prepare a business plan and report with recommendations.

Following MDH's verbal agreement to pursue a project related to his patent with TU, a Memorandum of Understanding (MOU) was prepared by the Development Office. The development office is the university's fundraising arm, working with foundations, corporations and individuals to build a successful university while meeting donors' needs. The MOU eventually proved to be an unsatisfactory vehicle for an agreement between the doctor and TU since it represents the transaction as a gift to the university and therefore beyond monitoring by the donor.

The university is a not for profit institution, and thus is generally not impacted by income tax laws on transactions related to the "purpose for which it is exempt." The doctor, however, may be able to impact his personal tax situation by the nature of the transaction and there would be a difference in his personal tax return dependant on whether the transaction is characterized as a contribution to the university or as a research contract performed by the university. Although tax implications are never discussed by the parties, the tax treatment of the transaction from MDH's perspective does change as the form of the agreement, the scope of the work to be done, monitoring mechanisms in the form of interplay between completion of work and progress payments, the agreement on ownership of any new intellectual property and the nature of the relationship are negotiated.

Rather than an MOU, MDH requested that a formal contract be written between himself and TU that would include specific deliverables (e.g. engineering drawings, test results, business plan, etc.). With these specifics in mind, CRM contacted the TU Research Office (RO) and began drafting a statement of work with the suggestions/assistance of MDH. It included a discussion on corrosion, a brief description of the work that would be performed, a schedule, and a list of deliverables. At the recommendation of the Research Officer, BRF from the College of Business was approached and asked to participate in the project by identifying an MBA student to write a business plan and supervise the work of that student. On April 24, 2007 a meeting was held at MDH's office with participation from the College of Business (BRF), College of Engineering (CRM), MDH, and the TU Research Office (ROM and ROK)

MDH pointed out that he felt many of his highly educated colleagues had innovative ideas but lacked the facilities to research and develop those ideas into usable products. He also noted the value to the university and its students if graduates (and/or students working on the project) were to subsequently become involved in a company that would market the products.

It was the hope of all involved that this project would demonstrate how the expertise of local universities could be brought to bear in developing and eventually marketing entrepreneurial ideas from their surrounding communities. BRF described it as a "chance for the university to create a process by which persons outside the university can present us with ideas and funds to scientifically validate their ideas. And, we are coupling scientific validation with guidance or contacts for commercializing the validated technology – using students to do the work with our guidance."

On the first of May a draft research agreement and preliminary budget were presented to MDH. Included in the budget was support for an engineering faculty & business faculty, an engineering graduate student & a business graduate student, technician & fabrication support, and equipment. At CRM's request, the support for the students became a separate item and that amount was requested up front to insure that when a graduate student came on board for the project, they would be assured sufficient funding until they graduated. The name washdown dispenser (or WDD) was selected for the invention.

A meeting was held with all parties to discuss the details. MDH presented the additional request that the agreement include: (1) provision for a back-up for each Principal Investigator (CRM and BRF), (2) Article 3 – Term and Termination be further clarified and include the return to him of any materials if TU terminates, (3) the milestones be defined by CRM and BRF, (4) in the event other patents are generated from this work they would be co-owned by TU and the sponsor, (5) the difference between co-inventor and co-owner be included, (6) while the students are working for the university they are covered by the university, (7) Article 12 – Export Control Regulations be deleted, (8) major milestones should be included in the business part with an estimate of dollars and dates and (9) product design would include a working prototype that aligns with the business plan.

At this point in time, most of the activity was between the RO and MDH's representative. Since the clock was ticking on the patent and since the summer term was about to begin, completion of the contract package was expedited so work could begin in May. The final composition of the project participants were: RO budget manager, Physician/Surgeon, College of Engineering Faculty, College of Business Faculty, College of Engineering graduate student and College of Business graduate student (both designated Fellowship Students), Laboratory Manager (Physicist/Electrical Engineer) and a laboratory technician.

The contract was signed and finalized by mid-May. ROM from the Research Office personally brought the contract to the involved parties for their signature. In addition MDH agreed to accept an appointment as research professor in the College of Engineering.

As the contract was signed, everyone involved was optimistic and filled with anticipation of good outcomes for all parties.

Appendix A to this case sets forth the milestones included in the final contract.

CURRENT CHALLENGES FACING THE ORGANIZATIONS

The engineering test results indicate that the washdown dispenser was successful in reducing corrosion and extending air conditioner life. However, the business analysis calls into question whether such a device would be of interest to the average home owner. Based on the market analyses, there appear to be market related problems with the product offering. These problems include:

1. Accelerated air conditioner corrosion exists in relatively small geographic areas – potential market is spread along coastal area
    a. The product will require direct sales effort to establish market entry
    b. The profit margins to the washdown dispenser company will be low
2. Customer awareness of "the problem" before the air conditioner fails is almost non existent
3. Customer willingness to pay for a solution at the time their air conditioner is replaced is low

MDH has received the final research report and results of the business analysis. It is unknown if attempts at commercialization will be made.

There is pressure on MDH to decide. Generally, a patent expires 20 years from the earliest filing date granted under 35 U.S. C. 111(a), 120, 121, 363, or 365(c) plus any calculated extension under 35 U.S. D. 154(b). However, the actual date a patent will expire is dependent on a number of factors, one of which is maintenance payments. These are due over the life of the patent at 3.5 years, 7.5 years and 11.5 years and less than 40% of patentees pay all three maintenance fees. Reasons not manageable by the patentee that result in more than 60% of patents granted not being in force for the full 20 years include causes of termination such as successful litigation challenging the issuance of the patent.

The dates for patent maintenance played a significant role in this project since MDH wished to determine if the patent was useful before the impending renewal date. In some regard this created an atmosphere of urgency which is not usually present in university research.

REFERENCES

Cohen, S. P. (2002). Negotiating skills for managers (p. 109). New York: McGraw-Hill.

Craver, C. B. (2009). Skills and values: Legal negotiating. LexisNexis Matthew Bender.

Fisher, R., Ury, W., & Patton, B. (1991). Getting to yes: Negotiation agreement without giving in. Penguin Books.

Golish, B. L., Besterfield-Sacre, M. E., & Shuman, L. J. (2008). Comparing Academic and Corporate Technology Development Processes. *Journal of Product Innovation Management*, *25*, 47–62.

Guthrie, C., & Orr, D. (2006). Anchoring, information, expertise, and negotiation: New insights from meta-analysis. Ohio State Journal on Dispute Resolution.

Klyver, K., Hindle, K., & Meyer, D. (2008). Influence of social network structure on entrepreneurship participation – A study of 20 national cultures. *The International Entrepreneurship and Management Journal*, *4*, 331–17. doi:10.1007/s11365-007-0053-0

Remington, M. J. (2005). The Bayh-Dole Act at twenty-five years: looking back, taking stock, acting for the future. *Journal of the Association of University Technology Managers*, *17*(1), 15–31.

Rothaermel, F. T., Agung, S. D., & Jiang, L. (2007). University entrepreneurship: a taxonomy of the literature. *Industrial and Corporate Change*, *16*(4), 691–791. doi:10.1093/icc/dtm023

Valentín, E. M. M. (2000). University-industry cooperation: a framework of benefits and obstacles. *Industry and Higher Education*, (June): 165–172.

ADDITIONAL READING

Ajamian, G. M., & Kown, P. A. *Technology stage-gate™: A structured process for managing high-risk new technology projects.* Retrieved August 23, 2009, from http://www.stevens.edu/cce/NEW/PDFs/TechStageGate_2.pdf

Association of University Technology Managers. (2009). *Technology transfer practice manual* (3rd ed.) AUTM. Retrieved from http://www.autm.net/TTP_Manual_Third_Edition/2866.htm

Cooper, R. G. (2009). How companies are reinventing their idea – to launch methodologies. *Research Technology Management*, *52*(2), 47–57.

Day, G. S. (2007). Is it real? can we win? is it worth doing? *Harvard Business Review*, *85*(12), 110–120.

Dillon, C. P. Economic evaluation of corrosion control measures. *Materials Protection*, *4*, 38-45.

Fontana, M. G., & Greene, N. D. (2000). Corrosion engineering. New York: McGraw-Hill.

Roberge, P. R. (2000). Handbook of corrosion engineering. New York: McGraw-Hill.

APPENDIX

Project Plan Attached to Contract

Development of a Washdown Dispenser for Air Conditioning Units

It is estimated that in the United States more than 4% of the annual gross domestic product is spent fighting corrosion or repairing the damage from corrosion. Approximately 15% is unnecessary if the appropriate preventive measures had been undertaken. This project will evaluate a proposed preventive measure for the alleviation of corrosion in air conditioning units.

Corrosion

Wet corrosion (as distinguished from dry corrosion) occurs when metals are exposed to moisture from the atmosphere or aqueous environments (such as the coast of Florida). There are several physical forms that can be taken by wet corrosion. For the purpose of this project, ones that will be considered are:

- General corrosion – which is non-localized and uniform
- Dealloying – where one constituent of the alloy is removed
- Galvanic – which occurs between dissimilar metals
- Pitting – which is accelerated localized attack
- Crevice – which occurs under debris, in lap joints etc?

Methods for preventing and controlling wet corrosion include:

- Coatings/passivation
- Alloy selection
- Design
- Anodic/cathodic protection
- Alteration of the environment by inhibition or sterilization
- Proper maintenance procedures.

The last two, alteration of the environment and proper maintenance procedures will be the focus of this project.

A/C Units

Any home owner will willingly testify to the propensity for corrosion on his home air conditioning unit (as will A/C maintenance personnel). Along with the 53% increase in the size of homes over the past thirty years, there has been an equally large (55%) increase in the use of air conditioning units. Currently space conditioning consumes 44% of the home's energy. So improvements in both the lifetime and the energy consumption of A/C units would be welcomed.

TU A/C Unit Project

The proposed project would involve developing the technology described in the U.S. Patent ("Air Conditioning system including liquid washdown dispenser and related methods") into a commercially viable product. It is recommended that this be a joint project between the College of Engineering (COE) and the College of Business (COB). A student from COE would pursue the technical research and development approach (thesis) while the MBA student would pursue the business plan/marketing approach (thesis).

Technical Research and Development Approach

The student will pursue a systematic approach that will include (but not necessarily be limited to):

1. Study of relevant corrosion mechanisms
2. Examination/evaluation of failed units
3. Literature search
4. Identification of relevant ASTM standards
5. Metallurgical evaluation
6. Heat transfer evaluation
7. Development of corrosion test program
    a. Components
    b. Individual units
    c. Test variables (exposure time, spray location, solution etc.)
    d. Power consumption and energy transfer
8. Test system design
    a. Timer
    b. Pump system
    c. Water reclamation
    d. Energy transfer
    e. Washdown location and characteristics
9. Product design
10. Publication of results

Business Plan Approach

The student will collaborate with the college of engineering student and faculty on his/her commercialization proposal which would include:

a. Evaluation of alternative pathways to commercialization
b. Market analysis
c. Industry analysis
d. Search for competing technologies, if any
e. Written business plan
f. Start-up company proposal

Estimated Time

Approximately one and one-half years.

Deliverables

College of Engineering thesis (MS) including system design.
College of Business thesis.
Publication.
An itemized equipment budget will be included in the test program and presented to the sponsor for final approval.

Formal Approach

Two **Fellowships** (College of Engineering and College of Business) will support students who are selected based on their qualification and ability to successfully complete the thesis and capstone projects. It is hoped that personal interactions among the students, faculty and sponsor will add further commitment by all to this project.

Desired Start Date

Summer term 2007.

Parties to the Case:

BRF-Business Researcher F
CRM – Corrosion Researcher M
MDH – Medical Doctor Inventor
MOU – Memorandum of Understanding
RO – Research Office
ROK – Research Office representative K
ROM – Research Office representative M
TU – Technological University
WDD – Washdown dispenser

Compilation of References

ACST - Prime Minister's Advisory Council on Science and Technology. (1999). *Public investments in university research: Reaping the benefits – Report of the expert panel on the commercialization of university research.* Industry Canada.

Adamson, W. (2004). *Finding the key elements of commercialisation – Process, skills and reward*, an independent White Paper prepared for the Australian Institute for Commercialisation, Retrieved October, 15, 2005, from: http://www.digitalinvestor.com.au/files/O2G65PO8WX/ADAMSON-Fixing-Key-Elements-Commercialisation.pdf.

Administration on Aging. (2002). A profile of older Americans. *Administration on Aging, U.S. Department of Health and Human Services.* Retrieved May 1, 2009, from http://www.aoa.gov/prof/Statistics/profile/2002profile.pdf

Agarwal, R., & Prasad, J. (1998). A conceptual and operational definition of personal innovativeness in the domain of information technology. *Information Systems Research, 9*(2), 204–215. doi:10.1287/isre.9.2.204

Agell, J. (2004). Why are small firms different? Managers' views. *The Scandinavian Journal of Economics, 106*(3), 437–452. doi:10.1111/j.0347-0520.2004.00371.x

Agrawal, A., & Henderson, R. (2002). Putting patents in context: Exploring knowledge transfer from MIT. *Management Science, 48*(1), 44–60. doi:10.1287/mnsc.48.1.44.14279

Ajzen, I. (1985). From intentions to action: A theory of planned behavior. In Kuhl, J., & Bechmann, J. (Eds.), *Action Control: From Cognition to Behavior* (pp. 11–39). New York: Springer Verlag.

Ajzen, I. (1991). The theory of planned behavior. *Organizational Behavior and Human Decision Processes, 50*, 179–211. doi:10.1016/0749-5978(91)90020-T

Al-Hakim, L. (2007). Information quality function deployment. In Al-Hakim, L. (Ed.), *Challenges of Managing Information Quality in Service Organizations* (pp. 26–50). Hershey, PA: IGI Global.

aliamoune-Lutz, M. (2003). An analysis of the determinants and effects of ICT diffusion in developing countries. Information Technology for Development, 10, 151–169.

Alter, S., & Sherer, S. A. (2004). A General, but readily adaptable model Of information system risk. *Communications of the Association for Information Systems*, (14): 1–28.

Altman, S., Shactman, D., & Eliat, E. (2006). Could U.S. hospitals go the way of U.S. airlines? *Health Affairs, 25*(1), 11–21. doi:10.1377/hlthaff.25.1.11

Al-Zamany, Y. Hoddell. S. E. J., & Savage, B. M. (2002). Self assessment and obstacles to their implementation in Yemen, TQM and change management. In S.K.M. Ho & J. Dalrymple (Ed.), *Proceedings of the 7th International Conference on ISO 9000 and TQM.* RMIT University, Melbourne, Australia.

American Foundation for the Blind. (2009). *What is normal vision? The aging eye- normal changes and their symptoms.* Retrieved May 2, 2009, from www.afb.org/seniorsite.asp?SectionID=63&DocumentID=3194.

American Medical Association. (2008). New AMA Guidelines on Medical Tourism.

Anderson, C., & Andersson, M. (2006). The long tail. *Wired Magazine, 12*(10).

Antonacopoulou, E. P., & FitzGerald, L. (1996). Reframing competency in management development. *Human Resource Management Journal, 6*, 27–46. doi:10.1111/j.1748-8583.1996.tb00395.x

Apte, U. S., & Karmarkar, U. M. (2007). Current research on managing in the information economy: Introductory note. In U. S. Apte & U. M. Karmarkar (Eds.), Managing in the Information Economy: Current Research Issues. New York: Springer Science+Business Media, LLC.

Argyres, N., & Liebeskind, J. (1998). Privatizing the intellectual commons: Universities and the commercialization of biotechnology. *Journal of Economic Behavior & Organization, 35*, 427–454. doi:10.1016/S0167-2681(98)00049-3

Argyris, C. (2004). *Reasons and Rationalizations: The Limits to Organizational Knowledge*. Oxford: Oxford University Press.

Arrington, M. (2008). Encyclopedia Britannica now free for bloggers. *TechCrunch*. Retrieved Aug 3, 2009 from http://www.techcrunch.com/2008/04/18/encyclopedia-britannica-now-free-for-bloggers/

Arvan, A. (1988). Those fabulous Japanese banks. *Bankers Monthly, 105*(1), 29–35.

Audretsch, D. (2001). The role of small firms in U.S. biotechnology clusters. *Small Business Economics, 17*(1-2), 3–15. doi:10.1023/A:1011140014334

Audretsch, D., & Lehmann, E. (2005). Do university policies make a difference? *Research Policy, 34*, 343–347. doi:10.1016/j.respol.2005.01.006

Auh, S., & Menguc, B. (2005). Top management team diversity and innovativeness: The moderating role of interfunctional coordination. *Industrial Marketing Management, 34*(3), 249–261. doi:10.1016/j.indmarman.2004.09.005

AUTM - The Association of University Technology Managers. (2003). *AUTM licensing survey: FY 2003 licensing summary*. Deerfield, IL: The Association of University Technology Managers.

AUTM - The Association of University Technology Managers. (2007). *The better world report part 2, technology transfer works: 100 innovations from academic research to real-world applications*. Deerfield, IL: The Association of University Technology Managers.

AUTM - The Association of University Technology Managers. (2007). *AUTM U.S. licensing survey: FY 2005 licensing summary*. Deerfield, IL: The Association of University Technology Managers.

AUTM - The Association of University Technology Managers. (2007). *AUTM Canadian licensing survey: FY 2005 licensing summary*. Deerfield, IL: The Association of University Technology Managers.

Avery, J. (1997). *Progress, Poverty and Population: Re-reading Condorcet, Godwin and Malthus*. Portland, Oregon: Frank Cass Publishers.

Bachelard, G. (1971). On Poetic Imagination and Reverie (C.Gaudin, Trans.). Dallas, TX: Spring.

Bacon, F. (1605). *The advancement oflearning*. Adamant Media Corporation.

Badr, E. A. (2007, April 1). *ENPO Vice President for IT Interview*.

Baida, Z., Liu, J., & Tan, Y.-H. (2007). Towards a methodology for designing e-government control procedures. In Electronic Government, 4646, 56-67. Berlin/Heidelberg: Springer.

Baida, Z., Rukanova, B., Wigand, R., & Tan, Y. H. (2007). Heineken shows benefits of customs collaboration. *Supply Chain Management Review, 11*(7), 11–12.

Bakic, O., & Hrabovski-Tomic, E. (2008). Informal education for aanagement in health tourism. *Tourism and Hospitability Management, 14*(1), 1–12.

Balaji, P., & Keniston, K. (2005, July). Tentative conclusions. Information and communications technologies for development: A comparative analysis of impacts and costs. Department of Information Technology, Government of India. Retrieved from http://www.iiitb.ac.in/Complete_report.pdf

Ballantine, J., & Levy, M. (1998). Evaluating information systems in small and medium-sized enterprises: issues and evidence. *European Journal of Information Systems, 7*(4), 241–251. doi:10.1057/palgrave.ejis.3000307

Ballou, D. P., & Pazer, H. L. (1995). Modelling data and process quality in multi-input, multi-output information systems. *Management Science, 31*(2), 150–162. doi:10.1287/mnsc.31.2.150

Bandura, A. (2001). Social cognitive theory: An agentic perspective. *Annual Review of Psychology, 52*, 1–26. doi:10.1146/annurev.psych.52.1.1

Banerjee, P. (2008). Biomedical Innovation in India: With a comparison to China and others. New Delhi, India: Har Anand.

Banerjee, P., & Chau, P. Y. K. (2004). An evaluative framework for analyzing e-government convergence capability in developing countries. *Electronic Government, 1*, 29–48. doi:10.1504/EG.2004.004135

Barnes, S. (2004). Wireless Support for Mobile Distributed Work: Taxonomy and Examples. In *Proc. 37th Hawaii International Conference on System Sciences* (pp 2-10), Big Island, Hawaii. Retrieved January 10, 2009, from http://csdl.computer.org/comp/proceedings/hicss/2004/2056/03/205630078a.pdf.

Bassanini, A. (2002). Growth, technology change, and ICT diffusion: Recent evidence from OECD countries. *Oxford Review of Economic Policy, 18*(3), 324–344. doi:10.1093/oxrep/18.3.324

Baumol, W. J., & Solow, R. (1998, Fall). Comments. *Issues in Science and Technology, 15*(1), 8–10.

Beck, U., Giddens, A., & Lash, S. (1994). *Reflexive modernization. Politics, tradition and aesthetics in the modern social order.* Cambridge: Polity Press.

Becker, S. A. (2004). A Study of Web Usability for Older Adults Seeking Online Health Resources. *ACM Transactions on Human-Computer Interaction, 11*(4), 387–406. doi:10.1145/1035575.1035578

Becker, S. A. (2005). Web Accessibility and Critical Issues Facing Older Adult Users. In M. Khosrow-Pour (Ed.), Encyclopedia of Information Science and Technology (pp. 3036-3041). Hershey, PA: IGI Global. Reprinted 2008, Encyclopedia of Information Science and Technology (2nd ed.) (pp. 4041-4046).

Bell, D. (1973). The coming of post-industrial society. *Business & Society Review/Innovation* (5).

Bell, W. (1996). An overview of future studies. In R. A. Slaughter (Ed.), The Knowledge Base of Future Studies (Vol. 1, Foundations, pp. 26-56). Hawthorne, Victoria: DDM Media Group.

Bellone, C., & Goerl, G. (1992). Reconciling public entrepreneurship and democracy. *Public Administration Review, 52*(2), 130–134. doi:10.2307/976466

Bies, W., & Zacharia, L. (2007). Medical tourism: Outsourcing surgery. *Mathematical and Computer Modelling, 46*, 1144–1159. doi:10.1016/j.mcm.2007.03.027

Bjorklund, D. F. (1995). *Information processing approaches: An introduction to cognitive development.* Washington, D.C.: Brooks-Cole.

Bjørn-Andersen, N., Razmerita, L. V., & Henriksen, H. Z. (2007). The streamlining of cross-border taxation using IT: The Danish eExport solution. In Makolm, J., & Orthofer, G. (Eds.), *E-Taxation: State & Perspectives: E-Government in the Field of Taxation: Scientific Basis, Implementation Strategies, Good Practice Examples* (pp. 195–206). Linz, Austria: Trauner Verlag.

Blomstrom, M., & Kokko, A. (1998). In G. B. Navaretti Foreign investment as a vehicle for international technology transfer. In G. Barba Navaretti, P. Dasgupta, K-G. Maler, & D. Siniscalco (Eds.), Creation and Transfer of Knowledge: Institutions and Incentives. New York: Springer Verlag.

Blouin, C. (2007). Trade policy and health: from conflicting interests to policy coherence. *Bulletin of the World Health Organization, 85*, 169–173. doi:10.2471/BLT.06.037413

Bobrowski, M., Marre, M., & Yankelevich, D. (2002). A Neat Approach for Data Quality Assessment. In Piattini, Calero & Genero (Ed.), Information and Database Quality (pp. 135-162), Hershey,PA: IGI Global.

Bok, D. (2003). *Universities in the marketplace: The commercialization of higher education.* Princeton, NJ: Princeton University Press.

Bookman, M. Z., & Bookman, K. R. (2007). Medical Tourism in Developing Countries. New York: Palgrave.

Borman, E. (2004). Health Tourism: Where healthcare, ethics, and the state collide. *British Medical Journal, 328*, 60–61. doi:10.1136/bmj.328.7431.60

Bottomore, T., & Nisbet, R. (Eds.). (1978). *A History of Sociological Analysis.* New York: Basic Books.

Boxall, P. (1996). The strategic HRM debate and the resource-based view of the firm. *Human Resource Management Journal, 6*, 59–70. doi:10.1111/j.1748-8583.1996.tb00412.x

Boyd, S. L., Hobbs, J. E., & Kerr, W. A. (2003). The Impact of Customs Procedures on Business to Consumer E-commerce in Food Products. *Supply Chain Management: An International Journal, 8*(3), 195–200. doi:10.1108/13598540310484591

Bozeman, B. (2000). Technology transfer and public policy: a review of research and theory. *Research Policy, 29*, 627–655. doi:10.1016/S0048-7333(99)00093-1

Breitkopf, A. (2007). Diebold machine was used to alter Wikipedia entry. *American Banker, 172*(159), 8–8.

Bridge, S., O'Neill, K., & Cromie, S. (2003). *Understanding enterprise, entrepreneurship and small business.* Basingstoke, UK: Palgrave Macmillan.

Brown, J. S., & Duguid, P. (2000). *The social life of information.* Boston: Harvard Business School Press.

Brown, S. (2008). Medical tourism: nations vie for health dollars. *Hospitals & Health Networks, 82*(12), 49.

Bruderl, J., Preisendorfer, P., & Ziegler, R. (1992). Survival chances of newly founded business organizations. *American Sociological Review, 57*, 227–242. doi:10.2307/2096207

Bryant, R. (2002). Despite turbulent times, Dead Sea draws medical tourism. *Dermatology Times, 23*(2), 8.

Bunnell, T. (2002, March). Multimedia utopia? A geographical critique of hgh-tech development in Malaysia's multimedia super corridor. *Antipode, 34*(2), 265. doi:10.1111/1467-8330.00238

Burn, J., & Robins, G. (2003). Moving towards e-government: A case study of organisational change process. *Logistics Information Management, 16*(1), 25–35. doi:10.1108/09576050310453714

Burns, T., & Stalker, G. M. (1961). *The management of innovation.* London: Tavistock.

Bygrave, W. D. (1989). The entrepreneurship paradigm (I): A philosophical look at its research methodologies. *Entrepreneurship Theory and Practice, 14*(1), 7–26.

Cairo, I. C. T. (2009). Retrieved March 20, 2009, from, www.cairoict.com.

Caldeira, M. M., & Ward, J. M. (2002). Understanding the successful adoption and use of IS/IT in SMEs: an explanation from Portuguese manufacturing industries. *Information Systems Journal, 12*(2), 121–152. doi:10.1046/j.1365-2575.2002.00119.x

Canada Foundation for Innovation. (2002). *Conference on Innovation and Commercialization of University Research* (Vol. 2). Canada Foundation for Innovation.

Carland, J., Hoy, F., Boulton, W., & Carland, J. (1984). Differentiating entrepreneurs from small business owners: A conceptualization. *Academy of Management Review, 9*(2), 354–359. doi:10.2307/258448

Carrera, P. (2006). Medical tourism. *Health Affairs, 25*(5), 1453. doi:10.1377/hlthaff.25.5.1453

Castonguay, G., & Brown, A. (1993). Plastic surgery tourism proving a boon for Costa Rica's surgeons. *Canadian Medical Association Journal, 148*(1), 74–76.

CED. (2009). *East Central Florida regional planning council's comprehensive economic development strat-*

egy. Retrieved September 4, 2009, from http://www.ecfrpc.org

Chadwick-Dias, A., Mcnulty, M., & Tullis, T. (2003). Web usability and age: How design changes can improve performance. *2003 Conference on Universal Usability*, Vancouver, British Columbia, Canada, 30-36. Chaparro, A., Bohan, M., Fernandez, J. E., Choi, S., D., & Kattel, B. (199). The impact of age on computer input device use: Psychophysical and physiological measures. *International Journal of Industrial Ergonomics*, *24*, 503–513.

Chaffey, D. (2004). E-business and e-commerce management (2nd ed.). Harlow, Essex: Pearson Education Ltd.

Chakrabarti, A. K., & Lester, R. K. (2002, August 20). Regional economic development comparative case studies in the US and Finland. In *Proceedings IEEE Conference on Engineering Management*. Cambridge, UK.

Chakravarthy, B. (1997). A new strategy framework for coping with turbulence. *Sloan Management Review*, (Winter): 69–82.

Charness, N. (2001). Aging and communication: Human factors issues. In N. Charness, D. C. Park, & B. A. Sabel (Eds.), Communication, Technology, and Aging: Opportunities and Challenges for the Future (pp. 1-29). New York: Springer Publishing Company.

Chase, S. E. (2005). Narrative inquiry. In Denzin, N. K., & Lincoln, Y. S. (Eds.), *Handbook of Qualitative Research* (3rd ed., pp. 651–679). Thousand Oaks, CA: Sage Publications.

Cheung, I., & Wilson, A. (2007). Arthroplasty tourism. *The Medical Journal of Australia*, *187*(11/12), 666–667.

Chomsky, N. (1996). *Language and problems of knowledge*. Mendocino, CA: MIT Press.

Christensen, C. (1997). *The innovator's dilemma: when new technologies cause great firms to fail*. Harvard Business School Press.

Christensen, C. M., & Shih, W. C. (11 Dec 2007). Successful Innovation: The Intersection of Theory and Practice Retrieved 30 Jan 2009, from http://www.hbs.edu/centennial/conversation/successful_innovation/

Churchill, N. C., & Lewis, V. L. (1983). The five stages of small business growth. *Harvard Business Review*, *61*(3), 30–49.

Clark, B. (1998). *Creating entrepreneurial universities*. Oxford, UK: International Association of Universities and Elsevier Science.

Clayman, B., & Holbrook, J. (2003). *The survival of university spin-offs and their relevance to regional development*. Canada Foundation for Innovation.

Clegg, S. R., Kornberger, M., & Rhodes, C. (2004). Noise, parasites and translation: Theory and practice in management consulting. *Management Learning*, *35*(1), 31–44. doi:10.1177/1350507604041163

Clough, P. T. (2000). Comments on setting criteria for experimental writing. *Qualitative Inquiry*, *6*, 278–291. doi:10.1177/107780040000600211

Cohen, W., & Levinthal, D. (1990). Absorptive capacity: a new perspective on learning and innovation. *Administrative Science Quarterly*, *35*, 128–152. doi:10.2307/2393553

Cohen, W., Nelson, R., & Walsh, J. (2002). Links and impacts: The influence of public research on industrial R&D. *Management Science*, *48*(1), 1–23. doi:10.1287/mnsc.48.1.1.14273

Cohen-Blankshtain, G., & Nijkamp, P. (2003, August). Still not there, but on our way: thinking of urban ICT policies in European cities. *Tijdschrift voor Economische en Sociale Geografie*, *94*(3), 390–400. doi:10.1111/1467-9663.00265

Collins, S., & Wakoh, H. (2000). Universities and technology transfer in Japan: Recent reforms in historical perspective. *The Journal of Technology Transfer*, *25*(2), 213–222. doi:10.1023/A:1007884925676

Colyvas, J., Crow, M., Gelijns, A., Mazzoleni, R., Nelson, R., Rosenberg, N., & Sampat, B. (2002). How do university inventions get into practice? *Management Science*, *48*(1), 61–72. doi:10.1287/mnsc.48.1.61.14272

Compeau, D. R., & Higgins, C. A. (1995). Computer self-efficacy: Development of a measure and initial test. *Management Information Systems Quarterly, 19*, 189–211. doi:10.2307/249688

Connell, D. (2006). *"Secrets" of the world's largest seed fund: How the United States Government uses its Small Business Innovation Research (SBIR) program and procurement budgets to support small technology firms.* Cambridge, UK: University of Cambridge, Centre for Business Research.

Connell, J. (2006). Medical tourism: Sea Sun Sand and… surgery. *Tourism Management, 27*(6), 1093–1100. doi:10.1016/j.tourman.2005.11.005

Corbett, J. (2005). Toresten Hägerstrand: Time Geography Retrieved 14 September 2005, from http://www.csiss.org/classics/content/29

Corley, K. (2002). Breaking away: an empirical examination of how organizational identity changes during a spin-off. Unpublished doctoral dissertation, The Pennsylvania State University, Pennsylvania.

Coughlin, J. F. (1999). Technology need of aging boomers. *Issues in Science and Technology*, (Fall): 53–60.

Cragg, P. B., & King, M. (1993). Small-firm computing: Motivators and inhibitors. *Management Information Systems Quarterly, 17*(1), 47–60. doi:10.2307/249509

Craver, C. B. (2009). Skills and values: Legal negotiating. LexisNexis Matthew Bender.

D'Cruz, C., & Ports, K. (2006). Space Coast innovation and technology commercialization outreach. In *Proceedings of the United States Association of Small Business and Entrepreneurship (USASBE),* Tucson, AZ.

Daft, R. L., & Lengel, R. H. (1986). Organizational information requirements, media richness and structural design. *Management Science, 32*, 554–571. doi:10.1287/mnsc.32.5.554

David, P. A. (1997). Rethinking technology transfers: Incentives, institutions and knowledge-based industrial development. In C. Feinstein & C. Howe (Eds.), Chinese Technology Transfer in the 1990s: Current Experience, Historical Problems and International Perspectives. Cheltenham, UK: Elgar.

Davis, F. D. (1986). *A technology acceptance model for empirically testing new end-user information systems: Theory and results.* Unpublished Doctoral dissertation, Massachusetts Institute of Technology.

Davis, F. D. (1989). Perceived usefulness, perceived ease of use, and user acceptance of information technology. *Management Information Systems Quarterly, 13*(3), 319–340. doi:10.2307/249008

Davis, F. D., Bagozzi, R. P., & Warshaw, P. R. (1989). User acceptance of computer technology: A comparison of two theoretical models. *Management Science, 35*(8), 982–1003. doi:10.1287/mnsc.35.8.982

De Sanctis, G., & Poole, M. S. (1994). Capturing the complexity in advanced technology use: Adaptive structuration theory. *Organization Science, 5*, 121–147. doi:10.1287/orsc.5.2.121

Debackere, K., & Veugelers, R. (2005). The role of academic technology transfer organizations in improving industry science links. *Research Policy, 34*, 321–342. doi:10.1016/j.respol.2004.12.003

Dedrick, J., & West, J. (2003). Why firms adopt open source platforms: A grounded theory of innovation and standards adoption. In J. L. King & K. Lyytinen (Eds.), *Proceedings of the Workshop on Standard Making: A Critical Research Frontier for Information Systems,* (pp. 236-257). Seattle, WA, USA.

Denton, D. K., & Wisdom, B. L. (1992). Shared vision. In Thompson, A. A. Jr, Fulmer, W. E., & Strickland, A. J. III, (Eds.), *Readings in strategic management* (pp. 52–56). Boston: Irwin.

Department of Health & Children. (2006, March). *Reducing the risk: A strategic approach - the report of the Task Force on Sudden Cardiac Death.* Public Report, Department of Health & Children, Ireland.

Dervin, B. (1992). From the mind's eye of the user: the sense-making qualitative-quantitative methodology. In Glazier, J. D., & Powell, R. R. (Eds.), *Qualitative research in information management* (pp. 61–84). Englewood, CO: Libraries Unlimited.

DEST Department of Education. Science and Training. (2003). Mapping Australian science and innovation: National and international linkages: Background paper. Canberra: Department of Education, Science and Training.

Docktor, R. (2004). Successful global ICT initiatives: Measuring results through an analysis of achieved goals, planning and readiness efforts, and stakeholder involvement. Presentation to the Council for Excellence in Government.

Domestic Policy Council - U.S. Domestic Policy Council Office of Science and Technology Policy. (2006). *American competitiveness initiative – Leading the world in innovation.* U.S. Domestic Policy Council Office of Science and Technology Policy.

Dosi, G. (1982). Technological paradigms and technological trajectories. *Research Policy, 11*(3), 147–162. doi:10.1016/0048-7333(82)90016-6

Drucker, P. F. (1985). *Innovation and entrepreneurship.* New York: Harper & Row.

Duarte, D. L., & Tennant Snyder, N. (2001). *Mastering virtual teams: Strategies, tools, and techniques that succeed.* San Francisco, CA: Jossey-Bass.

Duggan, E. W., & Reichgelt, H. (2006). *Measuring information systems delivery quality.* Hershey, PA: IGI Global.

Dunn, P. (2007). An American in Bangkok. *Hospitals & Health Networks, 81*(11), 43.

Dutton, J., & Duncan, R. (1983). *The creation of momentum for change through the process of organizational sensemaking.* Unpublished manuscript, New York University, New York.

Dutton, J., & Duncan, R. (1987). The influence of the strategic planning process on strategic change. *Strategic Management Journal, 8*(2), 103–116. doi:10.1002/smj.4250080202

Dytche, J., & Warren, A. (2007). *I99 Corridor Innovation Portal - summary: Farrell Center for Corporate Innovation and Entrepreneurship.* Pennsylvania: The Pennsylvania State University.

Dytche, J., & Warren, A. (2007). *Penn State KIZ Innovation Grant Team I-99 Innovation Network Portal: Farrell Center for Corporate Innovation and Entrepreneurship.* Pennsylvania: The Pennsylvania State University.

Easterly, W. (2001). The elusive quest for economic growth: Economists' adventures and misadventures in the tropics. Cambridge: MIT Press. *The Economist (2000), "Growth is good", The Economist, May 27*, 82.

Echt, K. V. (2002). Designing Web-based health information for older adults: Visual considerations and design directives. In R.W. Morrell (Ed.), Older Adults, Health Information, and the World Wide Web (pp. 61-88). Mahwah, NJ: Lawrence Erlbaum Associates, Inc.

Egypt Post. (2009). Retrieved March 25, 2009, http://egyptianpost.net/en/index.asp.

El-Labban, D. N. (2007, April 1). *ENPO Director for International Relations Interview.*

Ellet, W. (2007). *The Case Study Handbook.* Boston, Massachusetts: Harvard Business School Press.

Ellis, R. D., & Kurniawan, S. H. (2000). Increasing the usability of online information for older users: A case study in participatory design. *International Journal of Human-Computer Interaction, 12*(2), 263–276. doi:10.1207/S15327590IJHC1202_6

Ellul, J. (1964). *The Technological Society.* New York: Alfred A Knopf, Inc.

eMarketer.com. (2007).User-Generated Content: Will Web 2.0 Pay Its Way? *eMarketer.com.* Retrieved May 2, 2009, from http://www.emarketer.com/Reports/All/Emarketer_2000421.aspx?src=report2_home

English, L. P. (1999). *Improving data warehouse and business information quality: Methods for reducing costs and increasing profits.* New York: Wiley.

Etzkowitz, H., & Leydesdorff, L. (2001). *Universities and the global knowledge economy*. London: Continuum.

Evans, J. R., & Lindsay, W. M. (2005). *The management and control of quality* (6th ed.). Ohio: Thomson/South-Western.

Evans., J. & Dean, J. (2003). *Total quality management, Organization and strategy*. Ohio: Thomson/South-Western.

Faberman, R. (2002). Job flows and labor dynamics in the US Rust Belt. *Monthly Labor Review, 125*(9), 3–10.

Feinson, S. (2003, June). National innovation systems overview and country cases. Knowledge Flows and Knowledge Collectives: Understanding The Role of Science and Technology Policies in Development. In B. Bozeman, et al. (Eds.), Synthesis Report on the Findings of a Project for the Global Inclusion Program of the Rockefeller Foundation.

Feldman, M., Feller, I., Bercovitz, J., & Burton, R. (2002). Equity and the technology transfer strategies of American research universities. *Management Science, 48*(1), 105–121. doi:10.1287/mnsc.48.1.105.14276

Feller, I. (1997). Technology transfer from universities. In Smart, J. (Ed.), *Higher education: Handbook of theory and research* (*Vol. 12*, pp. 1–43). New York: Agathon Press.

Ferris, W. (2001). Australia chooses: Venture capital and a future Australia. *Australian Journal of Management, 26*, 45–64. doi:10.1177/031289620102601S03

Fields, J., & Casper, L. M. (2001). America's families and living arrangement population characteristics. P20-537. U.S. Bureau of the Census, U.S. Department of Commerce, Washington, D.C.

Fisher, D., & Atkinson-Grosjean, J. (2002). Brokers on the boundary: Academy-industry liaison in Canadian universities. *Higher Education, 44*, 449–467. doi:10.1023/A:1019842322513

Fisher, R., Ury, W., & Patton, B. (1991). Getting to yes: Negotiation agreement without giving in. Penguin Books.

Forgoione, D., & Smith, P. (2007). Medical tourism and Its impact on the US health care system. *Journal of Health Care Finance, 34*(1), 27–35.

Fox, W. M. (1995). Sociotechnical system principles and guidelines: past and present. *The Journal of Applied Behavioral Science, 31*, 91–105. doi:10.1177/0021886395311009

Franke, N., von Hippel, E., & Schreier, M. (2006). Finding commercially attractive user innovations: A test of Lead-user theory. *Journal of Product Innovation Management*, 301–315. doi:10.1111/j.1540-5885.2006.00203.x

Frieden, R. (2005). Lessons from broadband development in Canada, Japan, Korea and the United States. *Telecommunications Policy, 29*, 595–613. doi:10.1016/j.telpol.2005.06.002

Fussell, S. R., & Benimoff, I. (1995). Social and cognitive processes in interpersonal communication: Implications for advanced telecommunications technologies. *Human Factors, 37*, 228–250. doi:10.1518/001872095779064546

Gans, J., & Stern, S. (2003). *Assessing Australia's innovative capacity in the 21st century* (Working Paper 2003-16). Melbourne: University of Melbourne, Melbourne Business School.

GAO - U.S. General Accounting Office. (1990). *Case study evaluations (Transfer Paper 10.1.9)*. U.S. General Accounting Office.

Gao, J. Z., Shim, S., Mei, H., & Su, X. (2006). Engineering wireless based software systems and applications. Norwood, MA: Artech House.

Gatian, A. W., Brown, R. M., & Hicks, J. O. (1995). Organizational innovativeness, competitive strategy and investment success. *The Journal of Strategic Information Systems, 4*(1), 43. doi:10.1016/0963-8687(95)80014-H

Geiger, R., & Sá, C. (2008). *Tapping the riches of science; Universities and the promise of economic growth*. Cambridge, MA: Harvard University Press.

Gephart, R. P., Boje, D. M., & Thatchenkery, T. J. (1996). Postmodern management and the coming crises of organizational analysis. In Gephart, R. P. (Eds.), *Postmodern Management and Organization Theory* (pp. 1–20). Thousand Oaks, CA: Sage.

Giddens, A. (1984). *The constitution of society: Outline of the theory of structuration*. Cambridge, UK: Polity Press.

Giddens, A. (1991). *Modernity and self-identity. Self and society in the late modern age*. Stanford, CA: Stanford University Press.

Gillingham, A. (n.d.). *Bulk Mail Takes Notes and Goes Hi-Tech*.

Gioia, D., & Chittipeddi, K. (1991). Sensemaking and sensegiving in strategic change initiation. *Strategic Management Journal, 12*(6), 433–448. doi:10.1002/smj.4250120604

Gioia, D., & Thomas, J. (1996). Identity, image, and issue interpretation: Sensemaking during strategic change in academia. *Administrative Science Quarterly, 41*, 370–403. doi:10.2307/2393936

Golish, B. L., Besterfield-Sacre, M. E., & Shuman, L. J. (2008). Comparing Academic and Corporate Technology Development Processes. *Journal of Product Innovation Management, 25*, 47–62.

Goodhue, D. L., & Thompson, R. L. (1995). Task-technology fit and individual performance. *Management Information Systems Quarterly*, (June): 213–232. doi:10.2307/249689

Goodrich, J. (1993). Health-care tourism in the Caribbean. In D. Gayle & J. Goodrich (eds.) Tourism marketing and management in the Caribbean (pp. 122-128). New York: Routledge.

Goozner, M. (2004). *The $800 million pill: The truth behind the cost of new drugs*. California: University of California Press.

Graham, S. (2000, March). Symposium on cities and infrastructure networks: Constructing premium network spaces: reflections on infrastructure networks and contemporary urban development. *International Journal of Urban and Regional Research, 24*(1), 183. doi:10.1111/1468-2427.00242

Granberg, A., & Stankiewicz, R. (1981). The development of generic technologies - the cognitive aspects. In Grandstrand, O., & Sigurdson, J. (Eds.), *Technological and Industrial Policy in China and Europe* (pp. 196–224). Lund: Research Policy Institute.

Gray, H., & Poland, S. (2008). Medical Tourism: Crossing borders to access health care. *Kennedy Institute of Ethics Journal, 18*(2), 193–201.

Gray, H., & Sanzogni, L. (2004). Technology leapfrogging in Thailand: Issues for the support of ecommerce infrastructure. *Electronic Journal on Information Systems in Developing Countries, 16*(3), 1–26.

Greenberg, D. (2007). *Science for sale – The perils, rewards and delusions of campus capitalism*. Chicago, IL: The University of Chicago Press.

Greenhalgh, T., Robert, G., MacFarlane, F., Bate, P., & Kyriakidou, O. (2004). Diffusion of innovations in service organizations: Systematic review and recommendations. *The Milbank Quarterly, 82*(4), 581–629. doi:10.1111/j.0887-378X.2004.00325.x

Greer, C. R. (1995). *Strategy and human resources. A general managerial perspective*. Englewood Cliffs, NJ: Prentice-Hall.

Gregor, S., & Benbasat, I. (1999). Explanations from intelligent systems: Theoretical foundations and implications for practice. *Management Information Systems Quarterly, 23*, 497–527. doi:10.2307/249487

Grubesic, T. H., & Murray, A. T. (2004, Spring). Waiting for broadband: Local competition and the spatial distribution of advanced telecommunication services in the United States. *Growth and Change, 35*(2), 139–165. doi:10.1111/j.0017-4815.2004.00243.x

Guha, S., Grover, V., Kettinger, W. J., & Teng, J. T. C. (1997). Business process change and organizational performance: exploring an antecendent model. *Journal of Management Information Systems, 14*(1), 119–154.

Gunasekara, C. (2005). The role of universities in shaping regional agglomeration: case studies in the Australian setting. *International Journal of Technology Transfer & Commercialisation, 4*(4), 525–539.

Gupta, U. (2004). *The first venture capitalist – Georges Doriot on leadership, capital, & business organization.* Calgary, Canada: Gondolier.

Gustafson, L. T., & Reger, R. K. (1995). Using organizational identity to achieve stability and change in high velocity environments. *Academy of Management Journal, Best Papers Proceedings,* 464-468.

Guthrie, C., & Orr, D. (2006). Anchoring, information, expertise, and negotiation: New insights from meta-analysis. Ohio State Journal on Dispute Resolution.

Habermas, J. (1970). Knowledge and interest. In Emmet, D., & MacIntyre, A. (Eds.), *Sociological Theory and Philosophical Analysis* (pp. 36–54). London: Macmillan.

Habermas, J. (1974). *Theory and Practice*. London: Heinemann.

Hamel, G., & Prahalad. (1993). Strategy as stretch and leverage. *Harvard Business Review*, (March-April): 75–85.

Handy, C. (1995, May-June). Trust and the virtual organization. *Harvard Business Review*, *73*(3), 40–50.

Hanks, S. H., Watson, C. J., Jansen, E., & Chandler, G. N. (1993). Tightening the Life-Cycle Construct: A Taxonomic Study of Growth Stage Configurations in High-Technology Organizations. *Entrepreneurship: Theory and Practice*, *18*(2), 5–30.

Hanson, F. (2008). A revolution in healthcare: Medicine meets the marketplace. *Public Affairs Report*, *59*, 4.

Harman, G. (2002). Australian university-industry links: Researcher involvement, outputs, personal benefits and "withholding" behaviour. *Prometheus*, *20*(2), 143–158. doi:10.1080/08109020210137529

Hassan, S. T. G. (2008). Bringing Lead-user innovations to the market: Research and management implications. *SAM Advanced Management Journal*, 51-58.

Hatakenaka, S. (2005). *Development of third stream activity: Lessons from international experience*. Higher Education Policy Institute. Retrieved September 9, 2007, from: http://www.hepi.ac.uk/downloads/Developmentofthirdstreamfunding-SachiHatakenaka.pdf

Hawthorne, D. (2000). Possible implication of aging for interface designers. *Interacting with Computers*, *12*(5), 507–528. doi:10.1016/S0953-5438(99)00021-1

Hay, B. (2001). New tourism market: health and well being holidays. *Countryside Tourism*, *9*(2), 14–16.

Hayler, R., & Nichols, D. M. (2007). *Six Sigma for financial services, how leading companies are driving results with Lean, Six Sigma, and process improvement, profiles from global leaders including AMERICAN EXPRESS, BANK OF AMERICA, WACHOVIA, and LLOYDS TSB.* New York: McGraw-Hill.

He, W., Sengupta, M., Velkoff, V. A., & DeBarros, K. A. (2005). 65+ in the United States: 2005. Special report issued by the U.S. Department of Health and Human Services and the U.S. Department of Commerce, Washington, DC.

Helfert, M., Zellner, G., & Sousa, C. (2002). Data quality problems and proactive data quality management in data-warehouse-systems. In *Proceedings of BIT-World 2002*, Guyaquil, Ecuador.

Helm, B. (2005, December 14). Wikipedia: a work in progress. *Business Week Online*. Retrieved April 1, 2009, from http://www.businessweek.com/technology/content/dec2005/tc20051214_441708.htm.

Henderson, J. (2004). Healthcare tourism in Southeast Asia. *Tourism Review International*, *7*(3-4), 111–121.

Henderson, J. C., & Venkatraman, N. (1999). Strategic alignment: Leveraging information technology for transforming organizations. *IBM Systems Journal*, *38*(2/3), 472–484.

Henriksen, H. Z., & Rukanova, B. (2008, April 23-25). *Barriers and Drivers of eCustoms Implementation: Never Mind IT.* Paper presented at the 6th Eastern European eGovernment Days, Prague, Czech Republic.

Henriksen, H. Z., Rukanova, B., & Tan, Y.-H. (2008). Pacta Sunt Servanda but Where Is the Agreement? The Complicated Case of eCustoms. In Wimmer, M. A.,

Scholl, H. J., & Ferro, E. (Eds.), *EGOV 2008* (pp. 13–24). Berlin, Heidelberg: Springer-Verlag.

Herrick, D. (2007). Medical tourism: Global competition in health care. National Center for Policy Analysis

Herstatt, C., & Von Hippel, E. (1992). From experience: Developing new product concepts via the Lead-user method: a case study in a 'low tech' field. *Journal of Product Innovation Management, 9*(3), 213–221. doi:10.1016/0737-6782(92)90031-7

Hirsch, T., Forlizzi, J., Hyder, E., Goetz, J., Kurtz, C., & Stroback, J. (2000). The ELDer project: Social, emotional, and environmental factors in the design of eldercare technologies. In *Proceedings of CM Conference on Universal Usability,* Arlington, Virginia (pp. 72-79).

Hirschman, E. C. (1981). Innovativeness, novelty seeking, and consumer creativity. *The Journal of Consumer Research, 7*(4), 63–71.

Hobday, M., & Howard, R. (2007). Upgrading the technological capabilities of foreign transnational subsidiaries in developing countries: The case of electronics in Thailand. *Research Policy, 36*(9), 1335–1356. doi:10.1016/j.respol.2007.05.004

Holly, K. (4 Feb 2008). The politics of change. *BusinessWeek.* Retrieved from http://www.businessweek.com/innovate/content/feb2008/id2008024_250194.htm

Holt, B. J., & Morrell, R. W. (2002). Guidelines for Web site design for older adults: The ultimate influence of cognitive factors. In R. W. Morrell (Ed.), Older Adults, Health Information, and the World Wide Web (pp. 109-132). Mahwah, NJ: Lawrence Erlbaum Associates, Inc.

Holton, J. A. (2001). Building trust and collaboration in a virtual team. *Team Performance Management, 7*(3/4), 36–47. doi:10.1108/13527590110395621

Horowitz, M. D., & Rosensweig, J. A. (2007). Medical tourism - health care in the global economy. *Physician Executive, 33*(6), 24.

Houghton, L., & Kerr, D. V. (2004). *Understanding Feral Systems in Organisations: A case study of a SAP implementation that led to the creation of ad-hoc and unplanned systems in a large corporation.* Paper presented at the 9th Asia-Pacific Decision Sciences Institute Conference, Seoul, Korea.

Houghton, L., & Kerr, D. V. (2006). A study into the creation of feral information systems as a response to an ERP implementation within the supply chain of a large government-owned corporation. *International Journal of Internet and Enterprise Management, 4*(2), 135–147.

Howard, J. (2005). *The emerging business of knowledge transfer, creating value from intellectual products and services (Report of a Study Commissioned by the Department of Education, Science, and Training).* Canberra: Department of Education, Science and Training.

Howells, J. (2006). Intermediation and the role of intermediaries in innovation. *Research Policy, 35*(5), 715–728. doi:10.1016/j.respol.2006.03.005

Hsiao, F. S. T., & Hsiao, M.-Ch. W. (2003, February). Miracle growth in the twentieth century – International comparisons of East Asian development. *World Development,* 227–257. doi:10.1016/S0305-750X(02)00188-2

Huang, K.-T., Lee, Y. W., & Wang, R. Y. (1999). *Quality information and knowledge.* Upper Saddle River, NJ: Prentice Hall PTR.

Hult, G. T. M., & Hurley, R. F. (1998). Innovation, market orientation, and organisational learning: An integration and empirical examination. *Journal of Marketing, 62,* 42–54. doi:10.2307/1251742

Hult, G. T. M., Hurley, R. F., & Knight, G. A. (2004). Innovativeness: Its antecedents and impact on business performance. *Industrial Marketing Management, 33*(5), 429–438. doi:10.1016/j.indmarman.2003.08.015

Hutley, P. S. B., & Russell, P. A. (3rd ed.). (2005). An introduction to the Financial Services Reform Act, 2001. Australia: LexisNexis Butterworths.

IAB. (2007). IAB Internet Advertising Revenue Report. Industry survey conducted by PricewaterhouseCoopers and Sponsored by the Interactive Advertising Bureau (IAB). Retrieved from http://www.iab.net/media/file/IAB_PwC-2007Q2.pdf.

Ingebretsen, M. (2002). NASDAQ: A History of the Market that changed the world. USA: Forum, Random House.

Intarakumnerd, P. (2004, April). Thailand's national innovation system in transition. First Asialics International Conference on Innovation Systems and Clusters in Asia: Challenges and Regional Integration. National Science and Technology Development Agency, Bangkok Thailand.

Internet Outsider. (2007). *Time to update those Facebook revenue estimates.* Retrieved May 2, 2009, from www.internetoutsider.com/2007/07/time-to-update-.html.

Irwin, T. (2007).Web site click-through rates soar with human touch. *MediaPost Publications.* Retrieved May 2, 2009, from http://www.mediapost.com/publications/index.cfm?fa=Articles.showArticle&art_aid=65206.

Ismail, R., & Yussof, I. (2003). Labour market competitiveness and foreign direct investment: The case of Malaysia, Thailand and the Philippines. *Papers in Regional Science, 82,* 389–402. doi:10.1007/s10110-003-0170-2

ITAIDE. (2009). *Report on redesign of administrative processes, interoperability and standardization.* Retrieved from http://www.itaide.org

Janakiraman. R. (1992). *Interim report of the Janakiraman Committee to probe into the securities scam.* Government of India.

Jansen, K. (1995). The macroeconomic effects of direct foreign investment: The case of Thailand. *World Development, 23*(2), 193–210. doi:10.1016/0305-750X(94)00125-I

Jarzabkowski, P. (2008). Shaping strategy as a structuration process. *Academy of Management Journal, 51,* 621–650.

Javary, M., & Mansell, R. (2002). Emerging internet oligopolies: A political economy analysis. In Miller, E. S., & Samuels, W. J. (Eds.), *An Institutionalist Approach to Public Utilities Regulation* (pp. 162–201). East Lansing, MI: Michigan State University Press.

Jeyaraj, A., Rottman, J. W., & Lacity, M. C. (2004, December). *Understanding the Relationship between Organizational and Individual Adoption of IT Innovations: Literature Review and Analysis.* Paper presented at the Diffusion Interest Group in Information Technology, Washington D.C., USA.

Johnston, R. (1972). The Internal Structure of Technology. In Halmos, P., & Albrow, M. (Eds.), *The Sociological Review Monograph 18 - The Sociaology of Sciences* (pp. 117–130). J.H. Brookes Printers Limited, Keele.

Jomo, K. S. (2003). Growth and vulnerability before and after the Asian crisis: The fallacy of the universal model. In Martin, A., & Gunnarsson, C. (Eds.), *Development and structural change in Asia-Pacific: globalising miracles or end of a model?* (pp. 171–197). London: RoutledgeCurzon.

Jomo, K. S., Chung, C. Y., Folk, B. C., Ul-Haque, I., Phongpaichit, P., Simatupang, B., & Tateishi, M. (1997). *Southeast Asia's Misunderstood Miracle: Industrial Policy and Economic Development in Thailand, Malaysia, and Indonesia.* Boulder, CO: Westview.

Jomo, K. S., Rasiah, R., Alavi, R., & Gopal, J. (2003). Industrial policy and the emergence of internationally competitive manufacturing firms in Malaysia. In Jomo, K. S. (Ed.), *Manufacturing Competitiveness in Asia: How International Competitive National Firms and Industries Developed in East Asia* (pp. 106–172). London: RoutledgeCurzon.

JPC. (1992). Report of the Joint Parliamentary Committee to probe into the securities scam of 1992. Government of India.

Judkins, G. (2007). Persistence of the US - Mexico Border: Expansion of medical-tourism amid trade liberalization. *Journal of Latin American Geography, 6*(2), 11–32. doi:10.1353/lag.2007.0042

Kahai, S. S., Sosik, J. J., & Avolio, B. J. (1997). Effects of leadership style and problem structure on work group process and outcomes in an electronic meeting system environment. *Personnel Psychology, 50,* 121–146. doi:10.1111/j.1744-6570.1997.tb00903.x

Kahn, M. (1999). The silver lining of Rust Belt manufacturing decline. *Journal of Urban Economics*, *46*(3), 360–376. doi:10.1006/juec.1998.2127

Kaiser, S., & Müller-Seitz, G. (2008). Leveraging Lead-user knowledge in software development—the case of Weblog technology. *Industry and Innovation*, 199–221. doi:10.1080/13662710801954542

Kalakota, R., Oliva, R. A., & Donath, B. (1999). Move Over, E-Commerce. *Marketing Management*, *8*(3), 22–32.

Kam, W. P. (1999). Technological capability development by firms from East Asian NIEs: Possible lessons for Malaysia. In Jomo, K. S., & Felker, G. (Eds.), *Technology, Competitiveness, and the State: Malaysia's Industrial Technology Policies* (pp. 53–64). London: Routledge.

Kangas, B. (2007). Hope from abroad in the international medical travel of Yemeni patients. *Anthropology & Medicine*, *14*(3), 293–305. doi:10.1080/13648470701612646

Kaplan, R. S., & Norton, D. P. (1992). The balanced scorecard – Measures that drive performance. *Harvard Business Review*, *70*(1), 71–80.

Kaplan, R. S., & Norton, D. P. (1993). Putting the balanced scorecard to work. *Harvard Business Review*, *71*(5), 134–142.

Kaplan, R. S., & Norton, D. P. (1996). Using the balanced scorecard as a strategic management system. *Harvard Business Review*, *74*(1), 75–85.

KCA - Knowledge Commercialisation Australia. (2006). *Big book of ideas 2006*. Knowledge Commercialisation Australia.

Khalil, T. M. (1999). *Management of technology*. McGraw-Hill Science/Engineering/Math.

Kim, J., & Marschke, G. R. (2007). *How much US technological innovation begins in universities? Economic Commentary* (pp. 1–3). Cleveland: Federal Reserve Bank.

Kirp, D. (2003). *Shakespeare, Einstein, and the bottom line – The marketing of higher education*. Cambridge, MA: Harvard University Press.

Kirschner, A. (2006, June). What I know is. *PublishingTrends.Com*. Marketing Partners International. Retrieved March 31, 2009, from http://www.publishingtrends.com/copy/06/0606/0606WhatIKnowIsWiki.html.

Kirzner, I. (1984). The entrepreneurial process. In Kent, C. A. (Ed.), *The environment for entrepreneurship*. Lanham, MD: Lexington Books.

Kissoon, S. V. (2007). Continuous improvement teamwork in the Australian banking sector. In *Proceedings of the 5th ANZAM and 1st Asian Pacific Operations Management Symposium 2007*. RMIT University, Melbourne, Australia.

Kissoon, S. V. (2008). Toward the conceptual model of Continuous Improvement Teamwork: A participant observation study. In F. Zhao (Ed.), Information Technology Entrepreneurship and Innovation (250-276). Hershey, PA: IGI Global.

Kissoon, S. V. (2008). Ethnographic research cycle to evidence the Continuous Improvement Teamwork model. *Qualitative Research Journal*, *8*(2), 134–136. doi:10.3316/QRJ0802134

Klyver, K., Hindle, K., & Meyer, D. (2008). Influence of social network structure on entrepreneurship participation – A study of 20 national cultures. *The International Entrepreneurship and Management Journal*, *4*, 331–17. doi:10.1007/s11365-007-0053-0

Koestler, A. (1989). *The Sleepwalkers*. Arkana/Penguin.

Koh, W. T. H. (2006). Singapore's transition to innovation-based economic growth: infrastructure, institutions and government's role. *R & D Management*, *36*(2), 143–160. doi:10.1111/j.1467-9310.2006.00422.x

Kohpaiboon, A. (2006). Foreign direct investment and technology spillover: A cross-industry analysis of Thai manufacturing. *World Development*, *34*(3), 541–556. doi:10.1016/j.worlddev.2005.08.006

Kotorov, R. (2001). Virtual organization: Conceptual analysis of the limits of its decentralization. *Knowledge and Process Management*, *8*(1), 55–62. doi:10.1002/kpm.93

Kraut, R. E., Rice, R. E., Cool, C., & Fish, R. S. (1998). Varieties of social influence: The role of utility and norms in the success of a new communication medium. *Organization Science, 9*, 437–453. doi:10.1287/orsc.9.4.437

Kuhn, T. S. (1962). The Structure of Scientific Revolutions Chicago. University of Chicago Press. the problem of induction 1(1), 5.

Kuhn, T. S. (1996). *The structure of scientific revolutions*. Chicago: University of Chicago Press.

Kuhn, T. S., & Neurath, O. (1970). *The Structure of Scientific Revolutions: International Encyclopedia of Unified Science*. University of Chicago Press.

Kuiper, E. J. (2007). *Convergence by Cooperation in IT – The EU's Customs and Fiscalis Programmes*. Delft, The Netherlands: Delft University of Technology.

Kuratko, D. F. (2005). The emergence of entrepreneurship education: Developments, trends, and challenges. *Entrepreneurship Theory and Practice, 29*(5), 577–597. doi:10.1111/j.1540-6520.2005.00099.x

Kuratko, D. F., & Hodgetts, R. M. (2004). *Entrepreneurship: Theory, process, practice*. Mason, OH: South-Western College Publishers.

Labormarketinfo.com. (2009). State of Florida Agency for Workforce Innovation, Labor Market Statistics. Retrieved September 4, 2009, from http://www.labormarketinfo.com

Lakatos, I., & Musgrave, A. (1970). *Criticism and the Growth of Knowledge*. Cambridge Univ Press.

Lake, M. (2009). The art of creation in science: A consonant paradox. *Market Times, 19*, 278–197.

Lall, S. (1999). Technology policy and competitiveness in Malaysia. In Jomo, K. S., & Felker, G. (Eds.), *Technology, Competitiveness, and the State: Malaysia's Industrial Technology Policies* (pp. 148–179). London: Routledge.

Lambert, K. (2003). *Lambert review of business-university collaboration*. London: HMSO.

Lauriola, M., Levin, I. P., & Hart, S. S. (2007). Common and distinct factors in decision making under ambiguity and risk: A psychometric study of individual differences. *Organizational Behavior and Human Decision Processes, 104*, 130–149. doi:10.1016/j.obhdp.2007.04.001

Lee, P. M., & O'Neill, H. M. (2003). Ownership structures and R&D investments of US and Japanese firms: Agency and stewardship perspectives. *Academy of Management Journal, 46*, 195–211.

Lee, Y. (2000). The sustainability of university-industry research collaboration: An empirical assessment. *The Journal of Technology Transfer, 25*(2), 111–133. doi:10.1023/A:1007895322042

Lefebvre, E., & Lefebvre, L. A. (1992). Firm innovativeness and CEO characteristics in small manufacturing firms. *Journal of Engineering and Technology Management, 9*(3-4), 243–277. doi:10.1016/0923-4748(92)90018-Z

Lettl, C. H. (2006). Users' contributions to radical innovation: evidence from four cases in the field of medical equipment technology. *R & D Management*, 251–272. doi:10.1111/j.1467-9310.2006.00431.x

Levy, B., & Spiller, P. T. (1994). The institutional foundations of regulatory commitment: A comparative analysis of telecommunications regulation. *Journal of Law Economics and Organization, 10*(2), 201–246.

Levy, M., Powell, P., & Yetton, P. (1998). *SMEs and the gains from IS: from cost reduction to value added*. Paper presented at Joint Working Conference on Information Systems, Helsinki, Norway.

Li, R. (2003, October 27). Alignment of funding mechanisms with scientific opportunities. *Regional Forum on Research Business Models — Workshop Summary*, Berkeley, CA, (OSTP/NSTC Committee on Science).

Liefner, I., & Schiller, D. (2008). Academic capabilities in developing countries – A conceptual framework with empirical illustrations from Thailand. *Research Policy, 37*, 276–293. doi:10.1016/j.respol.2007.08.007

Lincoln, Y. S. (1997). Self, subject, audience, text: Living at the edge, writing in the margins. In Tierney, W. G., & Lincoln, Y. S. (Eds.), *Representation and the Text: Reframing the Narrative Voice* (pp. 37–55). Albany: State University of New York Press.

Lincoln, Y. S. (2000). Narrative authority vs perjured testimony: Courage, vulnerability, and truth. *Qualitative Studies in Education, 13*, 131–138. doi:10.1080/095183900235654

Lindmark, S. (2002). *Evolution of techno-economic systems: an investigation of the history of mobile communications*. Published Doctoral dissertation, Chalmers tekniska högskola, Göteborg.

Link, A., & Siegel, D. (2005). Generating science-based growth: An econometric analysis of the impact of organizational incentives on university–industry technology transfer. *European Journal of Finance, 11*(3), 169–181. doi:10.1080/1351847042000254211

Lipnack, J., & Stamps, J. (1997). *Virtual teams: Reaching across space, time and organizations with technology*. New York: John Wiley & Sons.

Litan, R., Mitchell, L., & Reedy, E. (2007). *Commercializing university innovations: Alternative approaches* (NBER Working Paper, 16 May 2007). Retrieved June 29, 2007, from: http://ssrn.com/abstract=976005.

Liu, L., & Chi, L. N. (2002). Evolutional data quality: A theory-specific view. In *Proceedings of 7th International Conference on Information Quality (ICIQ 2002)*, Cambridge, MA.

Lorincz, K., Malan, D., Fulford-Jones, T., Nawoj, A., Clavel, A., & Shnayder, V. (2004). *Sensor networks for emergency response: Challenges and opportunities. IEEE Pervasive Computing*. Oct/Dec.

Lucas, H. C. Jr, Wonseok, O., Bruce, W., & Weber, B. W. (2009). The defensive use of IT in a newly vulnerable market: The New York Stock Exchange, 1980–2007. *The Journal of Strategic Information Systems, 18*, 3–15. doi:10.1016/j.jsis.2009.01.003

Lüthje, C., & Herstatt, C. (2004). The Lead-user method: Theoretical-empirical foundation and practical implementation. *R & D Management, 34*(5), 549–564.

Maier, D., Muegeli, T., & Krejza, A. (2007). Customer investigation process at Credit Suisse: Meeting the rising demands of regulators. In Al-Hakim (Ed.), *Challenges of managing Information Quality in service Organizations* (pp. 52-76). Hershey, PA: IGI Global.

Malaysian Communications and Multimedia Commission and Ministry of Energy, Water and Communications. (2006). The national broadband plan: Enabling high speed broadband under MyICMS 886. Cyberjaya, Malaysia: Malaysian Communications and Multimedia Commission. Retrieved from http://www.mcmc.gov.my

Mansfield, E. (1991). Academic research and industrial innovation. *Research Policy, 20*, 1–12. doi:10.1016/0048-7333(91)90080-A

Mansfield, E. (1998). Academic research and industrial innovation: An update of empirical findings. *Research Policy, 26*, 773–776. doi:10.1016/S0048-7333(97)00043-7

Manz, C. C., & Stewart, G. L. (1997). Attaining flexible stability by integrating total quality management and socio-technical systems theory. *Organization Science, 8*, 59–70. doi:10.1287/orsc.8.1.59

Martin, M., Gruetzmacher, R., Lanham, R., & Brady, J. (2004). Assessing technology transfer and business development potential: Technology cluster analysis. *Economic Development Quarterly, 18*(2), 168–173. doi:10.1177/0891242403261088

Masuda, Y. (1980). *The information society as post-industrial society*. Tokyo, Japan: Institue for the Information Society.

Mazis, M., Ahtola, O., & Kippel, R. (1975). A comparison of four multi attribute models in the prediction of consumer attitudes. *The Journal of Consumer Research, 2*, 38–53. doi:10.1086/208614

Maznevski, M. L., & Chudoba, K. M. (2000). Bridging Space Over Time: Global Virtual Team Dynamics and Effectiveness. *Organization Science, 11*, 473–492. doi:10.1287/orsc.11.5.473.15200

McCallum, B. T., & Jacoby, P. F. (2007). Medical outsourcing: Reducing clients' health care risks. *Journal of Financial Planning, 20*(10), 60.

McGuckin, R., & Stiroh, K. (1998, Summer). Computers can accelerate productivity growth. *Issues in Science and Technology, 14*(4), 41–48.

McLuhan, M. (1964). *The medium is the message Understanding Media* (pp. 7–23). London: Routledge and Kegan Paul.

Melkas, H. (2007). Analyzing information quality in virtual networks of the services sector with qualitative interview data. In Al-Hakim, L. (Ed.), *Challenges of managing Information Quality in service Organizations* (pp. 187–212). Hershey, PA: IGI Global.

Mephokee, C., & Ruengsrichaiya, K. (2005, December). Information and communication technology (ICT) for development of small and medium-sized exporters in East Asia: Thailand. United Nations Publication, Comisión Económica para América Latina y el Caribe (CEPAL), Project Document.

Midgley, D. F., & Dowling, G. R. (1978, March). Innovativeness: the concept and its measurement. *The Journal of Consumer Research*, 4.

Miles, R. E., & Snow, C. C. (1992). Causes of failure in network organizations. *California Management Review, 34*(4), 53–72.

Miller, H. (1996). The multiple dimensions of information quality. *Information Systems Management, 13*(2), 79–82. doi:10.1080/10580539608906992

Mills, C. W. (2000). *The Sociological Imagination*. New York: Oxford University Press.

Milstein, A., & Smith, M. (2006). America's new refugees—Seeking affordable surgery offshore. *The New England Journal of Medicine, 355*(16), 1637–1640. doi:10.1056/NEJMp068190

Miner, J. B., & Raju, N. S. (2004). Risk propensity differences between managers and entrepreneurs and between low- and high-growth entrepreneurs: A reply in a more conservative vein. *The Journal of Applied Psychology, 89*(1), 3–13. doi:10.1037/0021-9010.89.1.3

Minniti, M., & Bygrave, W. D. (2004). *Global entrepreneurship monitor*. Kansas City, MO: Kauffman Center for Entrepreneurial Leadership.

Mintzberg, H. (1994). The fall and rise of strategic planning. *Harvard Business Review*, (January-February): 107–114.

Mintzberg, H., Raisinghani, D., & Théorêt, A. (1976). The structure of "unstructured" decision processes. *Administrative Science Quarterly, 21*, 246–275. doi:10.2307/2392045

Mirrer-Singer, P. (2007). Medical malpractice overseas: the legal uncertainty surrounding medical tourism. *Law and Contemporary Problems, 70*(2), 211–233.

Mitman, G. (2003). Hay fever holiday: Health, leisure, and place in gilded-age America. *Bulletin of the History of Medicine, 77*, 600–635. doi:10.1353/bhm.2003.0127

Mohannak, K. (1999). A national linkage program for technological innovation. *Prometheus, 17*(3), 323–336. doi:10.1080/08109029908632135

Moor, J. H. (1985). What is computer ethics? *Metaphilosophy, 16*(4), 266–275. doi:10.1111/j.1467-9973.1985.tb00173.x

Morrell, R. W., Dailey, S. R., Feldman, C., Mayhorn, C. B., Echt, K. V., Holt, B. J., & Podany, K. I. (2004). Older adults and information technology: A compendium of scientific research and Web site accessibility guidelines. National Institute on Aging, Bethesda, MD.

Morris, M. G., & Venkatesh, V. (2000). Age differences in technology adoption decisions: Implications for a changing work force. *Personnel Psychology, 53*, 365–401. doi:10.1111/j.1744-6570.2000.tb00206.x

Morrison, P. D., Roberts, J. H., & Midgley, D. F. (2004). The nature of Lead-users and measurement of leading edge status. *Research Policy, 33*(2), 351–362. doi:10.1016/j.respol.2003.09.007

Mowery, D., & Shane, S. (2002). Introduction to the special issue on university entrepreneurship and technology transfer. *Management Science*, *48*(1), v–ix. doi:10.1287/mnsc.48.1.0.14277

Mowery, D., Nelson, R., Sampat, B., & Ziedonis, A. (2001). The growth of patenting and licensing by U.S. universities: an assessment of the effects of the Bayh-Dole Act of 1980. *Research Policy*, *30*, 99–119. doi:10.1016/S0048-7333(99)00100-6

Narin, F., Hamilton, K., & Olivastro, D. (1997). The increasing linkage between U.S. technology and public science. *Research Policy*, *26*, 317–330. doi:10.1016/S0048-7333(97)00013-9

National Academy of Sciences. National Academy of Engineering, & Institute of Medicine. (2006). Rising above the gathering storm – Energizing and employing America for a brighter economic future. The National Academies.

National Institute on Aging. (2002). *Making your Web site senior-friendly: A checklist*. Retrieved May 2, 2009, from www.nlm.nih.gov/pubs/staffpubs/od/ocpl/agingchecklist.html.

NECTEC. (2003). *Thailand: Information and communication technology master plan (2002-2006)*. Bangkok, Malaysia: National Electronics and Computer Technology Center.

Nelson, R. (2001). Observations on the post-Bayh-Dole rise of patenting at American universities. *The Journal of Technology Transfer*, *26*(1-2), 13–19. doi:10.1023/A:1007875910066

Nelson, R., & Winter, S. (1982). *An evolutionary theory of economic change*. Harvard University Press.

Ngwenyama, O. K., & Lee, A. S. (1997). Communication richness in electronic mail: Critical social theory and the contextuality of meaning. *Management Information Systems Quarterly*, (June): 145–166. doi:10.2307/249417

Nielsen, J. (1999). Designing Web Usability: The Art of Simplicity. Indianapolis, IN: New Riders Publishing.

North, D. C. (1990). Institutions, Institutional Change and Economic Performance. Cambridge University Press.

O'Farrell, P., & Hitchens, D. (1988). Alternative theories of small-firm growth: a critical review. *Environment and Planning*, *20*(3), 1365–1383. doi:10.1068/a201365

O'Flynn, B., Angove, P., Barton, J., Gonzalez, A., O'Donoghue, J., & Herbert, J. (2006). Wireless biomonitor for ambient assisted living. In *Proceedings International Conference on Signals and Electronic Systems*, (ICSES'06), Lodz, Poland (pp. 257-260).

O'Reilly, T. (2005). What is Web 2.0: Design patterns and business models for the next generation of software? *O'Reilly Media*. Retrieved March 13, 2009, from http://www.oreillynet.com/pub/a/oreilly/tim/news/2005/09/20/what-is-web-20.html.

O'Shea, R., Allen, T., O'Gorman, C., & Roche, F. (2004). Universities and technology transfer: A review of academic entrepreneurship literature. *Irish Journal of Management*, *25*(2), 11–29.

Öberg, S. (2005, June). Hägerstrand and the remaking of Sweden. *Progress in Human Geography*, *29*(3), 340–349.

Opinion Research Corporation. (2006). Attitudes and beliefs about caregiving in the U.S. Findings of a national opinion survey. *Johnson and Johnson*. Retrieved May 1, 2009, from www.strengthforcaring.com/util/press/research/index.html.

Orlikowski, W. J. (1996). Improvising organizational transformation over time: A situated change perspective. *Information Systems Research*, *7*, 63–92. doi:10.1287/isre.7.1.63

Ormerod, P. (2005). *Why Most Things Fail: Evolution, Extinction and Economics*. London: Faber and Faber Limited.

Orwell, G. (1949). *Nineteen eighty-four*. London: Secker.

Owen-Smith, J., & Powell, W. (2001). To patent or not: Faculty decisions and institutional success at technology transfer. *The Journal of Technology Transfer, 26*(1-2), 99–114. doi:10.1023/A:1007892413701

Painter, M., & Wong, S.-F. (2007). The telecommunications regulatory regimes in Hong Kong and Singapore: When direct state intervention meets indirect policy instruments. *The Pacific Review, 20*(2), 173–195. doi:10.1080/09512740701306832

Palmintera, D., Bannon, J., Levin, M., & Pagan, A. (2000). Developing high technology communities: San Diego. Produced under contract to Office of Advocacy, U.S. Small Business Administration, by Innovation Associates, Inc., Reston, Virginia.

Pande, P. S., Newman, R. B., & Cavanagh, R. R. (2000). *The Six Sigma way, how GE, Motorola, and other top companies are honing the performance*. New York: McGraw-Hill.

Parker, C., Scott, C., Lacey, N., & Braithwaite, J. (2004). *Regulating law*. Oxford University Press.

Patil, R. H. (2006). Current state of the Indian capital market. *Economic and Political Weekly, 41*(11), 1001–1010.

Pau, L.-F. (1989). Failure diagnosis and performance monitoring (8th ed.). New York: Marcel Dekker.

Pava, C. (1986). Redesigning sociotechnical systems design: Concepts and methods for the 1990s. *The Journal of Applied Behavioral Science, 22*, 201–221. doi:10.1177/002188638602200303

Peine, A. (2008). Technological paradigms and complex technical systems—The case of Smart Homes. *Research Policy, 37*(3), 508–529. doi:10.1016/j.respol.2007.11.009

Pennings, G. (2002). Reproductive tourism as moral pluralism in motion. *Journal of Medical Ethics, 28*, 337–341. doi:10.1136/jme.28.6.337

Pennings, J., & Buitendam, A. (Eds.). (1987). *New Technology as Organizational Innovation*. Cambridge, Massachusetts: Ballinger.

Pesonen, M., Rossi, M., & Tuunainen, V.K. (2004, March). *Mobile Technology in Field Customer Service*. Presentation at Austin Mobility Roundtable, Austin, TX.

Peters, L., Rice, M., & Sundararajan, M. (2004). The role of incubators in the entrepreneurial process. *The Journal of Technology Transfer, 29*(1), 83–91. doi:10.1023/B:JOTT.0000011182.82350.df

Pfeffer, J., & Salancik, G. R. (1978). *The external control of organizations: A resource dependence perspective*. New York: Harper & Row.

Phillips, K. (2005, December 5). Live from... transcript: interview with John Seigenthaler and Jimmy Wales. *CNN*. Retrieved April 1, 2009, from http://transcripts.cnn.com/TRANSCRIPTS/0512/05/lol.02.html.

Pierce-Grove, R. (2007). *The Future of Social Networking: understanding market strategic and technological developments*. Datamonitor.com.

Planck, M. (1949). A scientific biography. In O. U. Press (Ed.), The Oxford dictionary (2004 ed., p. 596).

Plumlee, M. A. (2003). The effect of information complexity on analysts' use of that information. *Accounting Review, 78*, 275–296. doi:10.2308/accr.2003.78.1.275

Pollard, D. (2006). Innovation and technology transfer intermediaries: A systemic international study. *Advances in Interdisciplinary Studies of Work Teams, 12*, 137–174. doi:10.1016/S1572-0977(06)12006-3

Porat, M. U., & Rubin, M. R. (Eds.). (1977). *The Information Economy*. Washington, DC: US Department of Commerce/Office of Telecommunications Special Publication.

Porter, M. (1990). *The competitive advantage of nations*. New York: The Free Press.

Porter, M. (1998). Clusters and the new economics of competition. *Harvard Business Review, 76*, 77–90.

Porter, M. E. (1979). *How competitive forces shape strategy*. New York: Free Press.

Porter, M. E. (1996). What is strategy? *Harvard Business Review*, (November-December): 61–78.

Postman, N. (1992). *Technopoly: The Surrender of Culture to Technology*. New York: Knopf.

Press, E., & Washburn, J. (2000). The kept university. *Atlantic Monthly, 285*(3), 39–54.

Pressley, L., & McCallum, C. J. (2008, September/October). Putting the library in Wikipedia. *Online*. Retrieved July 15, 2009, from http://www.onlinemag.net

Productivity Commission. (2007). *Public support for science and innovation (Productivity Commission Report)*. Commonwealth of Australia.

Publishing Trends. (2007). Thomson teaches tech through TWikis. Retrieved April 25, 2009, from http://www.publishingtrends.com/copy/07/0702/0702Thomson.html.

Putnam, L. (2001). March/April). Distance teamwork: The realities of collaborating with virtual colleagues. *Online, 25*(2), 54–57.

Quinn, J. B. (1992). Managing strategic change. In A. A. Thompson, Jr., W. E. Fulmer & A. J. Strickland III (Eds.), Readings in Strategic Management (4th ed., pp. 19-42). Boston: Irwin. (Reprinted from Sloan Management Review, 21, 3-20).

Quinn, M. J. (2005). Ethics for the information age. Boston, MA: Pearson Addison Wesley.

Raisch, S., & Birkinshaw, J. (2008). Organizational ambidexterity: Antecedents, outcomes, and moderators. *Journal of Management, 34*, 375–409. doi:10.1177/0149206308316058

Ramasamy, B., Chakrabarty, A., & Cheah, M. (2004). Malaysia's leap into the future: an evaluation of the multimedia super corridor. *Technovation, 24*, 871–883. doi:10.1016/S0166-4972(03)00049-X

Rao, L. (2009). The French come calling for Wikipedia. Retrieved April 25, 2009, from http://www.techcrunch.com/tag/wikipedia/

Rasiah, R. (2003). Foreign ownership, technology and electronics exports from Malaysia and Thailand. *Journal of Asian Economics, 14*, 785–811. doi:10.1016/j.asieco.2003.10.006

Rasmusson, L., & Jansson, S. (1996). Simulated social control for secure internet commerce. In M. C. (Ed.), *Proceedings of the 1996 New Security Paradigms Workshop*. Retrieved April 1, 2009, from http://en.wikipedia.org/wiki/Soft_security.

Raus, M. (2009). *Value Assessment of Business-to-Government IT Innovations: a Case Study. 22nd Bled eConference eEnablement: Facilitating an Open, Effective and Representative eSociety*. Slovenia: Bled.

Raus, M., Flügge, B., & Boutellier, R. (2008). Innovation Steps in the Diffusion of e-Customs Solutions. In S. A. Chun, M. Janssen & J. R. Gil-Garcia (Eds.), *ACM International Conference Proceeding Series* (Vol. 289, pp. 315-324). Montréal, Canada: Digital Government Society of North America.

Raus, M., Flügge, B., & Boutellier, R. (2009). Electronic Customs Innovation: an Improvement of Governmental Infrastructure. *Government Information Quarterly, 26*(2). doi:10.1016/j.giq.2008.11.008

Raus, M., Kipp, A., & Boutellier, R. (2008). Diffusion of e-Government IT Innovation: a Case of Failure? In Cunningham, P., & Cunningham, M. (Eds.), *Collaboration and the Knowledge Economy: Issues, Applications, Case Studies*. Amsterdam: IOS Press.

Reamer, A., Icerman, L., & Youtie, J. (2003). *Technology transfer and commercialization: Their role in economic development*. Washington, DC: U.S. Department of Commerce, Economic Development Administration.

Remington, M. J. (2005). The Bayh-Dole Act at twenty-five years: looking back, taking stock, acting for the future. *Journal of the Association of University Technology Managers, 17*(1), 15–31.

Reynolds, P. D., Hay, M., & Camp, S. M. (1999). *Global entrepreneurship monitor*. Kansas City, MO: Kauffman Center for Entrepreneurial Leadership.

Rhea, S. (2008). Medical migration. *Modern Healthcare, 38*(18), 6.

Riche, Y., & Mackay, W. (2005). Peercare: Challenging the monitoring approach to care for the elderly. In *Proceedings of the British Human Computer Interaction 2005 Workshop on HCI and the Older Population*. Edinburgh, United Kingdom.

Riddle, C. (2004). *Commercialization strategies of Canadian universities & colleges: Challenges at the university/college-industry interface, including intellectual property policies (A Study for the Advisory Council on Science and Technology)*. Government of Canada.

Ringer, G. (2007). Healthy spaces, healing places – sharing experiences of wellness tourism in Oregon, USA. *Selective Tourism: The Journal for Tourist Theory and Practice*, *1*(1), 29–39.

Ritzer, G. (2005). Structuration theory. *Contemporary Sociology: A Journal of Reviews, 36*, 84-85.

Roger, W. (1999) *Postal service getting wired*. Retrieved May 31, 2009 from www.interactive-week.com

Rogers, E. M. (1962). *Diffusion of Innovations*. New York: The Free Press of Glencoe.

Rogers, E. M. (1983). *Diffusion of innovations*. New York: Free Press.

Rogers, E. M. (2003). *Diffusion of Innovations* (5th ed.). New York: Free Press/Simon & Schuster, Inc.

Rogers, E. M., & Kincaid, D. L. (1981). *Communication networks: Toward a new paradigm for research*. New York: Free Press.

Rogers, E., & Larsen, J. (1984). *Silicon valley fever: Growth of high-technology culture*. New York: Basic Books.

Rostow, W. W. (1960). *The stages of economic growth*. Cambridge, UK: Cambridge University Press.

Rotenberk, L. (2008). Medical tourism. As the world flattens, US hospitals expand their global reach. *Hospitals & Health Networks*, *82*(6), 14.

Rothaermel, F. T., Agung, S. D., & Jiang, L. (2007). University entrepreneurship: a taxonomy of the literature. *Industrial and Corporate Change*, *16*(4), 691–791. doi:10.1093/icc/dtm023

Rouse, W. B. (1999). Connectivity, creativity, and chaos: Challenges of loosely-structured organizations. *Information & Knowledge Systems Management*, *1*, 117–131.

Rousseau, D. (2006). Is there such a thing as "evidence-based management"? *Academy of Management Review*, *31*(2), 256–269.

Saunders, R. J. Jeremy J. Warford, and Björn Wellenius. (1994). Telecommunications and economic development (2nd ed.). Baltimore: Published for the World Bank by the Johns Hopkins University Press.

SBA. (1998). *The new American evolution: The role and impact of small firms*. Small Business Research Report. Retrieved September 12, 2008, from http://www.sba.gov/advo/

SBA. (2004). *Small business resources for faculty, student, and researchers*. Retrieved September 12, 2008, from http://www.sba.gov/advo/.

SBA. (2004). *Small Firms and Technology: Acquisitions, Inventor Movement, and Technology Transfer, Small Business Research Summary No. 233*. Retrieved September 12, 2008, from http://www.sba.gov/advo/

SBA. (2004). Top ten reasons to love small business. *Small Business Administration News Release, SBA 04-06 ADVO*. Retrieved September 12, 2008, from http://www.sba.gov/advo/.

Scevak, N. (2007, March 12). Wikipedia as a proxy for keyword analysis. *Bronte Media*. Retrieved April 26, 2009, from http://brontemedia.com/2007/03/12/wikipaedia-as-a-proxy-for-keyword-analysis/.

Schneider, K. G. (2007, September). Wikipedia's awkward adolescence. *Cio.com*. Retrieved July 16, 2009, from http://www.cio.com/article/141650/Wikipedia_s_Awkward_Adolescence.

Schneiderman, B. (1998). Designing the User Interface: Strategies for Effective Human-Computer Interaction (3rd Ed.). Boston, MA: Addison-Wesley.

Schreier, M., Oberhauser, S., & Prügl, R. (2007). Lead-users and the adoption and diffusion of new products: Insights from two extreme sports communities. *Marketing Letters*, 15–30. doi:10.1007/s11002-006-9009-3

SEBI. (2002, May 31). *Approach paper on implementation of STP in Indian markets*. SEBI. Retrieved from www.sebi.gov.in.

SEBI. (2004). *Revised scheme for implementing Straight-Through-Processing in India.* SEBI. Retrieved from www.sebi.gov.in.

Seidler, R., & Stelmach, G. (1996). Motor control. Encyclopedia of Gerontology: Age, Aging, and the Aged. San Diego, CA: Academic Press.

Seigenthaler, J. (2005, November 29). A false Wikipedia "biography". *USA Today*. Retrieved April 1, 2009, from http://www.usatoday.com/news/opinion/editorials/2005-11-29-wikipedia-edit_x.htm.

Semler, S. W. (1997). Systematic agreement: a theory of organizational alignment. *Human Resource Development Quarterly, 8*, 23–40. doi:10.1002/hrdq.3920080105

Shane, S. (2002). Selling university technology: Patterns from MIT. *Management Science, 48*(1), 122–137. doi:10.1287/mnsc.48.1.122.14281

Shane, S. (2004). Encouraging university entrepreneurship? The effect of the Bayh-Dole Act on university patenting in the United States. *Journal of Business Venturing, 19*, 127–151. doi:10.1016/S0883-9026(02)00114-3

Shane, S., & Stuart, T. (2002). Organizational endowments and the performance of university start-ups. *Management Science, 48*(1), 154–170. doi:10.1287/mnsc.48.1.154.14280

Shapira, P., Youtie, J., Yogeesvaran, K., & Zakiah, J. (2005, May). Knowledge economy measurement: Methods, results and insights from the Malaysian knowledge content study. Triple Helix 5 Conference - Panel Session on New Indicators for the Knowledge Economy, Turin, Italy.

Shari, I. (2003). Economic growth and social development in Malaysia, 1971-98: Does the state still matter in an era of economic globalisation? In Andersson, M., & Gunnarsson, C. (Eds.), *Development and structural change in Asia-Pacific: globalising miracles or end of a model?* (pp. 109–124). London: RoutledgeCurzon.

Shepherd, J. (2006, September). Transfers prove costly. [Higher Education Supplement]. *Times (London, England)*, 15.

Sheppard, B. H., Harwick, J., & Warshaw, P. R. (1988). The theory or reasoned action: A meta-analysis of past research with recommendations for modifications and future research. *The Journal of Consumer Research, 15*, 325–343. doi:10.1086/209170

Siegel, D., Waldman, D., & Link, A. (1999). *Assessing the impact of organizational practices on the productivity of university technology transfer offices: An exploratory study* (NBER Working Paper 7256). National Bureau of Economic Research.

Siegel, D., Waldman, D., Atwater, L., & Link, A. (2004). Toward a model of the effective transfer of scientific knowledge from academicians to practitioners: qualitative evidence from the commercialization of university technologies. *Journal of Engineering and Technology Management, 21*(1-2), 115–142. doi:10.1016/j.jengtecman.2003.12.006

Siegel, D., Westhead, P., & Wright, M. (2003). Science parks and the performance of new technology-based firms: A review of recent U.K. evidence and an agenda for future research. *Small Business Economics, 20*, 177–184. doi:10.1023/A:1022268100133

Sime, J. (2004). *The commercialisation of intellectual property from public sector research establishments: a discussion paper.* Unpublished paper based on a report to the Australian Institute for Commercialisation presented at the conclusion of a series of master classes conducted around Australia during 2003, Australian Institute for Commercialisation.

Simon, M., & Houghton, S. M. (2003). The relationship between overconfidence and the introduction of risky products: Evidence from a field study. *Academy of Management Journal, 46*, 139–149.

Sine, W. D., Mitsuhashi, H., & Kirsch, D. A. (2006). Revisiting Burns and Stalker: Formal structure and new venture performance in emerging economic sectors. *Academy of Management Journal, 49*, 121–132.

Si-young, H. (2007, June 11). Korea post utilizes cutting-edge IT. *The Korea Herald.*

Slaughter, S., & Leslie, L. (1997). *Academic capitalism – Politics, policies, and the entrepreneurial university.* Baltimore, MD: The Johns Hopkins University Press.

Smerd, J. (2008). Large companies hopping aboard medical tourism. *Workforce Management, 87*(11), 8.

Sobol, M., & Newell, K. (2003). Barriers to and measurements of the diffusion of technology from the university to industry. *Comparative Technology Transfer and Society, 1*(3), 255–278. doi:10.1353/ctt.2003.0032

Sohal, A. S., & Fitzpatrick, P. (2002). IT governance and management in large Australian organisations. *International Journal of Production Economics, 75*(1-2), 97–112. doi:10.1016/S0925-5273(01)00184-0

Speyer, M., Pohlmann, T., & Brown, K. (2006). *IT spending in the SMB sector.* Cambridge, MA: Forrester Research.

Spradley, J. P. (1980). *Participant observation.* USA: Thomson Learning Academic Resource Centre. Te`eni, D. (1993). Behavioural aspects of data production and their impact on data quality. *Journal of Database Management, 4*(2), 30–38.

Stacy, R. D. (1992). *Managing the unknowable. Strategic boundaries between order and chaos in organizations.* San Francisco: Jossey-Bass.

STA-Singapore Technologies Automobile. (1996). *Business improvement Hhandbook* (5th ed.). Singapore: Singapore Technologies.

Stefik, M., & Stefik, B. (2004). *Breakthrough: Stories and Strategies of Radical Innovation.* Cambridge, MA: MIT Press.

Stein, D. (Ed.). (2004). *Buying in or selling out? The commercialization of the American research university.* Piscataway, NJ: Rutgers University Press.

Steinmueller, W. E. (2001). ICTs and the possibilities for leapfrogging by developing countries. *International Labour Review, 120*(2), 193–210. doi:10.1111/j.1564-913X.2001.tb00220.x

Stephens, S. (2008). Healing the World: Reverse medical tourism. *Magazine of Physical Therapy, 16*(9), 26–29.

Stewart, W., Watson, W., Carland, J., & Carland, J. (1999). A proclivity for entrepreneurship A comparison of entrepreneurs, small business owners, and corporate managers. *Journal of Business Venturing, 14*(2), 189–214. doi:10.1016/S0883-9026(97)00070-0

Stoneburner, G., Feringa, A., & Goguen, A. (2002). *Risk management guide for information technology systems* (Tech. Rep. No. SP800-30). National Institute for Science and Technology.

Stones, R. (2005). *Structuration theory.* New York: Palgrave-Macmillan.

Storm, S., & Naastepad, C. W. M. (2005). Strategic factors in economic development: East Asian industrialization 1950–2003. *Development and Change, 36*(6), 1059–1094. doi:10.1111/j.0012-155X.2005.00450.x

Storrow, R. (2005). The handmaid's tale of fertility tourism: Passports and third parties in the religious regulation of assisted conception. *Texas Wesleyan Law Review, 12*, 189.

Strong, A. J., Walker, N., & Rogers, W. A. (2001). Searching the World Wide Web: Can older adults get what they need? In W.A. Rogers & A.D. Fisk (Eds.), Human Factors Interventions for the Health Care of Older Adults (pp. 255-269). Mahwah, NJ: Lawrence Erlbaum Associates, Inc.

Strover, S., & Berquist, L. (1999, November 22-24). Telecommunications infrastructure development: The evolving state and city role in the United States. Cities in the Global Information Society Conference. Newcastle upon Tyne.

Subramanian, A. (1996). Innovativeness: Redefining the concept. *Journal of Engineering and Technology Management, 13*(3-4), 223–243. doi:10.1016/S0923-4748(96)01007-7

Surowiecki, J. (2004). The wisdom of crowds: why the many are smarter than the few and how collective wisdom shapes business, economies, societies and nations. London: Little, Brown.

Survey: New Media (2006, April 20). The wiki principle. *Economist*. Retrieved March 31, 2009, from http://www.economist.com/surveys/displaystory.cfm?story_id=6794228.

Synnott, W. R. (1987). *The information weapon: Winning customers and markets with technology*. New York: John Wiley & Sons.

Taleb, N. N. (2007). *The Black Swan: The Impact of the Highly Improbable*. London: Penguine Books Ltd.

Tan, F. B., & Hunter, M. G. (2002). The repertory grid technique: A method for the study of cognition in information systems. *Management Information Systems Quarterly, 26*, 39–57. doi:10.2307/4132340

Tapscott, D., & Williams, A. (2008). Wikinomics: How mass collaboration changes everything. *Journal of Information Technology & Politics, 5*(2), 259–262. doi:10.1080/19331680802294487

Taylor, H., & Karlin, S. (1993). *Introduction to stochastic modeling*. London: Academic Press.

Taylor, S., & Todd, P. A. (1995). Understanding information technology usage: A test of competing models. *Information Systems Research, 6*, 144–176. doi:10.1287/isre.6.2.144

Teo, H. H., Tan, B. C. Y., & Wei, K. K. (1997). Organizational transformation using electronic data interchange: The case of TradeNet in Singapore. *Journal of Management Information Systems, 13*(4), 139–165.

Terry, D. J., Hogg, M. A., & White, K. M. (1999). The theory of planned behavior: Self-identity, social identity and group norms. *British Journal of Psychological Society, 38*, 225–244. doi:10.1348/014466699164149

Thiemjarus, S., Lo, B. P. L., & Yang, G.-Z. (2005). Body sensor network – a wireless sensor platform for pervasive healthcare monitoring, *Adjunct Proceedings of the 3rd International Conference on Pervasive Computing*, (PERVASIVE 2005) (pp. 77-80).

Thomas, J., Clark, S., & Gioia, D. (1993). Strategic sensemaking and organizational performance: Linkages among scanning, interpretation, action, and outcomes. *Academy of Management Journal, 36*(2), 239–270. doi:10.2307/256522

Thomke, S., & von Hippel, E. (2002). Customers as innovators: A new way to create value. *Harvard Business Review, 80*(4), 74–81.

Thull, J. (2005). *The Prime Solution*. Dearborn Trade Publishing.

Thursby, J., & Thursby, M. (2002). Who is selling the ivory tower? Sources of growth in university licensing. *Management Science, 48*(1), 90–104. doi:10.1287/mnsc.48.1.90.14271

Tihanyi, L., Johnson, R. A., Hoskisson, R. E., & Hitt, M. A. (2003). Institutional ownership differences and international diversification: The effects of boards of directors and technological opportunity. *Academy of Management Journal, 46*, 195–211.

Toffler, A. (1970). *Future Shock*. New York: Bantam Books.

Tornatzky, L., Waugaman, P. G., & Gray, D. (2002). *Innovation U: New university roles in a knowledge economy*. Southern Growth Policies Board.

Trist, E. (1971). New directions of hope. *Regional Studies, 13*, 439–451. doi:10.1080/09595237900185381

Tullverket. (2006). *Säkerhetsfragor I Tullverkets EDI-System* (1.0 ed., pp. 1-9).

Turner, L. (2007). Medical tourism: Family medicine and international health-related travel. *Canadian Family Physician Medecin de Famille Canadien, 53*, 1639–1641.

Tversky, A., & Kahneman, D. (1983). Extensional versus intuitive reasoning: The conjunction fallacy in probability judgment. *Psychological Review, 90*, 293–315. doi:10.1037/0033-295X.90.4.293

TWIKI. (2009). TWiki tutorial. Retrieved April 25, 2009, from http://twiki.org/cgi-bin/view/TWiki.TWikiTutorial.

TWIKI.NET. (2007). White paper: Bringing wikis to the enterprise. Retrieved April 25, 2009, from http://www.twiki.net/.

U.S. Census Bureau. (2009). U.S. Census Bureau State and County Quick Facts. Retrieved September 4, 2009, from http://quickfacts.census.gov

United Nations Economic Commission for Europe. (2005). *Recommendation and Guidelines on establishing a Single Window - Recommendation No. 33*. Geneva: UN/CEFACT.

Universal Postal Union. (2003). The role of postal services. WSIS Summit, Bern.

Urology Times. (2008). Medical Tourists receive HIFU treatment abroad. *Urology Times*, *36*(9), 12.

UVAPF - University of Virginia Patent Foundation. (2004). *Operating Manual.*

Valentín, E. M. M. (2000). University-industry cooperation: a framework of benefits and obstacles. *Industry and Higher Education*, (June): 165–172.

Van de Ven, A. H., Polley, D. E., Garud, R., & Venkataraman, S. (1999). *The Innovation Journey*. New York: Oxford University Press.

van der Smagt, T. (2000). Enhancing virtual teams: Social relations and communication technology. *Industrial Management + Data Systems, 100*(4), 148-156.

Van Hout, E. J. Th., & Bekkers, V. J. J. M. (2000). Patterns of virtual organization: The case of the National Clearinghouse for Geographic Information. *Information Infrastructure and Policy*, *6*, 197–207.

Venkatesh, V., & Davis, F. D. (2000). A theoretical extension of the Technology Acceptance Model: Four longitudinal field studies. *Management Science*, *46*(2), 186–204. doi:10.1287/mnsc.46.2.186.11926

Venkatraman, N. (1994). IT enabled business transformation. *Sloan Management Review*, *35*(2), 73–78.

Verspoor, M., & Lowie, W. (2003). Making sense of polysemous words. *Journal of Language Learning*, *53*, 547–586. doi:10.1111/1467-9922.00234

Vincenti, W. (1995). The Technical Shaping of Technology: Real-World Constraints and Technical Logic in Edison's Electrical Lighting System. *Social Studies of Science*, 553–574. doi:10.1177/030631295025003006

Vitale, M. (2004). *Commercialising Australian biotechnology*. Australian Business Foundation.

Volz, D. (2008). Reverse medical tourism. *Hospitals & Health Networks*, *82*(6), 12–14.

Von Hippel, E. (1988). *The Sources of Innovation*. Oxford: Oxford University Press.

Von Hippel, E. (2005). *Democratizing Innovation*. Cambridge, MA: MIT Press.

Von Hippel, E. (2007). Horizontal innovation networks - by and for users. *Industrial and Corporate Change*, 293–315. doi:10.1093/icc/dtm005

Von Hippel, E. (2008). Users as sources of invention. In Hall, B. H., & Rosenberg, N. (Eds.), *Handbook of Economics of Technological Change*. New York: Elsevier B.V. Press.

von Hippel, E., Thomke, S., & Sonnack, M. (1999). Creating breakthroughs at 3M. *Harvard Business Review*, *77*(5), 47–57.

Voss, H. (1996, July/August). Virtual organizations: The future is now. *Strategy and Leadership*, 12–16. doi:10.1108/eb054559

Wah, L. Y., & Narayanan, S. (1999). Technology utilization level and choice: The electronics and electrical sector in Penang, Malaysia. In Jomo, K. S., Felker, G., & Rasiah, R. (Eds.), *Industrial Technology Development in Malaysia* (pp. 107–124). London, New York: Routledge.

Walker, N., Millians, J., & Worden, A. (1996). Mouse accelerations and performance of older computer users. In *Proceedings of the Human Factors and Ergonomics Society 40th Annual Meeting,* Santa Monica, CA (pp. 151-154).

Wand, Y., & Wang, R. Y. (1996). Anchoring data quality dimensions in ontological foundations. *Communications of the ACM*, *39*(11), 86–95. doi:10.1145/240455.240479

Wang, N., Zhang, N., & Wang, M. (2006). Wireless sensors in agriculture and food industry: recent developments and future perspective. *Computers and Electronics in Agriculture, 50*(1), 1–14. doi:10.1016/j.compag.2005.09.003

Wang, R. Y., & Strong, D. M. (1996). Beyond accuracy: What data quality means to data consumers. *Journal of Management Information Systems, 12*(4), 5–33.

Wang, R. Y., Reddy, M. P., & Kon, H. B. (1995). Toward quality data: An attribute-based approach. *Decision Support Systems, 13*(3-4), 349–372. doi:10.1016/0167-9236(93)E0050-N

Warren, A., Hanke, R., & Trotzer, D. (2008). Models for university technology transfer: resolving conflicts between mission and methods and the dependency on geographic location. *Cambridge Journal of Regions. Economy and Society, 1*(2), 219–232.

Washburn, J. (2005). *University Inc. – The corporate corruption of higher education.* New York: Basic Books.

Wassenaar, A. (2000). *E-governmental value chain models-E-government from a business (modelling) perspective.*

Weber, B. W. (1999). Next-generation trading in futures markets: A comparison of open outcry and order matching systems. *Journal of Management Information Systems, 16*(2), 29–45.

Webster, J. (1998). Desktop video teleconferencing: Experiences of complete users, wary users, and nonusers. *Management Information Systems Quarterly,* (September): 257–286. doi:10.2307/249666

Webster's Dictionary. (1978). *Webster's new 20th century dictionary.* New York: Harper-Collins.

Wee, V. (2001, June). Imperatives for the k-economy: Challenges ahead. InfoSoc Malaysia Conference, Penang, Malaysia.

Weick, K. (1995). *Sensemaking in organizations.* Sage Publications, Inc.

Weick, K. E. (1993). The collapse of sensemaking in organizations: The Mann Gulch Disaster. *Administrative Science Quarterly, 38*(4), 628–652. doi:10.2307/2393339

Weick, K. E., & Sutcliffe, K. M. (2001). *Managing the unexpected: Assuring high performance in an age of complexity.* San Francisco, CA: Jossey-Bass.

Weill, P., & Ross, J. W. (2004). *IT Governance: How Top Performers Manage IT Decision Rights for Superior Results.* Boston, Massachusetts: Harvard Business School Press.

Weill, P., & Vitale, M. (2001). From place to space: migrating to e-business models. Cambridge, MA: Harvard Business School Press.

Wernerfelt, B. (1984). A Resource-based view of the firm. *Strategic Management Journal, 5*(2), 171–180. doi:10.1002/smj.4250050207

Westhead, P. (1997). R&D "inputs" and "outputs" of technology-based firms located on and off Science Parks, R&. *Mana, 27*(1), 45–62.

Wikimedia Foundation. (2009). *Home.* Retrieved April 22, 2009, from http://wikimediafoundation.org.

Wikimedia Foundation. (2009). *Gift policy.* Retrieved April 25, 2009, from http://wikimediafoundation.org/wiki/Gift_policy.

Wikipedia (2009). Wiki. *Knol.* Retrieved April 25, 2009, from http://en.wikipedia.org/wiki/Knol.

Wikipedia (2009). Wiki. *Wikipedia: The Free Encyclopedia.* Retrieved April 1, 2009, from http://en.wikipedia.org/wiki/Wiki.

Wikipedia (2009). Wikipedia: Biographies of living persons. *Wikipedia: The Free Encyclopedia.* Retrieved July 16, 2009, from http://en.wikipedia.org/wiki/Wikipedia:Biographies_of_living_persons.

Wikipedia (2009). Wiki. *Crowdsourcing.* Retrieved 10 July, 2009 from http://en.wikipedia.org/wiki/Crowdsourcing.

Wilde, G. J. S. (2001). *Target risk.* Toronto: PDE Publications.

Willcocks, L., Feeny, D., & Olson, N. (2006). Implementing Core IS Capabilities: Feeny-Willcocks IT Governance and Management Framework Revisited. *Eu-*

ropean Management Journal, 24(1), 28–37. doi:10.1016/j.emj.2005.12.005

Wilson, E. J. III, & Wong, K. (2003). African information revolution: A balance sheet. *Telecommunications Policy, 27*, 155–177. doi:10.1016/S0308-5961(02)00097-6

Wolfe, R. A. (1994). Organizational innovation: Review, critique and suggested research directions. *Journal of Management Studies, 31*, 405–427. doi:10.1111/j.1467-6486.1994.tb00624.x

Wong, P. (1999). National innovation systems for rapid technological catch-up: An analytical framework and a comparative analysis of Korea, Taiwan, and Singapore. Paper presented at the DRUID's Summer Conference, Rebild, Denmark.

Woodside, A. G. (2005). Firm orientations, innovativeness, and business performance: Advancing a system dynamics view following a comment on Hult, Hurley, and Knight's 2004 study. *Industrial Marketing Management, 34*(3), 275–279. doi:10.1016/j.indmarman.2004.10.001

Worden, A., Walker, N., Bharat, K., & Hudson, S. (1997). Making computers easier for older adults to use: Area cursors and sticky icons. In *Proceedings of the SIGCHI conference on Human factors in computing systems*, Atlanta, GA (pp. 266-271).

Workman, M. (2005). Expert decision support system use, disuse, and misuse: A study using the theory of planned behavior. *Journal of Computers in Human Behavior, 21*, 211–231. doi:10.1016/j.chb.2004.03.011

Workman, M. (2007). Advancements in technology: New opportunities to investigate factors contributing to differential technology and information use. *Journal of Management and Decision Making, 8*, 221–240.

Workman, M., & Bommer, W. (2004). Redesigning computer call center work: A longitudinal field experiment. *Journal of Organizational Behavior, 25*, 317–337. doi:10.1002/job.247

World Bank. (2008). *World Bank development indicators data base 2006*. Washington, DC: World Bank.

World Summit on Information Society. (2009). Retrieved February 10, 2009, from www.wsis-egypt.gov.eg.

Yin (2003). *Applications of case study research* (2nd ed.). Thousand Oaks, CA: Sage.

Yin, R. K. (1994). *Case Study Research: Design and Methods* (2nd ed.). Thousands Oaks, CA: Sage Publishing.

Zekri, N. (2006, April 16). Transformation of the post organization, *Al-Ahram Newspaper*.

Zemsky, R., Wegner, G., & Massy, W. (2006). *Remaking the American University – Market-smart and mission-centered*. Piscataway, NJ: Rutgers University Press.

Zhao, F. (2004). Commercialization of research: a case study of Australian universities. *Higher Education Research & Development, 23*(2), 223–236. doi:10.1080/0729436042000206672

About the Contributors

S. Ann Becker is an Associate Dean of Research in the College of Business, a University Professor of Management Information Systems and Computer Science, and President of the Women's Business Center at Florida Tech. She has a B.S. and an MBA from St. Cloud State University, and an M.S. and Ph.D. in Information Systems from the University of Maryland, College Park. Dr. Becker also holds a graduate certificate in Contract Management from Florida Tech. Dr. Becker has extensive experience in teaching, research, and consulting in electronic commerce, web usability, software engineering, database technologies, and contract management. She has published over 100 articles in these and related areas. She has received multiple research awards from government and industry sources.

Robert E. Niebuhr became dean of the College of Business at Florida Tech in August of 2007. Florida Tech currently has over 7,500 students and offers both undergraduate and graduate degrees in business. Dr. Niebuhr came to Florida Tech from Tennessee Technological University where he was the dean of business for six years. Prior to that, he served as a faculty member and administrator at Auburn University for twenty-four years. During that time, he was chairman of their management department, helped create their Executive MBA Program, and served as interim dean of the college for one year. Dr.Niebuhr's research deals with a variety of management topics and also examines the issues with different delivery approaches to student learning processes. He has been involved with creating and providing distance-based academic programs for over twenty years. Prior to his academic career, Dr. Niebuhr worked as a metallurgical engineer for ten years before pursuing his doctorate in management at the Ohio State University. His engineering degrees include a B.S. from the University of Cincinnati and an M.S. from The Ohio State University.

* * *

Roman Boutellier is since October 1, 2008, Vice President Human Resources and Infrastructure of ETH Zurich. He is professor and leads the Chair for Technology and Innovation Management at the Department of Management, Technology, and Economics (D-MTEC) at ETH Zurich since 2004. Since 1999 Prof. Dr. Boutellier is titular professor at the University of St. Gallen (HSG). His works appeared in R&D Management, Harvard Business Manager, ZFO and Drug Discovery Today. Roman Boutellier has held several leading positions in the industry, e.g. he was member of the management of Leica, Heerbrugg as well as CEO and delegate of the board of directors of the SIG Holding AG, Neuhausen. He is member of the board of directors of several Swiss large-scale enterprises. The focus of his research is the management of technology driven enterprises with a specific focus on innovation.

Brian Cameron is a Professor of Practice at the College of Information Sciences and Technology in The Pennsylvania State University. He has received his PhD from the Pennsylvania State University. He has strong technical research and teaching experiences including enterprise architecture, enterprise integration, information systems design and development, IT project/portfolio management, and service-oriented architecture and has conducted several studies, including a study on the Impact of Personality Type and Learning Style on Virtual IT Project Team Effectiveness.

Alan Collier is Research Fellow at the Centre for Entrepreneurship at the University of Otago, Dunedin, New Zealand and also teaches law at the Queensland University of Technology, Brisbane, Australia. He has qualifications and experience as an electrical engineer, lawyer and in management. His career highlights include starting entrepreneurial companies, working as a management consultant in technology-oriented assignments and practising as a lawyer in technology and intellectual property. His present research emphasis is university technology transfer and commercialization, which embraces management and legal scholarship.

Francisco C Cua CPA ACA PhD (Otago) MEntr (Otago) has more than thirty years of experience in accountancy, enterprise systems, and teaching. He has worked in Hong Kong, Shanghai, Beijing, and Manila as a consultant of Asian Development Bank's project and taken part in various projects as a business analyst, functional analyst, systems manager, and Oracle systems administrator in New Zealand. Diffusion of innovations, entrepreneurship, supply chain management, and business model are his research areas of interest.

C. Fausnaugh: Over a forty year career, Dr Fausnaugh has been accounting department manager in the home office of a fortune 500 company, founded and sold two companies, and taught at the university level in the United States, Australia and Singapore. Her research interests include technology transfer and company performance during the first seven years of operations. In addition to expertise in the financial operations, Dr. Fausnaugh has advised numerous companies in areas related to growth and financial management including managing cash flow during periods of start-up, rapid growth and contraction, taxation, development of corporate infrastructure, and investor relations. She is a Delaware CPA.

Jingwen He is a Research Assistant and a Graduate Student at the College of Information Sciences and Technology in The Pennsylvania State University. She holds her Bachelors in Engineering from Renmin University of China and a Master of Philosophy from Business School, City University of Hong Kong. Her current research focuses on risk management for start-ups and other entrepreneurial ventures.

Mareike Heinzen works as research associate at the Chair of Technology and Innovation Management at the Department of Management, Technology, and Economics (D-MTEC) at ETH Zurich and at the swiss CAR Group since 2008. She is PhD candidate in Management with focus on product development and innovation. Her research focuses on efficiency in the back-end of product development regarding routine innovation and continuous product improvement. Mareike Heinzen gained her MSc in industrial engineering at the Technical University of Karlsruhe, Germany and several years work experience with Daimler AG in Thailand, USA and Germany.

Sherif H. Kamel is Dean of the School of Business and Professor of MIS at the American University in Cairo. Previously, he was Associate Dean for executive education. He was director of the Management Center (2002-2008). Prior to joining the university, he was director of the Regional IT Institute (1992-2001) and co-established and managed the training department of the Cabinet of Egypt Information and Decision Support Centre (1987-1992). In 1996, he was a co-founding member of the Internet Society of Egypt. His research and teaching interests include management of information technology, information technology transfer to developing nations, electronic business, human resources development, and decision support systems. Kamel is the author of many publications in IS and management books and journals. Kamel serves on the editorial and advisory board of a number of IS journals and is the associate editor of the Journal of Cases on Information Technology, Journal of IT for Development and the Electronic Journal of IS in Developing Countries. He served as VP for communications in the Executive Council of the Information Resources Management Association (2000-2007). He was appointed as member of the board of trustees of the Information Technology Institute since 2005 and the Sadat Academy for Management Sciences (2006-2007). He was the Chairman of the Chevening Association in Egypt (2004-2009). He is an Eisenhower Fellow (2005) and a member of the Eisenhower Fellowships Alumni Advisory Council since 2008. He serves as the co-chair of the ICT core committee of the American Chamber of Commerce in Egypt since 2008. He holds a PhD from London School of Economics and Political Science, an MBA and a BA in Business Administration from The American University in Cairo.

Arvind Karunakaran is a Research Assistant and a Graduate Student at the College of Information Sciences and Technology in The Pennsylvania State University. He has received his Bachelors in Computer Science and Engineering from Anna University, India. His research interest is at the intersection of collaborative sensemaking, organizational design and identity.

Bob Keimer is an Adjunct Instructor in the Florida Tech College of Business where he teaches a new business start-up course. He also teaches in the Florida Tech Online MBA Program where he has helped over 150 MBA students develop professional business plans. Mr. Keimer is a founding investor and EVP of Business Development at Airgonomix, LLC, a new high-tech business started in 2008. Prior to Airgonomix, Mr. Keimer has held various senior operating management positions at a publicly traded distribution company. Mr. Keimer holds bachelors and masters degrees from Columbia University.

Lara Khansa is Assistant Professor in the Department of Business Information Technology, Pamplin College of Business, at Virginia Polytechnic Institute and State University. She received the Ph.D. in Information Systems, M.S. in Computer Engineering, and MBA in Finance and Investment Banking from the University of Wisconsin-Madison, and the B.E. in Computer and Communication Engineering from the American University of Beirut. Her primary research interests include the economics of information security and the resulting implications on the IT industry landscape. Dr. Khansa has published in Computers and Security, the European Journal of Operational Research, and the Communications of the ACM.

Suryadeo Vinay Kissoon has about 15 years practical work experience in general management. Corporate member and chartered quality professional with Chartered Quality Institute (U.K).Accredited trainer for training in field of quality management and finance in industry and tertiary education. Previously worked as operations manager in a major manufacturing company for about 12 years administering about 150 employees. He has been working in some major Australian organizations in fields of banking and finance, customer service and railway industries. Five nominations and quality awards nationally and internationally as operations manager. One nomination in quality award and one employee excellence award in Australia. Best paper award by a Ph.d candidate in 2007 in the international Association of Qualitative Research conference. Author has been involved as part-time consultancy and lecturer for nearly ten years in fields of HRM, quality management, service quality, financial management, strategic management, marketing management, supply chain management, logistic system analysis and production management. Author has been working extensively with quality circles (PDCA), kaizen teams (application of Six Sigma DMAIC methodology), workplace improvement teams, process improvement team and virtual teams. Author has also been working on TQM, Continuous Improvement, ISO 9000, EMS, HACCP, Information Quality Systems and some other quality management principles and methodologies. He has also been involved in research project works and written about 20 papers in various fields of management.

Jeffrey P. Landry is a professor in the School of Computer and Information Sciences at the University of South Alabama, where he teaches courses in information systems strategy and policy, project management, and programming. Dr. Landry is developing risk assessments of voting systems for federal elections; has conducted risk management for a national, online certification exam; has published on trust in the IS context; and completed a doctoral dissertation on software process innovations, all following a ten-year software industry career.

Divakaran Liginlal is an Associate Teaching Professor of Information Systems at Carnegie Mellon University. Lal has won numerous awards for teaching including the Mabel Chipman Outstanding Faculty Teaching Award from the University of Wisconsin-Madison and the University of Arizona Foundation Award for Meritorious Teaching. His research in information security and decision support systems has received support from Microsoft, CISCO, and HP among others. He has published in Computers and Security, IEEE Trans. on Systems, Man, and Cybernetics, IEEE Trans. on Knowledge and Data Engineering, the European Journal of Operational Research, Fuzzy Sets and Systems, Decision Support Systems, and CACM.

Nicholas C. Maynard is a Policy Researcher at the RAND Corporation where he focuses on information technology, S&T policy, and economic development. His current projects include technology acquisition, economic development policies, and best practices for R&D management. Dr. Maynard has led research teams to develop strategic plans for technology development initiatives as well as performed several national case studies, and developed a plan for cross-border technology centers. Dr. Maynard also led a multi-year effort to benchmark the US national innovation system for European Commission, comparing the US system against its peers in Europe and the Americas. Dr. Maynard received his BA and MA from the University of Chicago in Political Science and he completed a Public Policy PhD at University of North Carolina at Chapel Hill. His dissertation research on national technology strategies was supported through a National Science Foundation grant.

M. H. McCay graduated in Materials Science and Metallurgical Engineering. She began her career at NASA Marshall Space Flight Center where she pursued materials processing in space and failure analysis. Later, as a Payload Specialist Astronaut and a Principal Investigator, she conducted materials processing in space experiments on two Spacelab missions. After leaving NASA, Dr. McCay became a professor of engineering science at the University of Tennessee where she studied gravitational effects on solidification and laser material interactions. Among other awards, Dr. McCay received the NASA Exceptional Scientific Achievement Medal and the UT Chancellor's Award for Creativity and Research. She has fourteen patents.

Tim Muth is the Director of Student and Program Assessment in the College of Business at Florida Tech. Mr. Muth teaches Finance and International Business. He coordinates and oversees Florida Tech's College of Business AACSB accreditation efforts. Mr. Muth has held executive management positions, with extensive international business experience, at Harris Corporation, Intersil Corporation, and Conexant Systems. His management positions included Vice President, Business Development (Strategic Alliance and Acquisitions); Vice President, Marketing; Vice President, Supply Management; and Director, Financial Planning. Mr. Muth is active in the local community serving on boards and executive committees.

Desai Narasimhalu is a Professor of Information Systems Practice at the School of Information Systems of the Singapore Management University and holds the university level responsibility as the Director of the Institute of Innovation and Entrepreneurship. Desai has more than thirty five years of innovation and innovation management experience and over ten years of new business coaching experience. He has been responsible for organizational innovation, ideation management, intellectual property management, business development, and licensing, legal and corporate communications at different times in his career. He is a member of the Board of Advisors of the International Society of Professional Innovation Management and sits on the editorial boards of several international journals.

John O'Donoghue is a lecturer in Business Information Systems, University College Cork, Ireland. He received his PhD from the Department of Computer Science, University College Cork. His research interests include pervasive data management, data and information quality, health informatics and medical based information systems. His research focus examines the paradigm shift from centralised decision making networks, to remote autonomous devices. In particular, how to collect, correlate and disseminate this new information pool in an intelligent manner to help support our medical decision support systems (DSS).

Brian O'Flaherty is a lecturer is Business Information Systems, University College Cork, Ireland. He originally studied Computer Science and received his PhD in Management Science from Strathclyde University, Glasgow, Scotland. This research interests include teaching innovating practises, new product development and the role of mentors in experiential learning. He has developed many courses in technology entrepreneurship and commercialisation for technology graduates at both postgraduate and PhD level.

L-F Pau is Professor of Mobile Communications and Media at the Copenhagen Business School and at the Rotterdam School of Management. He was recently until an accident: CTO of L.M. Ericsson's Network Systems division with worldwide responsibilities, which he joined from a prior position as CTO for Digital Equipment / Hewlett Packard Europe. He was earlier or in parallel on the faculties of

Danish Technical University, Ecole Nationale Supérieure des Télécommunications (Paris), M.I.T , and University of Tokyo. He is a Fellow of IEEE (USA), BCS (UK), JSPS (Japan).

Sandeep Purao is an Associate Professor of Information Sciences and Technology in The Pennsylvania State University. He has the following experience: He has received his PhD from the University of Wisconsin-Milwaukee. His research focuses on various aspects of information system design and development. His current research projects include risk management for small and medium business, integrating workflow patterns into design, reuse-based design, flexible information system design, empirical investigations of individual design behaviors, and design theory.

Marta Raus works as research associate at the Chair of Technology and Innovation Management at the Department of Management, Technology, and Economics (D-MTEC) at ETH Zurich and at SAP Research Lab Zurich since 2006. She is PhD candidate in Management with focus on Business Innovation at the University of St. Gallen (HSG). Her research focuses on diffusion of IT innovations and value assessment models in the field of e-government at European level. Marta Raus gained her MSc in managerial and production sciences at ETH Zurich with emphasis on integrated product development and technology and innovation management.

Biswatosh Saha is affiliated with the Strategic Management group of Indian Institute of Management Calcutta as Assistant Professor. His research interests lie broadly in the field of entrepreneurship and innovation, especially on institutional dimension of innovation. He also has an interest in understanding networks of business and knowledge. He is on the Editorial Board of AI & Society (Springer, UK) and Decision (IIMC, India). He has also reviewed for several other journals, including Water Policy, Metamorphosis and others.

Michael Workman received his Ph.D. from Georgia State University, and has over 50 reviewed research publications. He came to academic life in 2001 with nearly thirty years of experience in the computer industry where he began as a software engineer, then moved into management. Reflecting on problems he faced, his research area investigates how to exploit technologies, tasks, and human factors to improve how well people work. Prior to coming to Florida Tech, Michael was an assistant and associate professor of information systems at the Florida State University.

Fang Zhao is an Associate Professor of Management at the School of Business & Management, American University of Sharjah in the UAE. Dr Zhao has done extensive research into innovation management and e-business management areas. She is the founding Editor-in-Chief of the International Journal of e-Business Management (ISSN: 1835-5412) at http://www.rmitpublishing.com.au/ijebm.html. She has authored/edited and published 3 research-based books in the past few years and authored and co-authored over 50 other peer-reviewed publications internationally. Dr Zhao has provided consulting services to Siemens (Australia) in managing technology innovations and the Department of Justice of Victoria government (Australia) in knowledge management.

Index

Symbols

A

B

G

H

I

J

K

L

M

Q

R

S

T

U